The Borderlands of Science

By the Same Author

Denying History: Who Says the Holocaust Never Happened and Why Do They Say It? (with Alex Grobman; University of California Press, 2000)

How We Believe: The Search for God in an Age of Science (W. H. Freeman, 1999)

Why People Believe Weird Things: Pseudoscience, Supersitition, and Other Confusions of Our Time (Foreword by Stephen Jay Gould, W. H. Freeman, 1997)

Endzeittaumel: Propheten, Prognosen, Propaganda (Edited by Michael Shermer, Benno Maidhof-Christig, Lee Traynor. Berlin: IBDK Verlag. 1998. German only.)

Argumente und Kritik: Skeptisches Jahrbuch. Rassismus, die Leugnung des Holocaust, AIDS ohne HIV und andere fragwürdige Behauptungen (Ed. with Benno Maidhof-Christig and Lee Traynor; Berlin: IBDK Verlag. 1996. German only.)

Mathemagics (with Art Benjamin; Contemporary Books, 1993)

Teach Your Child Math (with Art Benjamin; Contemporary Books, 1991)

Teach Your Child Science (Contemporary Books, 1989)

Race Across America (WRS Publishing, 1993)

The Woman Cyclist (with Elaine Mariolle; Contemporary Books, 1988)

Cycling: Endurance and Speed (Contemporary Books, 1986)

Sport Cycling (Contemporary Books, 1984)

THE BORDERLANDS OF SCIENCE

Where Sense Meets Nonsense

MICHAEL SHERMER

OXFORD
UNIVERSITY PRESS
2001

OXFORD
UNIVERSITY PRESS

Oxford   New York
Athens   Auckland   Bangkok   Bogotá   Buenos Aires   Calcutta
Cape Town   Chennai   Dar es Salaam   Delhi   Florence   Hong Kong   Istanbul
Karachi   Kuala Lumpur   Madrid   Melbourne   Mexico City   Mumbai
Nairobi   Paris   São Paulo   Shanghai   Singapore   Taipei   Tokyo   Toronto   Warsaw

and associated companies in
Berlin   Ibadan

Copyright © 2001 by Michael Shermer

Author contact:
Skeptics Society
2761 N. Marengo Ave.
Altadena, CA 91001
Phone: 626/794-3119
FAX: 626/794-1301
E-mail: skepticmag@aol.com
WEB Page: www.skeptic.com

Published by Oxford University Press, Inc.
198 Madison Avenue, New York, New York 10016

Oxford is a registered trademark of Oxford University Press

Library of Congress Cataloging-in-Publication Data

Shermer, Michael.
    The borderlands of science : where sense meets nonsense / by Michael Shermer.
        p. cm.
    Includes bibliographical references and index.
    ISBN 0-19-514326-4
    1. Science—Miscellanea.   2. Belief and doubt.   3. Parapsychology and science.
    4. Skepticism.   I. Title.

Q173.S56 2001
500—dc21   00-063673

ISBN 0-19-514326-4
1 3 5 7 9 8 6 4 2

Printed in the United States of
America on acid-free paper

Plac'd in this isthmus of a middle state
A being darkly wise and rudely great
With too much knowledge for the sceptic side,
With too much weakness for the stoic pride,
He hangs between; in doubt to act or rest;
In doubt to deem himself a god or beast;
In doubt his Mind or Body to prefer;
Born but to die, and reas'ning but to err;
. . . .

Created half to rise, and half to fall,
Great lord of all things, yet a prey to all;
Sole judge of Truth, in endless error hurl'd;
The glory, jest and riddle of the world.
—Alexander Pope, "Essay on Man"

CONTENTS

Part III: Borderlands History

INTRODUCTION
BLURRY LINES
AND FUZZY SETS
The Boundary Detection Problem in the Borderlands of Science

IN LATE SEPTEMBER OF 1999 I went to Stonehenge, the magnificent Druidical stones laid out in the countryside of southern England. Well, sort of. I traveled to Stonehenge . . . in my mind . . . as part of an experiment on a phenomenon called "remote viewing," the belief that one can, in the words of my remote viewing instructor—Dr. Wayne Carr of the Western Institute of Remote Viewing in Reno, Nevada—"experience, feel, see, and describe, detailed and accurate information on any event, person, being, place, process or object that has ever existed, does exist, or will exist." According to Carr:

> Historically, remote viewing was developed at Stanford Research Institute for the army and the Defense Intelligence Agency. It was used in a secret espionage program for twenty years. This is why few people had heard of remote viewing until about three years ago when the government went public on "Nightline." Protocols have now been refined to allow trained remote viewers consistent detailed accuracy. Remote viewing could be considered a distant cousin to some other psychic disciplines, with the main difference being the extremely high and consistent accuracy. A single remote viewing usually takes about an hour or more. During this time, one can become "bilocated" and have strong "target" contact with all of one's senses. A target can be in the past, present or future. This is not some kind of "psychic network"; rather it is a serious scientific technique for exploration.[1]

Since I am a social scientist and historian of science who studies fringe and borderland claims to determine if they are scientific, pseudoscientific, or non-scientific, and I had seen the *Nightline* report on the CIA's twenty-year experimental program in remote viewing (originally set up to discover, for example, the location of hidden Soviet military bases), I wanted to try it myself. I signed up for Dr. Carr's weekend seminar in remote viewing—touted as "Professional Targeting Services, Corporate Business & Private Consulting & Target Contracting, Guaranteed Quality"—and joined a dozen other hopefuls who were going to be taught how to discover, according to the brochure, "the location and condition of a missing person, child or object, future potential markets in a certain area, the cause of an event or disaster, possible medical diagnostic considerations, personal family history and events, unsolved cases or mysteries, the effects of a personal decision, the location of mineral or petroleum deposits," and much more.[2]

As its name implies, remote viewing involves sitting in a room and "viewing" something remotely, that is, outside of the normal range of one's senses. Some claims for remote viewing's powers are modest, others not so. Science writer Jim Schnabel produced the first full-length volume on remote viewing (by a nonparticipant), tracing the U.S. government's involvement with some of the world's most famous "psychics," including Russell Targ, Hal Puthoff, Uri Geller, Ed Dames, and Joe McMoneagle.[3] Schnabel's tome recounts endless anecdotes, usually confirmed with additional anecdotes by believing eyewitnesses who were themselves in remote viewing, including:

> —A part-time Christmas-tree salesman remote-viewed his way into the heart of a super-secret National Security Agency installation buried in the West Virginia mountains.
> —The same psychic described previously unknown details of a high-tech Soviet military research facility—details that were later confirmed by spy satellite.
> —An Army remote viewer was the first in the U.S. intelligence community to describe the Soviets' new *Typhoon*-class submarine—while it was still indoors, under construction.
> —A woman in Ohio psychically found the location of a crashed Soviet bomber in the jungles of Zaire, helping a CIA team to recover the wreckage before the Soviets got there—and earning praise from President Jimmy Carter: "She went into a trace. And while she was in the trance, she gave us some latitude and longitude figures. We focused our satellite cameras on that point, and the plane was there."[4]

Shortly we shall examine the problem with remote viewing protocols that lead to the mistaken belief that the number of "hits" by remote viewers is above chance. Experimental psychologist Ray Hyman, professionally trained and experienced in proper experimenter protocols and the only outside observer allowed to review the raw data from the CIA's remote viewing experiments, concluded rather definitively: "By both scientific and parapsychological standards . . . the case for remote viewing is not just very weak, but virtually nonexistent. It seems that the preeminent position that remote viewing occupies in the minds of many proponents results from the highly exaggerated claims made for the early experiments, as well as the subjectively compelling, but illusory correspondence that experimenters and participants find between components of the descriptions and the target sites."[5] And as we shall see, these claims for the power of remote viewing are conservative in comparison to what has been claimed for it in recent years, even compared to the following observation by one of the government's top remote viewers, Fern Gauvin:

> The biggest concern is—will I be invaded by evil spirits? Maybe, but I can protect myself . . . Some other people call it, Okay, "cover yourself with the white light," and so on. All that is good intention. And if I have good intention—I don't care if you [a seductive evil spirit] are a whore on Fourteenth Street, I don't want anything to do with you—then you don't stand a chance, I don't care what the price. It's because I don't want to. I think that goes a long way in this line of work.[6]

Such discussions, many this absurd, went on at taxpayers' expense for twenty years under the cloak of national security. And this is comparatively conservative material. For my weekly radio show *Science Talk* on NPR affiliate KPCC in Southern California, I once spent an hour talking about remote viewing with one of its champions of the 1990s, Courtney Brown, a political science professor at Emory University (although I was not allowed to introduce him as such because of a contractual agreement between Brown and Emory that he not mention his affiliation when discussing remote viewing). For Brown, locating crashed planes and missing persons is child's play. He's after bigger targets that include, according to his 1996 book *Cosmic Voyage: A Scientific Discovery of Extraterrestrials Visiting Earth*, Martians and aliens from other planets, multidimensional beings from other galaxies, spiritual leaders such as Jesus and the Buddha, and even God (who, he says, actually resides within each of us). According to Brown, he has even had conversations with

Jesus about life on earth and the multidimensional life to come. Yet, over and over throughout his books and in my interview with him, Brown insists that he is a scientist and that remote viewing is solid science, as good as anything to be found in the social sciences. In fact, Brown has renamed the phenomenon "Scientific Remote Viewing," or SRV for short, and in his 1999 sequel, *Cosmic Explorers*, he reviews the detailed procedures for proper data collection, identifying target coordinates, SRV protocols, and the classification of categories of remote-viewing data. That claim places this phenomenon squarely in the seat of testable knowledge. And as we shall see, there are some serious flaws in the protocols of remote viewing that lead it to fail all tests.

But failed tests aside, the outrageousness of the claims alone should sound our skeptical alarms. The following passages from *Cosmic Explorers* could have been written by a pulp science fiction writer for 1950s B movies instead of a tenured college professor at a major American university (note the scientistic language and cachet of data-speak):

> Apparently, Buddha and the Galactic Federation are deeply involved in an intense struggle that conveys the sense of a major conflict, perhaps a war. I do not know from the data in this session if the struggle is exactly the same as that associated with the renegade Reptilians, but I suspect that the two conflicts are related.[7]
>
> In my interpretation of these data, it appears that the agenda of the Reptilian extraterrestrials is to use the genetic stock of humanity to create a new race that is partially human and partially Reptilian. There is no indication in the data of this session to suggest what the Reptilians plan on doing with the remainder of humanity.[8]
>
> My reading of these data suggest that the Galactic Federation will defend our right to evolve as a species, to make our own mistakes, and to learn from our own hardships. In essence, their agenda is to leave us alone, to let us find our own way in life. They respect our freedom to learn, to grow, and to err. And I suspect they will be waiting for us with eager anticipation of our abilities to contribute to an expanding Galactic civilization when we once again rise off the surface of this planet, wiser, more loving, and with a deep inner desire to explore, and to serve our gradually maturing universe.[9]

With this background, then, imagine the anticipation I experienced on the eve of trying remote viewing myself. Since we were neophytes, Dr. Carr explained that we could not expect to find, say, the location of Jimmy Hoffa's body or who killed Jon Benet Ramsey our first day, let alone talk to the Buddha. We had to learn the basics first. On the podium at the front of the room Carr had placed in an opaque envelope a photograph of a famous site somewhere in the world. Our task was to remotely view the contents of that envelope. Carr explained that we could do so not just by attempting to view the contents of the envelope in our minds, but to actually go to this place remotely, to "see" it in our "mind's eye."

To do so we began with a series of short remote viewing "templates" that consisted of a list of descriptive terms followed by an "ideogram," or picture of what it is we were viewing. This was not necessarily the target, Carr continued. In fact, it most likely was *not* the target, but with a series of these descriptive lists and ideogrammatic drawings, we would approximate the target and perhaps, eventually, even nail it down precisely. We're beginners, he reminded us. This is a serious science that takes correspondingly serious practice. We began in "Stage 1" with descriptive words. *Primitive Descriptors* include such terms as "hard, soft, semi-hard, semi-soft, wet, mushy," etc. *Intermediate Descriptors* include "natural, man made, biological, movement, energetics," etc. *Advanced Descriptors* include "structure, subjects, dry land, city, motion, mountain, water, wetland, sand, ice, hills," etc. In "Stage 2" we moved to more detailed descriptions (all the while writing them down and making sketches), such as: *Textures*—"smooth, soft, shiny, rough, matted, sharp," etc., *Temps*—"warm, cool, hot, freezing, frigid," etc., *Dimensionals*—"high, low, tall, towering, deep, flat, wide, open, thick, narrow," etc., and *Energetics*—"vibrating, pulsing, humming, vibration, movement, energy, penetrating, emanating, squeezing, pushing, pulling, attracting," etc.[10] Since we were instructed to allow our imaginations to follow the descriptive terms, this last list of dimensional descriptors led me to the remotely viewed object in FIGURE 1.

In my "Session SummaryPage" that followed the pages presented in FIGURE 1 I wrote: "I started off with something sexual and arousing, as if it were two people, but then switched to a statue, guessed *The Kiss,* then at 500 feet above [we were instructed to move around and above our target] it looked like people at a monument of some kind, perhaps a park in London, Hyde Park with statues, or perhaps at a movie theater. Very nebulous."

We continued to refine our remotely viewed targets, and after about an hour of this Carr was ready to reveal the content of the photograph in the envelope. Before he did this, however, he moved about the room, carefully examining each person's numerous sketches and descriptions. Some he responded to quite favorably, others he explained that as beginners we could not all expect to do

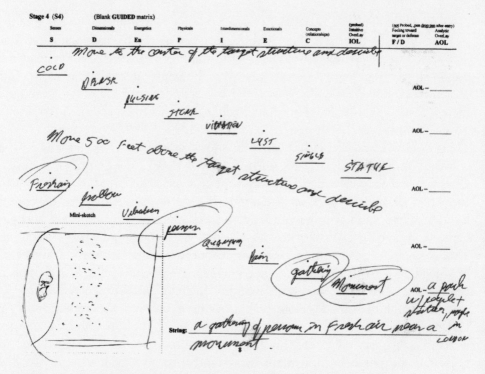

Figure 1. The Author's Remote Viewing Experimental Results

well at first. He seemed especially excited by my drawing and description. Had I mastered remote viewing my first time out?

The target, it turns out, was Stonehenge. I wasn't even close. Or was I? Carr proclaimed that I had great potential as a remote viewer because I had gotten a stone statue in England, which he felt was remarkably similar to Stonehenge. Herein lies the first problem in remote viewing experiments—determining what constitutes a "hit." The answer depends on how much wiggle room is allowed. Operational definitions and advanced selective criteria that are basic protocols in social science research, are all but missing in remote viewing research, or are constructed in such a way as to give the experimenter subjective leeway in determining whether a trial is considered a hit or a miss. Since all of the remote viewing experiments that I know of have been conducted by believers in the phenomenon, this calls into question their protocol criteria.

There was, however, one gentleman in our group whose results needed to be cut no slack at all. He drew a picture of big stones in a circle, and wrote at

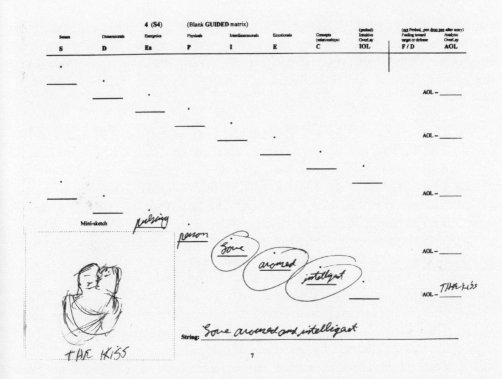

the bottom of his drawing "Stonehenge." A hit! There was no subjective interpretation needed for this one. I was befuddled until I discovered that this man is a good friend of Carr's who had, in fact, driven to San Francisco (where the seminar was being held) from his home in Reno earlier that morning. When Carr later asked me to explain the Stonehenge hit I simply said I thought that Carr probably told him the target ahead of time. Surprisingly, and tellingly, he did not abjure my charge. A real test, I explained, would be if no one knew what the target was and then we tried a remote-viewing experiment. This was the next step. I had brought my own sealed opaque envelope with a photograph of a target, so now we would see how well Carr, his expert remote viewer buddy, and another remote-viewer that Carr described as one of the best in the world, would do under controlled conditions.

To select a "good" target for remote viewing, I was given a list of "detailed instructions for target preparation" that explained "it is important that you exactly follow the following procedure in a NONHURRIED relaxed way: You need your PILE of individual target pages (a target page is a photo and description of an individual target) AND a separate CLEARED TABLE or desk that can REMAIN CLEARED (except for the selected one target page) during the viewing." I carefully followed the nineteen different steps as if they were a ritual of sympathetic magic—an appropriate analogy for this process—and read through the sheet labeled "Ideal Remote Viewing Target Characteristics." These include:

1. They inherently are OF INTEREST and inherently ATTRACT AT-TENTION (not boring) Good: Pyramids at Giza, Old Faithful, etc. Bad: a pair of scissors in Dr. Carr's desk, an eraser.
2. They ALWAYS are NAILED DOWN IN TIME AND SPACE, as much as possible, using TIME AND SPACE QUALIFIERS such as THE PLACE, CITY, HEIGHT, DISTANCE, ACTIVITY, COUN-TRY, PERSON(S), TIME, THE DATE. Good: The largest Pyramid at Giza. Bad: The Egyptian Pyramids.
3. If the target is an EVENT, then it should have "/event" written after it. Good: The First Human Heart transplant, done by . . . at . . . hos-pital/EVENT (date). Bad: The first heart transplant.
4. They are DISCRETE, not vague or open ended. Good: Arch de Tri-umph/Paris/France. Bad: A Roman bridge.
5. You can POINT AT the actual physical target or event. Good: The Empire State Building. Bad: The Stock Market Crash in 1982.
6. They have their OWN DISCRETE BOUNDARIES in time and space. They have more than conceptual boundaries such as state lines or coun-try borders. Good: Alcatraz Island. Bad: The State of Nebraska.
7. There is a good "FIGURE-GROUND" CONTRAST between the tar-get activity or object and the background activity or objects. Good: Mt. Shasta. Bad: The middle of the Pacific Ocean.[11]

This list goes on and on, each step suggesting good and bad targets. Here we see a second major flaw in remote-viewing research—forced choices. Ma-gicians will immediately recognize this for what it is. Many card tricks, for example, involve very careful instructions that force the subject into a situation that either insures the magician will pick the right card, or reduce the choices to a minimum. For example, think of a two-digit number between 50 and 100 where both digits are even numbers (thereby eliminating the 50s, 70s, and 90s and all the odd numbers in the 60s and 80s), like 62 or 82, but not where both digits are the same like 66 or 88 (thereby suggesting that the subject not select *those* numbers, leaving just a few two-digit numbers from which to select). The illusion is that you have made a free choice. The reality is that the magician has made the choice for you. The purpose of this remote-viewing target characteristics list is to reduce the number of potential targets to, basically, famous buildings, monuments, and locations.

To make this an objective test of remote viewing, therefore, I had to get around this trick, which I did by selecting a target unlike anything suggested in the list, but one that elsewhere in Carr's literature he said he had targeted before: galaxies. As I was sitting at my desk thinking about this problem I looked up and noticed on the wall of my office the Hubble Space Telescope's Wide Field and Planetary Camera 2 photograph taken of the tiny slice of sky 1/140th the apparent size of the full moon near the handle of the Big Dipper, that consists of literally thousands of galaxies. Since Carr said that they could remotely view galaxies, and this is one of the most famous and widely publicized photographs of galaxies ever taken (it graces countless magazine and book covers), I reasoned that this should be a fair target.

A third major problem in remote viewing experiments, and one related to the target selection list, is what types of drawings people make. When producing sketches amateur artists use just a handful of design elements—mostly lines and curves—to attempt to depict, however crudely, the object under gaze. Thus, a handful of sketched lines and curves on a page could be interpreted as almost anything, especially when the list of potential targets is limited to buildings, monuments, and natural objects with striking and recognizable features. In other words, there are only so many variations on a theme and with even a modicum of subjective wiggle room, almost any set of lines and curves could be interpreted as a hit.

The test began with Carr's two colleagues (Carr begged off the formal test) working on their drawings and word lists for over an hour. Each had generated over a dozen pages of sketches and descriptions. When both remote viewers were finished Carr demanded to know the target. "No, no," I explained. "The purpose of this test was for *you* to tell *me* what the target is." Carr then stammered through a disclaimer about how hard remote viewing is, how subjective and nebulous it can sometimes be, how this was not a true controlled test, and so forth. "But," I responded, "your friend here just nailed down Stonehenge by drawing, description, and name. No subjectivity there. No waffling. If this really works he should be able to tell me right now what is in that envelope." This was followed by several minutes of speculative fishing through all the different drawings, explaining that the target could be *this*, it could be *that*, etc. As the time passed they were squirming in their seats in what was clearly a state of high anxiety. They again asked me what was in the envelope. Again I responded that the onus was on them to tell me what was in the envelope. This went on for another few minutes until I decided to end the suspense.

"Before I open the envelope let me tell you what you are going to do when I reveal the contents. You are going to look through all those dozens of drawings, select the one that comes closest to what is in this photograph, and announce that you got it." To my utter amazement Carr explained that, indeed, this is how remote viewing experiments work! I explained to him that in science it has to work the other way around. This is a fourth problem in remote viewing research—the confirmation bias and the hindsight bias. In cognitive psychology and critical thinking studies, it is well known that subjects only look for confirmatory evidence and ignore disconfirmatory evidence of their preconceived beliefs, and they look back with hindsight to explain how they arrived at their beliefs in a justificatory mode. This is not allowed in science.

With this brief lesson in the philosophy of science over, I opened the envelope and revealed the target. Without missing a beat Carr immediately riffled through the sheets of paper strewn about the table, pulled out a sketch that was described as a "ferris wheel," and announced that this was, in fact, a galaxy! It was at this point that I knew that remote viewing is not normal science or even borderlands science. It is pseudoscience, which I defined in my book *Why People Believe Weird Things* as "claims presented so that they appear scientific even though they lack supporting evidence and plausibility." How did I determine what constitutes pseudoscience? Through a series of questions that I ask about all claims that I investigate for *Skeptic* magazine, a science publication that I edit, and for a television series on the Fox Family Channel called *Exploring the Unknown*, for which I am a cohost and coproducer, and a segment for which we filmed this remote viewing experiment. In exploring the unknown we often find ourselves in the borderlands of knowledge—in that fuzzy area between orthodoxy and heresy—and thus a consideration of some specific claims can help us learn where to draw the boundary line between science and pseudoscience, or between science and nonscience.

EXPLORING THE UNKNOWN

As even casual purveyors of the little screen know, the Fox television network is not known for sticking closely to a truth-in-advertising policy when it comes to its so-called "reality programming." If their alien autopsy film wasn't farcical enough, they followed that two years later with another special, this one debunking their own autopsy footage, which itself was a bait and switch—the "secrets revealed" was actually on a different alien film altogether, one not even aired in the original show! But then, this is the same network that gave us the world's deadliest animals, the most dangerous car chases, the powers of the paranormal "put to the test" by a boxing commentator, and, as a metahoax, a special about machines that seek revenge on their owners, including an angry automobile that purposely drove its owner off a cliff!

Reality programming, in reality, means low production costs (other people's video footage is vastly cheaper than union camera crews) coupled to good ratings ("if it bleeds, it leads"), resulting in robust profits. Television, to be blunt, is a series of commercials with blank spots in between that have to be filled with programming compelling enough to keep the viewer watching until the next commercial. "Don't go away," "stay tuned," and "when we return" are carefully crafted phrases that mean "don't touch the clicker." Clicker phobia lies hidden just beneath the surface of the television business. No show segment should be longer than seven or eight minutes—the perceived attention span of the American public. Interviews are chopped up into sound bites—nothing more than three to four sentences. Background music must be upbeat. Edit cuts are quick and abbreviated—no long, slow pans across mountain ranges or lakes as one might see in a Ken Burns PBS documentary. "Long" segments—fourteen to fifteen minutes—are chopped up into two-parters where, at the end of the first part, teasers of what is coming up in the second part keeps your fingers away from the clicker.

Television is a business and television executives are in business to make money, plain and simple. It's the American way. So let's not unfairly target Fox. When NBC aired a "documentary" hosted by Charlton Heston in which it was claimed that the Egyptian pyramids were actually built by an ancient civilization some ten thousand years ago, we should not be surprised that not a single archaeologist, scientist, or skeptic of any academic or mainstream credibility appeared on the show to present even an iota of dissent. This is because the show was not a documentary. It was what I call an *entertainmentary*—an entertainment show gussied up to look like a documentary. NBC is not alone. In 1993 CBS aired an entertainmentary produced by Sun International Pictures entitled *The Incredible Discovery of Noah's Ark*. The show's producer, David Balsiger, explained the timing philosophy: "What happens is that we attempt to keep as many interviewees in as possible, so we have to shorten their pieces. Maybe they were speaking for a minute, they get shortened to thirty seconds. A sentence or two is cut off the end or somewhere, not to change their point of view or anything, but to let them make the longest point they are making in a shorter period of time."[12] Had Sun Pictures spent a little more time in actually listening to what the interviewees said, perhaps they might not have been taken in by George Jammal, a Long Beach, California actor who convinced the producers he had a genuine piece of the ark—actually a hunk of wood he knocked out of a railroad tie near his home that he subsequently soaked in teriyaki sauce and other spices on his stove. Any archaeologist would have spotted the hoax in a second, but none were consulted. Balsiger responded to the sting in anger, particularly at the media attention it generated: "There is something wrong with the ethics of the news media when they glorify the acts of humanist hoaxers who intentionally and successfully deceive 40 million TV viewers; and then blame the show producer and CBS for not discovering their elaborate hoax."[13] Elaborate? One would have thought that even without consulting experts they might have spotted Jammal's clues that it was all a setup, such as the names of his phantom assistant "Mr. Asholian," his bogus Polish friend "Vladimir Sobitchsky," and his nonexistent son-in-law "Allis Buls Hitian."[14] As the good book warns, there are none so blind as those who will not see.

Bashing television and bemoaning programming is a favorite pastime of scientists and skeptics, and I've not shied away from launching my fair share of salvos against the little screen. But in the spirit of lighting a candle instead of cursing the dark, since the founding of the Skeptics Society and *Skeptic* magazine in 1992 I have been shopping around a skeptical reality show. To nearly every producer I met on nearly every show I appeared on I talked up my idea of a series in which both the believers' *and* the skeptics' points of view would be presented. In 1994 and 1995, I made several appearances on NBC's daytime paranormal show called *The Other Side* (hosted by an amiable stand-up comedian and one-time minister—such multifarious career combinations are common in the always-uncertain entertainment business), and got to know the producers. A few years later I made a formal pitch to their production company (networks rarely produce their own shows—they almost always purchase them or hire them to be produced by independent production companies of which there are dozens in Southern California), but it didn't fly.

Several years later, however, one of the executives of this particular production company moved to the newly formed Fox Family Channel (Rupert Murdoch, the owner of Fox, purchased the Family Channel from Pat Robertson, and as part of the deal Robertson got to keep his nightly 700 Club show, which, ironically, airs immediately after *Exploring the Unknown*!). This executive liked my show treatment and asked me to pitch it again to Fox Family, which I enthusiastically did. After months of negotiations (television contracts are complex enough to require the services of entertainment attorneys, a legal specialty of sizable proportions in Hollywood), the deal was sealed, and the company that produced over 200 episodes of the paranormal series, *Sightings,* was selected to produce what came to be called *Exploring the Unknown.* (This maximally equivocal name was chosen so as not to tip our hand to either viewers or potential guests as to the skeptical nature of the show—imagine the response of potential guests to a call from a show researcher that begins "Hello, we'd like to know if you would appear on our Fox Family show called *Debunking the Unknown.*")

Working on this show has been a wonderful educational experience, not only for learning how television series are produced, but in the actual investigations we have conducted. The show is a direct extension of the work we do at the Skeptics Society and *Skeptic* magazine, but with a budget of a couple of hundred thousand dollars per episode (normal for cable, cheap by major network standards) we can do a lot more than we have ever been able to afford to do through the society. And we can reach a lot more people. *Skeptic* magazine, for example, is distributed to nearly every bookstore and a majority of newsstands in America and has a respectable circulation of 40,000 people—an order of magnitude larger than most scientific journals and an order of magnitude, or two, smaller than the largest of magazines. And my books, by book publishing standards, have done well. *Why People Believe Weird Things,* for example, sold about 30,000 copies in hardback and is at about 50,000 in paperback at the time of this writing. My publisher, W. H. Freeman, is very pleased with these numbers that are, like *Skeptic,* an order of magnitude larger than most books that are published, and an order of magnitude, or two, smaller than the very best of the best-selling books on the market.

But compare these figures to the ratings of our little television series on an average cable channel. Airing Friday nights at 10 P.M.—not an especially good time slot—we typically get a .7 or .8 share, which translates to 700,000 to 800,000 homes that watch our show on any given week! That's an order of magnitude larger than both my magazine and my books, and these ratings, by television standards, are paltry compared to some shows, such as *Who Wants to be a Millionaire?,* typically watched by over twenty-five million people every night. The first series cycle of seven hours of *Exploring the Unknown* ran seven times, which translates to roughly five million people who watched the show. The simple and powerful fact is, if you want to reach a lot of people with your message you've got to do it through television.

Given this reality, I have worked hard to translate the message from my magazine and books into the most powerful medium of communication in history. What is that message? There is a method known as science that can help us answer questions, solve mysteries, and understand our cosmos, our world, and ourselves. Science cannot solve *all* mysteries (thus, *Solved Mysteries* was rejected as a potential show title), but it can solve a lot more mysteries than most people realize, although most television producers know it. (Yes, most people that work in the television business realize that most of the claims presented on these paranormal shows are utter nonsense. They know but they don't care because they are in the business of selling commercials, not ideas. Those of us in the ideas business must face this reality and work around it.) I am grateful to the good folks at Fox Family for giving us the green light not only to explore the unknown, but to explain it fully and, where appropriate, debunk it thoroughly (although always in a polite manner so as not to embarrass the guests). Interestingly, the researchers and producers for our production company were also glad to have their hands untied and be allowed to actually reveal or explain mysteries. It turns out that most of them have known all along that many of these claims are bunk, but they were restricted by the networks from saying so on the air. (Network executives approve all show ideas before production begins, and often read through, edit, and have final approval over voice-over scripts.) So, for *Exploring the Unknown,* we have been free to tell it like it is, to give the full explanation if we have one, and to say whether something makes sense or is nonsense.

THE BOUNDARY PROBLEM AND ITS FUZZY SOLUTION

Here's the rub: how do we know if a claim is sensical or nonsensical? How do we tell the difference between science and pseudoscience, or between science and nonscience? Can we always and clearly distinguish between reality and fantasy, fact and fiction? The opening line of every episode of *Exploring the Unknown,* dramatically read by actor Mitch Pileggi (who plays FBI Assistant Director Skinner on *X-Files,* a show that itself explores these themes in a dramatic format, albeit with far less skepticism) is: "Things are not always what they seem when you are exploring the unknown." Things are not always what they seem because we do not live in a black and white world of unambiguous yeses and noes. We are faced here with a "boundary problem"—where do we draw the boundary between orthodoxy and heresy, between orthodox science and heretical science, or between science and pseudoscience, science and nonscience, and between science and nonsense?

The boundary is the line of demarcation, or the border to be drawn between these geographies of knowledge, these countries of claims. The problem with this geographical/political analogy is that it does not fully hold. Where rivers and mountain ranges, and deserts and seas help geographers and politicians demarcate (however artificially) the boundaries between geographical areas and countries (necessarily cleanly drawn for legal reasons and sometimes right down the middle of a featureless landscape), knowledge sets are fuzzier and the lines between them are blurry. It is not always, or even usually clear where to draw the line. Whether a particular claim should be put into the set labeled science or pseudoscience will depend on both the claim and the definition of the set. Here fuzzy logic, as opposed to Aristotelian logic, may help us resolve this classic problem for philosophers of science.

Aristotelian logic says that A is A. A cannot be non-A. A male is defined by a set of characteristics—XY chromosomes, a penis and testicles, high levels of testosterone, a deep voice, beard and body hair, and so forth—and thus so defined cannot also be a nonmale. Yet even this classic and simple example runs aground in the fuzzy borderlands between the sets male and nonmale. Granted, most individuals falling into these two sets are clearly and distinctly either male or nonmale (female). But there are individuals who are not clearly in one or the other set and who, in fact, may even be represented by a third set called transgender. There are also hermaphrodites. There are males with an XXY genetic condition (Klinefelter's syndrome) that makes them sterile and significantly more feminine in physical appearance. On the other extreme there are XYY "supermales" who allegedly exhibit high levels of violence and aggression.[15] There are some males whose levels of testosterone are so low that their bodies are soft, their skin smooth and hairless, and their voices effeminate. Correspondingly, there are females whose levels of testosterone are so high that they do not qualify as females as defined by the International Olympic Committee's gender criteria, where a simple chromosomal check for an XX or XY will not suffice for their competitive definitions. (For example, in the bicycle Race Across America, which I cofounded, raced in five times, and directed or codirected for thirteen years, we used the IOC drug lab at UCLA to test for drug and steroid use. One year we had a close call when our female winner tested dangerously close to male levels of testosterone, which would have disqualified her as a female in the race. She was not taking testosterone; her levels were just naturally high.) And these examples only include physical definitions of maleness. There are behavioral examples as well, such as males who cross-dress as females and enjoy playing the role of female more than male. Such social and psychological factors blur the boundaries even more.

The fuzzy logic solution to this problem is to avoid the binary sets altogether and assign subjects a fuzzy fraction. USC engineering professor and fuzzy logic pioneering guru Bart Kosko uses the color of the sky as an example.[16] Aristotelian logic demands that it must be either blue or nonblue, but not both. Yet the sky cannot properly be characterized as either-or. By fuzzy logic reasoning, depending on the time of day and the patch of sky to be evaluated, a fuzzy fraction is a more accurate description. At dawn on the sunrise horizon the sky might be .1 blue and .9 nonblue (or, say, .9 orange). At noon overhead the sky might be .9 blue and .1 nonblue (or, say, .1 orange). At dusk on the sunset horizon the sky might be .2 blue and .8 nonblue (or, say, .8 orange). Likewise, most males could be assigned a fuzzy fraction of, say, .9 or .8. But, depending on the criteria used in a definition of maleness, we all know men who would be better classified as .7 or .6 males, and even a few who would be best described as .2 or .1 males.

When we move away from such simple sets as skies and males and into much more complex and socially influenced phenomena as knowledge claims, the sets overlap considerably more, the borderlands between them are wider and fuzzier, and the boundary lines of demarcation are much more difficult to draw. Fuzzy logic is critical to our understanding of how the world works, particularly in assigning fuzzy fractions not only to the knowledge sets and their inhabitants, but to the degrees of certainty we hold about those individuals and claims. Here we find ourselves in a very familiar area of science known as probabilities and statistics. In the social sciences, for example, we say that we reject the null hypothesis at the .05 level of confidence (where we are 95 percent certain that the effect we found was not due to chance), or at the .01 level of confidence (where we are 99 percent certain), or even at the .0001 level of confidence (where the odds of the effect being due to chance are only one in ten thousand). This is fuzzy logic at its best, and such fuzzy thinking (in the good sense) will help us solve the boundary problem in science.

In my book *Why People Believe Weird Things*, I discussed the problem of defining what constitutes a "weird thing." After all, one individual's weird thing may be another's cherished belief. Defining weirdness is like defining art, or pornography—I know it when I see it from a deep and vast experience of studying it, but a definitive definition is difficult to formulate. I can define neither weirdness nor the boundary between weirdness and nonweirdness with semantic precision in a single definition that would cover all phenomena, because of the variation and complexity of the claims being made and the diversity of the knowledge sets in which the claims may fall. It simply is not fair to the claims or the claimants to shoehorn them to a single categorical definition. Nevertheless, we can get our minds around this problem by examining in detail a number of specific claims and attempt to glean some principles from these examples of what we might look for in attempting to draw the boundary. We did this with remote viewing and it became clear that this claim is definitely not science. We will shortly examine another claim where the boundary is not so clear.

In my book *Denying History* (coauthored with Alex Grobman), I outline some questions one might ask about a claim in an attempt to distinguish between legitimate historical revisionism and illegitimate historical denial (in this case, Holocaust denial). I called this a Denial Detection list.[17] We can use the same list to detect pseudoscience, nonscience, or just plane nonsense. That is, this set of questions, when applied to any knowledge claim, can help us determine where to draw the boundary between fuzzy sets, or what sort of fuzzy fraction to assign a particular claim. In his book *The Demon-Haunted World*, Carl Sagan presented what he called his Baloney Detection Kit.[18] Since in *The Borderlands of Science* I will be dealing with a lot of claims that cannot be fairly categorized as baloney, with deference to Carl let's call this the Boundary Detection Kit.

THE BOUNDARY DETECTION KIT

Like any kit to be properly built and used, one must read the instructions carefully to receive the full benefit of the product. The Boundary Detection Kit requires the user to examine each claim in great detail, and to get to know the subject deeply enough to have a good feel for how to answer these questions. In so doing there is an implicit commitment to be honest and fair, and to not go into the investigation with a prearranged verdict in mind. This is difficult to do, of course, since none of us comes to the data with unvarnished thoughts free of theory. Science is theory laden. We all bring to the table a set of preconceptions born from the paradigms in which we were trained or raised.

Nevertheless, we can rise above our biases, if not to an Archimedean point of unsullied objectivity, at least to a level at which the claimant under investigation might feel he or she got a fair shake. In fact, a principle of fairness in our Boundary Detection Kit might be to ask this question—what I call the *fairness question*—before all others: *If I were to ask the holders of the claim if they feel that they and their beliefs were fairly treated, how would they respond?* Where possible, in fact, why not ask them? I have done so on a number of occasions and to my considerable surprise I discovered that I had not been fair in my analysis, particularly in truncating someone's beliefs to a handful of simplified tenets that could be more easily analyzed (and, usually, debunked). This is sometimes called the "straw man" fallacy in logic, where one sets up a straw man that can be easily toppled but does not represent anyone's actual position. I find that I learn a lot more in the process when I bear in mind the fairness question. In many cases questioning the belief holder is not practical, but the fairness question still works as a hypothetical standard toward which to aim.

Given these caveats, here are ten useful questions to ask in determining the validity of a claim:

1. *How reliable is the source of the claim?* People like Holocaust denier David Irving appear quite reliable as they cite facts and figures; but often, when examined more closely, these details are distorted, taken out of context, or occasionally even fabricated. Scientists are usually reliable. Pseudoscientists are often not reliable. This is a matter of degree, of course, since everyone makes mistakes. As Daniel Kevles showed so effectively in his book *The Baltimore Affair,* in investigating possible scientific fraud there is a boundary problem in detecting a fraudulent signal within the background noise of mistakes and sloppiness that is a normal part of the scientific process.[19] The investigation of a particular set of research notes of Thereza Imanishi-Kari (a collaborator of Nobel laureate David Baltimore) by an independent committee established by Congress to investigate potential fraud, revealed a surprising number of mistakes. But as a historian of science Kevles knows that, in reality, science is messier than most outsiders realize. On top of this, research in molecular biology is far more complex than, say, particle physics. Molecular biological experiments are complicated by the fact that individual cells and viruses are far more variable than, say, hydrogen atoms. The question then becomes: can a distinction be made between intentional and unintentional distortion of the data and interpretations? This was, in fact, a central point of discussion in the famous Holocaust denial trial of early 2000, where Deborah Lipstadt's attorneys and expert witnesses attempted to show that the errors and omissions in David Irving's numerous books on the Nazis and World War II were not just the product of normal scholarship sloppiness, but the intentional distortion of the historical record by Irving. What they showed (and the judge agreed in ruling in favor of Lipstadt) was that Irving's mistakes were almost always in the direction of exonerating Hitler and the Nazis.

2. *Does this source often make similar claims?* Extremists, deniers, and pseu-

doscientists have a habit of going well beyond the facts, so when one individual makes numerous such claims it is a sign that they are more than just revisionists or iconoclasts. Again, this is a matter of quantitative scaling, since some great thinkers often go beyond the data in their creative speculations. Cornell scientist Thomas Gold is notorious for his radical ideas, but he has been right often enough that other scientists listen to what he has to say, and those same scientists are also testing these ideas for their validity. Gold's book, *The Deep Hot Biosphere*, for example, proposes the heretical idea that oil is not a fossil fuel at all, but the by-product of a massive subterranean colony of bacteria living in rocks.[20] Hardly any earth scientists I have spoken with take this thesis seriously, yet they do not consider Gold a crank. Why? Because he plays by the rules of the game of science. What we are looking for here is a pattern of fringe thinking that consistently ignores or distorts data not for creative purposes, but for ideological agendas.

3. *Have the claims been verified by another source?* Typically, nonscientists and pseudoscientists will make statements that are unverified, or verified by a source within their own belief circle. We must ask who is checking the claims, and even who is checking the checkers. The biggest problem with the cold fusion debacle, for example, was not that Stanley Pons and Martin Fleischman were wrong; it was that they announced their spectacular discovery before it was verified by other laboratories (at a press conference no less), and, worse, when cold fusion was not verified anywhere they continued to cling to their belief in the phenomenon despite the lack of evidence. Pons and Fleischman abandoned the rules of science, and in the process their science became their faith. Science writer Gary Taubes called this "bad science."[21] Physicist Robert Park calls it "voodoo science."[22] By whatever name, outside verification is crucial to good science.

4. *How does this fit with what we know about the world and how it works?* An extraordinary claim must be placed into a larger context to see how and where it fits. When deniers construct elaborate conspiracy theories about how the Jews have concocted the Holocaust story in order to extract reparations from Germany and support for Israel out of America, they are naive or deceptive about how modern political systems work. German reparations were calculated based on survivors, not victims; and America supports Israel primarily for selfish economic and political reasons, not out of altruism, guilt, or sympathy.[23] When pseudoarchaeologists claim that the Pyramids and the Sphinx were built over 10,000 years ago by an advanced race of humans (because the Egyptians could not have moved those heavy stone blocks and because the Sphinx shows signs

of water weathering that could not have happened after the end of the last ice age), they are not presenting any context for that earlier civilization.[24] Where are the rest of the artifacts of those people? Where are their works of art, their weapons, their clothing, their tools, their trash? This is simply not how history or archaeology works.[25]

5. *Has anyone, including and especially the claimant, gone out of the way to disprove the claim, or has only confirmatory evidence been sought?* This is the confirmation bias, or the tendency to seek confirmatory evidence and reject or ignore disconfirmatory evidence.[26] The confirmation bias is powerful and pervasive, and is almost impossible for any of us to avoid. It is why the methods of science that emphasize checking and rechecking, verification and replication, and especially attempts to falsify a claim, are so critical. David Irving's books are classic examples of an ideology in search of facts. When it comes to the Holocaust he rarely attempts to falsify or disprove his interpretations (although he does do so with great alacrity with most other aspects of the war). Disconfirmatory evidence for the Holocaust story are eagerly embraced (e.g., Nazi survivors who deny it, trivial anomalies in the physical evidence), whereas the disconfirming evidence that abounds for most of his claims are adroitly evaded. In like manner, there is so much evidence against cold fusion that all but a handful of die-hard physicists, chemists, and hopelessly optimistic futurists long ago gave up conducting further research, yet the purveyors of "infinite energy" (there is even a magazine of this title) cling to the slimmest of experimental results and blithely sweep the disconfirming evidence under the rug of conspiracy theories where, for example, oil and electrical conglomerates are said to be preventing the positive evidence from reaching the American public.[27]

6. *In the absence of clearly defined proof, does the preponderance of evidence converge to the claimant's conclusion, or a different one?* Deniers do not look for evidence that converges to a conclusion; they look for evidence that fits their ideology. In examining the various eyewitness accounts of the gassing of prisoners at Auschwitz, for example, a consistent core of a story develops among them, to the point where we now have a fairly good understanding of what happened. Deniers, on the other hand, take the minor discrepancies in the eyewitness stories and treat these as anomalies that disconfirm the theory. To the contrary, and it seems counterintuitive at first, these variations in minutia *confirm* the theory by virtue of the fact that no one remembers the details of the past perfectly and, of course, specific events are similar only in generalities, not in specifics, which will vary depending on conditions. UFOlogists suffer the same fallacy in their continued focus on a handful of unexplained (or poorly ex-

plained) atmospheric anomalies and visual misperceptions by uninformed eye-witnesses, while conveniently ignoring the fact that the vast majority (I estimate 90 to 95 percent) of UFO sightings are fully explicable with prosaic answers.[28]

7. *Is the claimant employing the accepted rules of reason and tools of research, or have these been abandoned in favor of others that lead to the desired conclusion?* Most deniers do not even *know* the accepted rules of scholarship, let alone employ them fairly. But those who do know, or should know—like Mark Weber, Robert Faurisson, and David Irving—seemingly abandon them in the service of their ideologies. Here I am not just talking about citing sources in articles published in scholarly looking publications like the *Journal of Historical Review*, or thick books with dozens of pages of references in the bibliography. I am talking about the *honest* employment of these tools where, in the quiet solitude of examining a particular document or translating a certain word or phrase, one has done one's best to consider the historical content and context. Creationists—whom I prefer to call evolution deniers—are especially subject to this problem, along with a lack of convergent thinking. Creationists (mainly the young-earth creationists) do not study the history of life. In fact, they have no interest in the history of life whatsoever since they already know the history as it is laid down in the book of Genesis. No one fossil, no one piece of biological or paleontological evidence has "evolution" written on it; instead there is a convergence of evidence from tens of thousands of evidentiary bits that adds up to a story of the evolution of life. Creationists must not only ignore this convergence, they have to abandon the rules of science, which isn't difficult for them since most of them, in fact, are not practicing scientists. The only reason creationists read scientific journals at all is to either find flaws in the theory of evolution or to find ways to fit scientific ideas into their religious doctrines.[29]

8. *Has the claimant provided a different explanation for the observed phenomena, or is it strictly a process of denying the existing explanation?* Deniers usually have no new theory of history to offer, but concentrate instead on knocking down the accepted doctrines of the field. This is a classic debate strategy—criticize your opponent and never affirm what you believe in order to avoid criticism. But this stratagem is unacceptable in science and scientific history. Revisionism may involve legitimate critiques of the existing paradigm, or offer a replacement with a new paradigm, but denial rarely amounts to more than simply attacks on the status quo. Creationists' only "theory" to replace evolution is "God did it."[30] Proponents of the Pyramids as being built by pre-Egyptians offer no evidence of just who these people are, and instead just pick at anomalies in the work of Egyptian archaeologists. Critics of the Big Bang ignore the convergence of

evidence of this cosmological model, focus on the few flaws in the accepted model, and have yet to offer a viable cosmological alternative that carries a preponderance of evidence in favor of it.

9. *If the claimant has proffered a new explanation, does it account for as many phenomena as the old explanation?* Occasionally new theories of history are offered (e.g., extreme Afrocentrists and radical Feminists), but rarely do these theories account for as much of the past as the one they hope to replace. It is in these details of the past that disconfirming evidence can be found in the form of unexplained events. If the Holocaust did not occur, then what happened to the millions of Jews unaccounted for after the war? If the Holocaust did not happen, then how do deniers explain all those references to the "ausrotten" (extermination) of the Jews? They do not explain them. They ignore them, rationalize them, or deny them. Similarly, the HIV-AIDS skeptics argue that lifestyle (drug use or promiscuity, coupled to a correlation with a naturally-weakened immune system), not HIV, causes AIDS. Yet, to make this argument they must ignore the convergence of evidence in support of HIV as the causal vector in AIDS, and simultaneously ignore such blatant evidence as the significant correlation between the rise in AIDS among hemophiliacs just after HIV was inadvertently introduced into the blood supply. On top of this, their alternative theory does not explain nearly as much of the data as the HIV theory.[31]

10. *Do the claimants' personal beliefs and biases drive the conclusions, or vice versa?* All of us are biased. All scientists and historians hold social, political, and ideological beliefs that could potentially slant their interpretations of the data. The question then becomes: how do those biases and beliefs affect the research? It is true that even the most well-intentioned scientists and historians may find themselves searching for facts to fit their preconceptions. But at some point, usually during the peer-review system (either informally, when one finds colleagues to read a manuscript before publication submission, or formally when the manuscript is read and critiqued by colleagues, or publicly after publication), such biases and beliefs are rooted out, or the paper or book is rejected for publication. This is why one should not work in a vacuum. Intellect stumbles and falters without critical feedback. If you don't catch your biases in your research, someone else will.

With this Boundary Detection Kit we can expand the fuzzy logic heuristic into three sets that I will call *normal science, borderlands science,* and *nonscience* — a trinary system instead of the restrictive two-set binary system. Here are some examples from my experience of asking these questions in the process of studying in considerable detail a number of claims that fuzzily fall into one of these three categories, along with my own subjectively assigned fuzzy fractions (.9 highest, .1 lowest, in relation to their level of scientific validity).

Normal science. On the science side of the boundary:

—Heliocentrism, .9
—Evolution, .9
—Quantum mechanics, .9
—Big Bang cosmology, .9
—Plate tectonics, .9
—Neurophysiology of brain functions, .8
—Punctuated equilibrium, .7
—Sociobiology/evolutionary psychology, .5
—Chaos and Complexity theory, .4
—Intelligence and intelligence testing, .3

Nonscience. On the nonscience, pseudoscience, or nonsense side of the boundary:

—Creationism, .1
—Holocaust revisionism, .1
—Remote viewing, .1
—Astrology, .1
—Bible Code, .1
—alien abductions, .1
—Big Foot, .1
—UFOs, .1
—Freudian psychoanalytic theory, .1
—Recovered memories, .1

Borderlands science. In the borderlands between normal science and nonscience:

—Superstring theory, .7
—Inflationary cosmology, .6
—Theories of consciousness, .5
—Grand theories of economics (objectivism, socialism, etc.), .5
—SETI, .5
—Hypnosis, .5
—Chiropractic, .4

—Acupuncture, .3
—Cryonics, .2
—Omega Point Theory, .1

Since these categories and fractional evaluations are fuzzy it is possible for them to be moved and reevaluated with changing evidence. Indeed, all of the normal science claims were at one time in either the nonscience or borderlands science categories. How they moved from nonscience to borderlands science, or from the borderlands to normal science (or how some normal science claims slipped back into the borderlands or even into nonscience), is one of the most important aspects of the study of the history and philosophy of science.

SETI, or the Search for Extraterrestrial Intelligence, for example, is not pseudoscience because it is not claiming to have found anything (or anyone) yet, is conducted by professional scientists who publish their findings in peer-reviewed journals, it polices its own claims and does not hesitate to debunk the occasional signals found in the data, and it fits well within our understanding of the history and structure of the cosmos and the evolution of life. But SETI is not normal science either because its central theme has yet to surface as reality. Thus far no aliens have phoned in and, as much as I support the search, this still seems to me to belong in the borderlands. UFOlogy, by contrast, is non-science (and sometimes pseudoscience) pure and simple. Proponents do not play by the rules of science, do not publish in peer-reviewed journals, ignore the 90–95 percent of sightings that are fully explicable, focus on anomalies, are not self-policing, and depend heavily on conspiratorial theorizing about government cover-ups, hidden spacecraft, and aliens holed up in Nevada caves.

Likewise, superstring theory and inflationary cosmology are at the top of borderlands science, soon to be either bumped up into full-scale normal science or abandoned altogether, depending on the evidence that is now starting to come in for these previously untested ideas. What makes them borderlands science instead of pseudoscience or nonscience is that the practitioners in the field are professional scientists who publish in peer-reviewed journals and are trying to discover ways to test their theories. By contrast, creationists who devise cosmologies that they think will best fit the book of Genesis are typically not professional scientists, do not publish in peer-reviewed journals, and have no interest in testing their theories, except against what they believe to be the divine words of God.

Theories of consciousness are borderlands science, whereas psychoanalytic theories are pseudoscience, because the former are being tested and are grounded in sound facts of neurophysiology, whereas the latter have been tested, have failed the tests again and again, and are grounded in discredited nineteenth-century theories of the mind. Similarly, recovered memory theory is bunk because we now understand that memory is not like a videotape that one can rewind and play back, and that the very process of "recovering" a memory itself contaminates it. But hypnosis, by contrast, is tapping into something else in the brain, and there may very well be sound scientific evidence in support of some of its claims, so to that end we will wrap up this treatise on blurry lines and fuzzy boundaries by exploring in detail this borderlands science.

EXPLORING THE BORDERLANDS

Academic philosophy of science often finds itself bogged down in the thickets of symbolic logic, hypothetical scenarios, and theoretical speculations that lack correspondence to the real world. This is why I provided brief examples for each of the ten questions in the Boundary Detection Kit, and offered specific cases of normal science, borderlands science, and nonscience. To take one of these examples in detail, I recount an investigation we conducted for *Exploring the Unknown*, as further consideration of the boundary problem in science.

On Saturday, May 13, 2000, I was hypnotized by James Mapes, hypnotherapist and motivational speaker, for an episode segment on hypnosis. The theoretical question at hand is this: is hypnosis an altered state of consciousness, or is it nothing more than fantasy role-playing on the part of the subject in tacit agreement with the hypnotist? This is borderlands science because, on the one hand, we have some remarkable experimental results that suggest an altered state of consciousness is real, yet, on the other hand, despite over a century of research on the subject, scientists have been unable to agree on what is really going on in the hypnotic trance. Hypnosis skeptics argue that there is nothing the hypnotized subject does in a hypnotic trance that an unhypnotized subject could not do through either outright faking or (more likely) through an intense state of role-playing and fantasy enactment directed by the hypnotist.[32] In other words, an actor could duplicate anything an allegedly hypnotized subject could do such that an outside observer would be unable to distinguish between the two. The stage magician Kreskin, in fact, has a standing offer of $100,000 to anyone who could pass this test. No one has collected thus far.

Hypnosis believers, by contrast, point to research of Stanford University experimental psychologist Ernst Hilgard, and his discovery of the so-called "hidden observer." Hilgard's experimental protocol was to dip subjects' arms into a bucket of ice-cold water—water so cold that within minutes subjects reported high levels of pain. When hypnotized, the subjects were told that the water would cause no discomfort, and, in this state, they indeed reported low levels of pain. But when these same subjects were later asked to evaluate the level of pain they experienced when in the earlier hypnotized state, they reported the same levels of pain as the first groups of nonhypnotized subjects. In other words, during the hypnotic state in which one part of their brain reported a low level of pain, another part recorded the higher level of pain. That part of the brain Hilgard called the "hidden observer." This hidden observer is dissociated from the other part of the brain that is in the hypnotic altered state, and thus this is known as a dissociation theory of hypnosis, which Hilgard describes as a "multiplicity of functional systems that are hierarchically organized but can become dissociated from one another."[33]

Critics argued that Hilgard instructed his subjects to create a "hidden observer," itself a concept highly metaphorical, like the premodern notion of the homunculus—the little man inside the sperm cell that simply unfolds into a full-grown man. Is the hidden observer nothing more than a nonexistent homunculus? Here is what Hilgard told his subjects:

> When I place my hand on your shoulder (after you are hypnotized) I shall be able to talk to a hidden part of you that knows things are going on in your body, things that are unknown to the part of you to which I am now talking. The part to which I am now talking will not know what you are telling me or even that you are talking. . . . You will remember that there is a part of you that knows many things that are going on that may be hidden from either your normal consciousness or the hypnotized part of you.[34]

Did Hilgard simply invent the "hidden observer," then plant that idea into subjects' minds, who then obliged him accordingly? Maybe, but it seems unlikely—impossible more like it—for a subject to consciously create a dissociated state of mind. But even if this were possible, does that not make the case for the "hidden observer" as something real? Shortly before we filmed our piece on hypnosis, I attended the Western Psychological Association conference in Portland, Oregon, at which one of the world's leading neuroscientists spoke, Dr. Richard Thompson from the University of Southern California. Thompson arguably knows as much as or more about the brain than anyone in the world, so I presented this problem to him. His lecture, by chance, was on breakthroughs in brain studies as a result of brain imaging techniques (where different parts of the brain "light up" when active). To my delight Thompson showed how Hilgard's "hidden observer" experiments were replicated using neuroimaging techniques, in which it was demonstrated that the part of the brain active in nonhypnotized subjects experiencing pain is the same part of the brain active in hypnotized subjects who report feeling no pain, but in fact, later report the same levels of pain. "*There* is the hidden observer," Thompson dramatically proclaimed. At last, a neurological correlation for Hilgard's metaphorical concept. The hidden observer would appear to be real and the dissociation theory of hypnosis is further supported by experimental evidence.

With this theoretical debate in mind we gathered about forty people at a studio in Glendale, California for an afternoon of hypnosis. This was not the first time I had experienced hypnosis. In 1982, I had gone through a series of sessions with a trained hypnotherapist name Gina Kuras, a friend of mine from graduate school days in an experimental psychology program at California State University, Fullerton. Gina now made a living, in part, from hypnotherapy, and I was preparing for the first ever 3,000-mile, nonstop transcontinental bicycle Race Across America. Four of us were to compete around the clock, riding 24-hours a day for ten straight days, with sleep breaks and rest stops optional. Since ultra-marathon cycling is clearly more than just a physical sport, I figured I would need all the psychological help I could get. So, in addition to training over 500 miles a week, utilizing a specialized diet, and trying out any and all alternative new age therapies for health and fitness, I went through a series of hypnosis sessions with Gina to learn to control pain and stay focused.

I wasn't very good at it at first, but as all hypnotists know, some people are highly suggestible and make excellent subjects, while others are less suggestible and have a difficult time slipping into the altered state. For whatever reason, I fall into this latter category—far too self-analytical and aware of my surroundings to let myself slip into the psychological bonds with the hypnotist. But after several sessions I did get into it and was able to be hypnotized. My final session before the race, in fact, was filmed by ABC's *Wide World of Sports* for an "up close and personal" they were preparing on each of the riders as part of their coverage of the event. I was so "under" that I gave Gina a scare when I did not come out of the hypnotic trance upon her usual command of "3, 2, 1, awake." For a moment she feared she might have shut down my brain permanently, and on national television no less!

Twenty years later I was uncertain whether I would be able to get back into the state of hypnosis, or if my logical, scientific state of mind would cause me to be too aware of my external surroundings and internal states of mind to let myself go. That was, in fact, the case. I came close to being hypnotized and tried vigilantly to let myself go, following Mapes's suggestions as far as I could take them, but I don't think I was hypnotized. By contrast, however, there were half a dozen of our forty subjects who were most definitely in an altered state. While I am quite skeptical of some of the tasks Mapes put them through, such as age regression (which does strike me as nothing more than fantasy role-playing), his most dramatic demonstration was to plant the suggestion that a certain single-digit number does not exist, such that when the subjects counted their fingers all of them skipped that number and ended with their pinkie as finger number eleven. The hypnotically-implanted suggestion caused a number to disappear for these subjects. Yet when brought out of the hypnotic altered state of consciousness, every one of them not only remembered the number, they recalled that they could not previously say the number and that it was, like the tip-of-the-tongue phenomenon, a most peculiar and frustrating experience. In other words, the hidden observer knew the number and was aware of the other part of the brain's inability to say it, but the dissociated brain module in the hypnotized state was unable to do so.

So intrigued was I by this that we dismissed all of the subjects but one, a young woman named Jocelyn who works as a legal secretary in a Los Angeles-based law firm. Jocelyn seemed especially suggestible and emotionally involved in the hypnotic session, so Mapes and I sat back down with her for a private hypnosis session, all taped for the show. Even though over an hour had passed by the time the crew relit the set for this segment, Mapes was able to command her back into a hypnotic state in seconds with a touch to the back of her neck and the command "sleep." I instructed Mapes to suggest to Jocelyn that she would be unable to say the number eight. He did so, and when I asked Jocelyn to count on her fingers to ten she obliged us as before (but with a different number) with the count "1, 2, 3, 4, 5, 6, 7, 9, 10, 11," followed by a puzzled look at her "eleventh" finger. I asked her how many fingers she has. "Ten," she answered confidently. I asked her to count again, this time starting on the other hand. Same result, followed by the same confused look. I then explained to Jocelyn about the "hidden observer" and that a part of her brain did, in fact, know the number eight. I repeated the number over and over. I explained that the number eight comes after seven and before nine. I told her that Mapes had simply planted that suggestion, and that she was now to override that sugges-tion at my command and count straight through to ten. So she did: "1, 2, 3, 4, 5, 6, 7, 9, 10, 11." I couldn't believe it. I had Mapes bring her back out of the hypnotic state and asked Jocelyn to count to ten, which she did flawlessly. Did she remember that Mapes had told her she could not say the number eight? She did. Her hidden observer knew all along.

Still skeptical, I tried another experiment. I had Mapes give Jocelyn the hypnotic suggestion that she could not say the color "black" such that the color and concept "black" does not exist for her. Now awake and hypnotized, I asked Jocelyn the color of my black shoes. No answer. I asked her the color of the black microphone I was holding in front of her. No answer. I asked her to name a number of other colors. No problem. Once again I explained to her about the "hidden observer" that actually knew the color, and that she was to override Mapes's suggestion and now tell me the color of the microphone. No answer (and a very puzzled look). I then spelled out "b-l-u-e" and asked Jocelyn to tell me what it spells. "Blue," she said confidently. "Now say this word: 'b-l-a-c-k'," I commanded. I could see her lips start to move and her mouth trying to form the word, but she could not. " 'B-l-a-c-k' spells 'black'," I said. " 'Bl-ack'," I parsed my lips slowly and with emphasis on each syllable. "Say it, 'bl-ack'." She couldn't. "Try 'bl-ue'," I suggested. "Blue," she immediately answered. "Good, now listen, 'b-l-a-c-k' spells 'black,' or 'bl-ack,' 'black.' Say it. 'Bl-ack', 'black'." All I got was a puzzled stare.

Back down for one more trial. This time I instructed Mapes to tell Jocelyn that the back of my right hand was red hot like a stove. I first asked Jocelyn to touch my left hand. No problem. Then the right. She got her fingers to within about an inch of my hand when she suddenly pulled back with a frightened look on her face. "What's the matter?" I inquired. "Your hand is really hot." "No it's not," I explained as I freely touched the back of my right hand with my left. "As you can see I can touch my hand with no problem. And so can you. Mapes simply told you that my hand is hot, but as you can see it's not hot. So, go ahead, touch my hand." Again Jocelyn reached out to within an inch of my hand before pulling back, eyes saucer-wide and face scrunched up in fear. I grabbed her hand, pulled it toward my hand, and commanded "Jocelyn, listen. My hand is not hot. Your hand will not be burned." As I got her fingers to within an inch of my hand she yanked hers back in horror and shot me a look as if I had just physically assaulted her.

When the session was over Jocelyn, now alert, conscious, and not in a hypnotic state, recalled in vivid detail her inability to say the number "eight," the color "black," and to touch my hand. Her hidden observer was well aware of everything that had transpired, yet in this dissociated condition it could do nothing about it. To one part of the brain, the fantasy was more real than the reality is for the hidden observer. Why should this be? No one knows. How does this work? No one knows. What is the neurophysiology of hypnosis? No one knows. This is why hypnosis is a borderlands science. Clearly there is something going on here that demands an explanation. This is neither pseudoscience nor nonscience, and it certainly is not nonsense. But what is it? We don't know.

Hypnosis is emblematic of a larger borderlands science problem of consciousness itself. What does it mean to be conscious, as opposed to unconscious? Where does the conscious "self" go when the brain goes into an unconscious state like sleep, or an altered state of consciousness like hypnosis? No one knows. The brain consists of a number of modules specifically adapted for generalized functions such as language, vision, hearing, and balance, and a number of specific functions such as speech patterns, face recognition, motion detection, and so forth. Yet none of us "feels" like we consist of a myriad of modules. We all have a sense of self, a *single* sense of self. An I. A me. Where is this self located? What brain module coordinates all the other modules to create a sense of a single self? I presented these questions to the neuroscientist Richard Thompson. His response: "No one knows."

This is borderlands science at its best, and neuroscientists hope that sometime in the twenty-first century this great problem will be solved and become a part of normal science.

THE CROOKED TIMBER OF SCIENCE

In the early nineteenth century the German philosopher Immanuel Kant made this observation of history and the human condition: "Out of the crooked timber of humanity no straight thing was ever made." The substrate of *The Borderlands of Science: Where Orthodoxy Meets Heresy* is how human emotions, biases, preferences, and especially culture shape the process of how we explore our world (science), our past (history), and ourselves (biography), and how even though scientists and the methods of science themselves are inexorably intertwined with their social and cultural surroundings, we still have before us the best method ever devised for understanding reality. Let us use science, whether in its normal or borderlands mode, to maximize our knowledge and wisdom.

This book is about the fuzzy borderlands of science, and throughout we shall be exploring the boundary problem between orthodoxy and heresy in science in general, and between normal science and nonscience, revolutionary science, radical science, pseudoscience, protoscience, and nonsense in particular. In **Part I: Borderlands Theories**, we begin in **chapter 1** with a discussion of what I call our "knowledge filters"—the lenses through which we look at the world and how the color of those lenses very much influences what we think we see. Despite the limitations our knowledge filters set for us, however, I show how and why science is still the best method we have for understanding our world, our past, and ourselves, and that reality must take precedence in the search for truth. **Chapter 2** reviews in detail a number of "theories of everything," as I call these ideas generated by people on the margins of science that attempt to produce a unifying theory that explains the complexity of the world in a single principle. I start with theories of everything in 1950 as described by Martin Gardner in his classic work *In the Name of Science*, then compare these half a century later to theories of everything in 2000, to show that the theories—mostly in the fuzzy sets of pseudoscience or borderlands science—have changed in their particulars, but not in their general goal of explaining everything. In **chapter 3** we consider cloning and genetic engineering as a test of the moral limits set on science by society, and see what happens when a borderlands science, like genetics, mutates into a normal science, like genetic engineering, society lays down restrictions based on what it considers "normal." Once again we see how and why science does not and should not operate in a cultural vacuum. **Chapter 4** touches on one of the most sensitive of all borderlands sciences—racial differences and what they really mean—as a splendid example of how complex biological and social phenomena can so parsimoniously (and unjustly) be reduced to a handful of principles (and, occasionally, social policy). The science (and sometimes pseudoscience) of black-white differences is very much culturally bound and time-dependent to a particular culture and era. **Chapter 5** considers a history and sociology of the development of the theory of punctuated equilibrium which, at the time it was presented in the early 1970s, was considered a radical (even heretical) revision of Darwinian gradualism as the best description of the evolution of life on earth. Is punctuated equilibrium a new paradigm? Was there a paradigm shift from gradualism to punctuated equilibrium? Has punctuated equilibrium made the transition from borderlands science to normal science?

In **Part II: Borderlands People**, we begin in **chapter 6** with a deep look into the sociology and psychology behind the resistance to the first great revolution in science—the Copernican revolution—and why social and psychological forces had as much to do with its rejection and eventual acceptance as did evidence. The Copernican revolution is the quintessential case study in the shift from protoscience to borderlands science to normal science, and how this shift occurred because of both scientific and cultural factors, which are especially illuminated through Frank Sulloway's social science model of scientific orthodoxy and heresy. **Chapters 7 and 8** present a biographical analysis of one of the great borderlands explorers of the nineteenth century, Alfred Russel Wallace. In Wallace we see the boundaries and limits of science in particular and the borderlands of knowledge in general, as he is not only the codiscoverer of the primary mechanism of evolution—natural selection—and the creator of the entire branch of normal science called biogeography, but he also championed such pseudoscientific and nonscientific causes as spiritualism, land nationalism, antivivisection, and many others. **Chapter 9** turns to the lives of two of the greatest and mythic heroes of the modern world—Charles Darwin and Sigmund Freud—by comparing how the two men differed in their perceptions as to whether they were revolutionaries or not, and how and why one has stood the test of time and come down to us as one of the greatest scientists in history, while the other has become nothing more than a historical curiosity as to how such pseudoscientific ideas could have such a powerful impact on culture. **Chapter 10** uses the life of a contemporary scientist, Carl Sagan, to show how throughout his career he hovered in and around the borderlands of science in search of answers to the deepest questions of humanity, while all the time trying to find that exquisite balance between being open-minded enough to recognize radical new ideas when they come along, but not being so open-minded that quackery and nonsense are also eagerly accepted. Sagan was masterful in striking that balance, and in this chapter I attempt to show how and why he did so by applying theories of history, sociology, and psychology to biography, itself a borderlands science.

In **Part III: Borderlands History**, we begin in **chapter 11** by exploring how the study of the past can be done using the scientific methods employed by other historical sciences, such as cosmology, historical geology, paleontology, and archaeology and that, in fact, it must do so in order to avoid the quagmire of deconstructionism and postmodernism that has infected the field. This idea of a scientific history, although championed in the nineteenth century, is today considered heretical and not at all a normal part of what historians do when they "do history." Why don't they? If the methods of science are so effective at answering questions and solving mysteries, why don't historians employ these methods to questions and mysteries about the past? Specifically, this chapter explores what I call the "beautiful people myth," in which native peoples are thought to have lived in blissful harmony with nature and each other before the arrival of Dead White European Males (DWEMs). Here I employ several sciences and their methods to answer questions about the past, and argue that this should be a normal part of historical science. **Chapter 12** applies modern cognitive science to the study of historical figures and in the process debunks the "miraculous" nature of genius and shows how the exceptional geniuses in history are really only quantitatively different from the rest of us. It is an interesting study in how quantitative differences can accumulate into qualitative differences in what we might think of metaphorically as the borderlands between normalcy and genius. **Chapter 13** examines in detail the great priority dispute over who really deserves credit for the theory of evolution (Darwin or Wallace) and what this dispute tells us about how science, especially revolutionary science, really works. The story is more complicated than is usually told, and the outcome more amiable than often witnessed in the annals of scientific disputes. Finally, **chapter 14** is a splendid case study in the backward shift from science to pseudoscience through one of the greatest scientific flimflams of the twentieth century—the Piltdown hoax. Here we see how the data never just speak for themselves and how scientists mold information to fit their expectations—in this case the belief that humans evolved a big brain first and thus one should find in the fossil record a humanlike skull atop an apelike jaw, which was precisely what was "discovered" at Piltdown. Normal science for four decades before it was exposed as a hoax in the 1950s, Piltdown serves as a lesson in scientific humility before the data, and of the self-policing nature of scientists themselves.

Part of the self-policing process includes the acknowledgment of sources of data and influence. To that end there are a number of individuals who have greatly shaped me in the overall development of my ideas about science, as well as directly influenced the development of this book. First without equal is my colleague, friend, and confidant Frank Sulloway, who put the science back into social science for me and reawakened my passion for psychology.

My job as publisher of *Skeptic* magazine is intimately linked to my authorship of this and my earlier books, and to that end I owe a significant debt of gratitude to the magazine's board members: Richard Abanes, David Alexander, Steve Allen, Arthur Benjamin, Roger Bingham, Napoleon Chagnon, K. C. Cole, Jared Diamond, Clayton J. Drees, Mark Edward, George Fischbeck, Greg Forbes, Stephen Jay Gould, John Gribbin, Steve Harris, William Jarvis, Penn Jillette, Gerald Larue, Jeffrey Lehman, William McComas, John Mosley, Richard Olson, Donald Prothero, James Randi, Vincent Sarich, Eugenie Scott, Nancy Segal, Elie Shneour, Jay Stuart Snelson, Carol Tavris, Teller, and Stuart Vyse. As always I acknowledge the support of the Skeptics Society and *Skeptic* magazine provided by Dan Kevles, David Baltimore, Alison Winter, Susan Davis, and Chris Harcourt at the California Institute of Technology; Larry Mantle, Ilsa Setziol, Jackie Oclaray, and Linda Othenin-Girard at KPCC 89.3 FM radio in

Pasadena; Linda Urban at Vroman's bookstore in Pasadena; as well as those who help at every level of our organization, including editors and staff: Yolanda Anderson, Stephen Asma, Jaime Botero, Jason Bowes, Jean Paul Buquet, Adam Caldwell, Bonnie Callahan, Tim Callahan, Cliff Caplan, Randy Cassingham, Amanda Chesworth, Shoshana Cohen, John Coulter, Brad Davies, Janet Dreyer, Bob Friedhoffer, Jerry Friedman, Gene Friedman, Nick Gerlich, Sheila Gibson, Michael Gilmore, Tyson Gilmore, Greg Hart, Andrew Harter, Lisa Hoffart, Laurie Johanson, Terry Kirker, Diane Knudtson, Joe Lee, Bernard Leikind, Betty McCollister, Tom McDonough, Sara Meric, Tom McIver, Frank Miele, Dave Patton, Rouven Schaefer, Brian Siano, Tanja Sterrmann, Lee Traynor, and Harry Ziel.

I am especially grateful for the additional input provided by my agents Katinka Matson and John Brockman, my editor Kirk Jensen, production manager Ruth Mannes, Brian Hughes at Oxford, as well as all the good folks in sales and production at this prestigious publishing house. Bruce Mazet's support for my research and work is always deeply appreciated and I am always stimulated by his gentle reminders that there are many theories in mainstream science that perhaps should be in the borderlands, and maybe not even there. Gerry Ohrstrom's support, friendship and stimulating conversation have generated much light. I am also thankful for the art director of *Skeptic* magazine, Pat Linse, who deserves a lot more credit and public recognition than she is given due primarily to the division of labor between us where one job is high profile and the other low, but neither more nor less important. The preparation of the illustrations for this book was done entirely by her and I am fortunate that she understands that humans are the most visual of all the primates.

Finally, I thank my best friend and life partner Kim . . . for everything.

PART I
BORDERLANDS
THEORIES

Science is a very human form of knowledge. We are always at the brink of the known, we always feel forward for what is to be hoped. Every judgment in science stands on the edge of error, and is personal. Science is a tribute to what we can know although we are fallible. In the end the words were said by Oliver Cromwell: "I beseech you, in the bowels of Christ, think it possible you may be mistaken." I owe it as a scientist to my friend Leo Szilard, I owe it as a human being to the many members of my family who died at Auschwitz, to stand here by the pond as a survivor and a witness. We have to cure ourselves of the itch for absolute knowledge and power. We have to close the distance between the push-button order and the human act. We have to touch people.

—Jacob Bronowski, final paragraphs, chapter 11
"Knowledge or Certainty," *The Ascent of Man*, 1973

I

THE KNOWLEDGE FILTER

Reality Must Take Precedence in the Search for Truth

> Martha: Truth and illusion George; you don't know the difference.
> George: No; but we must carry on as though we did.
> Martha: Amen.
>
> — Edward Albee, *Who's Afraid of Virginia Woolf?*

WITHIN HOURS OF THE TRAGIC DEATH of Princess Diana, theories about what *really* happened to her began to proliferate via the Internet. One posting admonished readers: "Anyone who doesn't know that the order to murder Diana came from the Hanover/Windsor power structure is lacking an understanding of world affairs." Another explained: "A car wreck can very easily be engineered." The conspirators, it seems, were "acting on papal orders and financed by Du Pont." Within days the BBC reported that Libyan leader Muammar Qaddafi told his followers in a televised speech that the "accident" was a combined French and British conspiracy "because they were annoyed that an Arab man might marry a British princess."

By the time of the funeral, dozens of theories were ripe for the picking, including: Diana was murdered, but Dodi was an innocent victim; the driver was a "Manchurian Candidate" programmed to self-destruct at the right moment; Trevor Rees-Jones was in on the murder as a secret service agent; either M.I.5, the British Domestic Intelligence Agency, or M.I.6, the British Secret Service, orchestrated the murder; Diana's murder came about because of her stand against land mines—the arms industry was not going to put up with this attack on one of their major profit centers; Diana was three months pregnant and the monarchy was not going to allow any half-Arab child to become a part of the British Empire. The most bizarre conspiracy theory of all revealed that Diana was supposed to move to America and marry Bill Clinton, with Hillary either murdered or stepping out of the picture. Son William (the heir) would remain in England to become king, while second son Harry (the spare) moves to America, becoming a U.S. senator. By controlling both American and British banking and politics, the Rockefellers (who orchestrated the conspiracy), would rule the world and finally better their archrivals, the Rothschilds. The problem, we are informed, is that Princess Di wouldn't marry Clinton, so she needed to be eliminated. Perhaps he was too preoccupied with Monica to notice Diana.

The mythologist Joseph Campbell once observed: "Why should it be that whenever men have looked for something solid on which to found their lives, they have chosen not the facts in which the world abounds, but the myths of an immemorial imagination."[1] That's not quite right. Thinking is a combination of imagination and facts (and pseudo "factoids"). For example, what evidence do these cyber-conspiratorialists offer for their theories? They begin with the "fact" that everyone knows that the Rothschilds and Rockefellers have been competing to take over the world. Then they note that in July, 1996, Amschel Rothschild was "murdered" in Paris, on the anniversary of the "murder" of John D. Rockefeller III. In February 1997, Pamela Harriman, U.S. ambassador to France and a major financial supporter of Al Gore (controlled by the Rothschilds), was "murdered" in Paris. Di was "murdered" on a sacred section of the Paris highway called Pont de l'Alma, from which is derived the word "Pontiff," or "Pope," and is also an ancient site dating back to the time of the Merovingian kings in the sixth and seventh centuries. Before this, Pont de l'Alma had been a pagan sacrificial site. One translation has it meaning "bridge of the soul," another as "passage of nourishment." Since all true European royalty is descended from the Merovingians, themselves descendants from Christ (see, for example, the book *The Hanover Plot* by Hugh Schonfield), the tunnel in which Princess Di was murdered is spectacularly well connected to history. Oh, and don't forget, TWA Flight 800 was Paris-bound before it was "shot down," and among its victims were 60 French nationals and eight members of the French Secret Police.

TRUTH AND ILLUSION

Okay, so most of us don't buy into such wild speculations about cabalistic plots to take over the world, but why not? Because we have a *knowledge filter* that, unlike George in *Who's Afraid of Virginia Woolf?*, helps us discriminate between truth and illusion. Most of the time the knowledge filter works quite well. We *can* tell the difference between truth and illusion, and when we can't there is usually a good reason for it—a magician is trying to fool us, or we choose to be fooled. Think of the knowledge filter as a mental module that screens incoming ideas for their veracity. It works by comparing new facts and ideas with what we already know from previous experiences.

Society too has knowledge filters. Newspapers, magazines, radio, and television networks have journalistic standards and ethics that ask the same sorts of questions our personal knowledge filters ask. You rarely see an unopposed guest on *Nightline*, for example. There is always another perspective to a story, a point-counterpoint to most assertions. Medicine and science have a built-in knowledge filter called the peer-review system. In order to get a paper published in a medical or scientific journal it must be read by several of your colleagues. Rarely is a paper immediately accepted for publication. Most are rejected on first submission, and those that are published usually go through numerous resubmissions, or are published in journals of lower reputation. Errors are weeded out, faulty reasoning is exposed, inappropriate conclusions are rebuffed. And since the reviewers remain anonymous, the critiques can sometimes be quite harsh. This is no place for the thin-skinned.

But after centuries of constructing knowledge filters in the various fields of information dispersal, something has gone terribly wrong. Ideas are doing end-runs around the normal channels of communication, via what promises to be the most powerful source of knowledge diffusion in history—the Internet. Good ideas, bad ideas, interesting ideas, and wacky ideas stream through cyberspace into our computers at breathtaking speed. On one level that's good. Recall what the printing press did for the accelerating rate of knowledge growth. In religion it made everyone his or her own priest; in science it made everyone his or her own scholar. The Protestant Reformation and Scientific Revolution were the result. The telegraph, telephone, radio, and television had similar impacts. The problem is that it takes time for the knowledge filters to be implemented. There are no standards on the Internet, no peer review, and no editors fact-checking a story before publication. Matt Drudge's Internet gossip column is a case in point. This "Walter Winchell" of the Web is considered by many to be the king of cybergossip—post first and ask questions later. Sometimes he scoops the big boys with breaking stories that turn out true, other times he receives letters from attorneys threatening libel suits. Drudge summed it up with a statement to *USA Today*: "I don't give a damn what the bureau chief's going to think. I don't have one." And that's the point.

The result of this new everyone-their-own publisher is a mind-blurring pot-pourri of factoids and theory-mongering for the choosing. But how do we choose? I don't have time to check out all the sources and evidence for these ideas, do you? How do you know the government isn't hiding the bodies of aliens from another planet, or that the CIA did not smuggle drugs into the streets of Los Angeles, or that a secret branch of the government didn't invent AIDS to decimate the gay and black populations? After all, the government has lied to us so often in the past (and who knows how much more they haven't admitted) that it sometimes seems like anything is possible. Maybe Kurt Cobain was murdered. Maybe the Du Pont family maneuvered Congress to make mari-juana illegal for fear that cannabis would supplant many of their own manufac-tured chemicals. Maybe the KKK really does run Snapple (see the "slave" ship on its label). I'm skeptical of all of these claims, but short of conducting my own investigations into each and every case, how can I know for sure?

HOW THE KNOWLEDGE FILTER WORKS

Let's begin with a simple example of how one's knowledge filter works where truth and illusion overlap—dreams. When we are asleep the knowledge filter is off and dreams seem as real as our waking experiences. When we first wake up and our minds are in a fog, the line between truth and illusion is blurry. But with time it clears and we can reflect with amusement about what once seemed so real. We can discriminate between truth and illusion with dreams because our knowledge filter compares them with reality. For some people, however, their knowledge filters never engage and their dreams *become* reality, as in a significant percentage of alien abduction claims.

While we are awake our knowledge filters are hard at work comparing new images in the world with old images in memory. When you do a doubletake of someone's face, your knowledge filter is making lightning-fast comparisons of specific features of that face with all the faces in your memory. The knowledge filter then declares "match" or "no match." Ideas are treated much the same way. Recall the last time you encountered a get-rich scheme that seemed too good to be true. It probably was, but until you get burned the knowledge filter has no database from which to compare. Thus, most of us fall for such schemes at least once. Someone recently sent me a plan on how I can make beaucoup bucks in the Asian stock market through a company called "Financial Astrology." It seems that the forecasts of a professional "Financial Astrologer" were 71 percent and 74 percent accurate for the past two quarters respectively. For "only" $395 I can get her next picks. So why don't I mail my check or credit card number in their postage-paid return envelope? Because my knowledge filter has heard of schemes like this before—predict the rise or fall of eight stocks that will generate a possible 256 different outcomes (2^8). Mail these 256 outcomes to a large database, keeping track of who gets what combination. Assuming a stock has an equal chance of going up or down, for every group of 256 recipients, one person on average will get a letter with all predictions correct, 7 people will have 7 of the 8 correct and 28 will have 6 out of 8 correct. Mail letters to those who have received only your correct picks and ask them to send in their money. Clever, huh?

TRUTH AND ILLUSION IN MEDICINE

Things get more complicated with medical and scientific ideas. Answers to our knowledge filter's questions are not always obvious. The facts do not just speak for themselves. Experts disagree. How are nonexperts to know what to think? Is coffee bad for you or not? Do breast implants cause degenerative tissue disease or don't they? Should we use air bags or shouldn't we? Is the hole in the ozone caused by pollution or isn't it? Is there a greenhouse effect warming the earth, or is this warming trend just part of the ongoing variation in global temperatures? Just what is the carrying capacity of the earth—are we already past it and on our way to doomsday, or are we nowhere near it and able to handle another ten billion souls?

If traditional medicine controversies befog our knowledge filters, then alternative medicine claims cause them to cloud over altogether. If you want to experience the alternative medical movement at its epicenter attend a Whole Life Expo. Cures for everything from AIDS and cancer to baldness and impotency are offered, along with massages, chiropractic adjustments, acupuncture, acupressure, iridology, yoga, dowsing, channeling, aura readings, homeopathy, hypnosis, herbalism, aromatherapy, oxygen therapy, past-life regression therapy, and even future life progression therapy. Seminars at a recent Los Angeles convention included "How to Improve Your Eyesight Without Glasses, Contact Lenses, or Surgery," "Healing with Sound," "Your Aura Colors—What They

Say About You," "How Your Past Lives Are Influencing You Now," "Channeled Insights From the Dolphins," and "Cosmic Orgasm for Enlightenment," where you will "learn how to have orgasms so intense, so cosmic that with each climax you come back a new, better, happier person." Beam me up Scotty!

Do people really go for this stuff? They do, to the tune of billions of dollars a year. Why? Because traditional medicine, as marvelous as it is, is still incapable of curing AIDS, cancer, and many other life-threatening diseases. Traditional medicine doesn't offer "whole" human experiences involving one's psychological being or "spiritual" needs. Even when successful, many traditional medical practices are harsh, expensive, and delivered like automotive repair. Alternative services proliferate when the mainstream products fail to meet the needs of customers. No one knows this better than Deepak Chopra, a traditional medical doctor turned alternative medical guru (and author, novelist, poet, screenplay writer, lecturer, CD-ROM producer, and regular talk show guest). His success may be attributed to any number of factors—his medical credentials, his seeming polymathic mastery of such abstruse sciences as quantum mechanics, his Indian accent, his marketing acumen—the most important being that he meets an apparently unfulfilled need in millions of people that is not being met by traditional medicine. Take a trip to his Center for Well Being, nestled in a cozy corner of old town La Jolla that hugs the jagged cliffs overlooking the Pacific Ocean, and you'll see what I mean. Wholesomely attractive women in California casual wear greet you at the welcome desk. An organic juice and salad bar invites you to first pass through the bookstore that is nothing short of a Chopra Shrine offering all manner of nostrums not to be found on the shelves of your local pharmacy. Incense and massage oil call forth primal memories of smells and feelings that shut out real-world stresses and anxieties. Books on love and life and lust bring a sense that here is to be found Something Special. I don't recall getting these same feelings at my local Kaiser. When was the last time your HMO told you "Bring your mind into balance and your body will follow." Can Western medicine claim "the 5,000 year old healing traditions of Ayurveda which address the full range of human experience"? Have you ever seen a medical brochure with anything resembling the following offer?: "This mind body connection holds the potential for not only freedom from disease, but a higher state of health. By enlivening our inner healer, balance, wholeness, and well being can be restored." This is feel-good medicine. Large, multistoried hospitals with computers, instruments, and faceless physicians who grudgingly give you eight minutes of their time (the average doctor visit) amounts to feel-bad medicine.

Something is amiss here. A recent survey of the nation's 126 medical schools found that 34 of them offer course electives in alternative medicine. In 1991, the National Institutes of Health established an Office of Alternative Medicine to test such claims. Why don't we have an Office of Alternative Airlines, where they test planes with one wing? Because regular airlines' success rate is so remarkably high compared to other forms of transportation that there is no public demand for it. Modern medicine cannot claim such success rates. Frankly, there is no way I would ever go to Doc Chopra before I would go to Doc Kaiser, and all of the alternative medical claims I have taken the time to investigate turned out to be total bunk. But I can understand why those who have been let down by the medical establishment, or who face certain death from a disease on which their doctors simply put a doomsday clock are tempted by such alluring offers. Our personal knowledge filters are simply not equipped to deal with such complex medical claims. That's why we need more and better science.

PUTTING THE SCIENCE KNOWLEDGE FILTER TO THE TEST

The best knowledge filter ever invented is science. Flawed as it is at times, the methods developed over the past four centuries were specifically designed to help us avoid errors in our thinking. As an example of how the science knowledge filter works in a very simple and straight-forward manner (as a demonstration, not as a controlled experiment), on Monday, November 9, 1998, James Randi and I tested a Chinese psychic healer named Dr. Kam Yuen from Shaolin West International (in Canoga Park, California), an "Institute of Martial Arts and Natural Medicine." According to his card, Yuen is a doctor of "Chinese Energetic Medicine, Chiropractic, Homeopathic Medicine," as well as a "Nutritional Consultant."

Dr. Yuen's organization contacted Randi in order to be tested for the million-dollar challenge the James Randi Educational Foundation offers. The test was arranged through the NBC television show *Extra!*. Randi was the principal investigator in the experiment. Dr. Fleishman was the attending physician who would monitor the patients for pain, while my role was to monitor Dr. Fleishman and the other people involved to ensure that proper experimental controls were employed.

The claim was that Dr. Yuen can heal people, in a matter of seconds, of intense pain and illness of virtually any kind. He stands or sits in front of the person, stares at them intensely, waves his hands and fingers around in a ersatz-Kung Fu style for a few seconds, and, chango presto, the healing is completed. Patients, we are told, suddenly feel better. How do you test such a claim?

With the help of a glitzy health club on Los Angeles' trendy west side, *Extra!* arranged to find, with the assistance of Dr. Fleishman, five people who had noticeable and constant pain of a kind that could be easily recognized if there were any changes. Two alternates were requested. One was supplied and utilized. Each of the six subjects (five for the test and one alternate) were screened before the test by Dr. Fleishman and myself, with Randi and cameras present. Each subject selected a number that they then attached to their clothing.

#1. Mary had lower back pain caused by scoliosis. She had pain to the touch that went down her right leg. On a scale of 1–10, she rated her pain as a 4 or 5.

#2. Gary had neuropathy-caused pain in his feet, especially the left foot, causing numbness in his toes and pain and a burning sensation in the ankle. He rated his pain as a 4.

#3. Nadine had carpel tunnel syndrome that causes a tingling sensation and numbness in the fingers after about five seconds of pressure applied to the wrist.

#4. Paula had an inflamed tendon and adjacent nerve that, when pressed, caused pain rated as a 7 to the touch.

#5. Don had severe pain in his right knee, which he rated as a 10 when Dr. Fleishman pressed on a particularly tender spot.

#6. (Alternate #1). Miranda had lower back and hip pain, tender to the touch, which she rated as a 5.

For the test, Dr. Yuen was brought in and introduced to the five subjects seated in front of him. He sat in a chair about five feet away. Each of them informed Dr. Yuen of their problem and pain. The subjects were seated very close to each other—only about an inch apart, but he claimed it would not matter and that he could heal each person individually. The subjects were numbered 1–5 from left to right. All five were blindfolded using a Randi-approved blindfold called the "Mindfold Relaxation Mask," that prevented the subjects from knowing which one Dr. Yuen would be healing.

Dr. Yuen then selected a number from an envelope to determine which subject he would heal for that trial. By chance alone he then would have a 20 percent probability of a match between his healing attempt and a blindfolded subject's reporting that they felt better. Of course, this protocol was not as tight as we would have liked it to be, since it was entirely possible that more or less than one each time would report a change, and that the change could be better or worse. But Dr. Yuen made it quite clear that he could isolate a single patient, cause a reduction in pain, and that he could do this five out of five times. So that was the test we were running as a preliminary to try for the million dollars.

Trial One. Mary, #1, reported a reduction in her lower back pain, from a 4/5 to a 2, and Nadine, #3, reported a dramatic improvement in the numbness in her fingers, from five seconds to the onset of symptoms in the pre-exam, to 30 seconds in this trial, and the numbness was significantly less. Dr. Yuen had selected Paula, #4, to work on. She reported no change at all and her pain remained the same at 7. This ended the formal test since Dr. Yuen claimed he could get five out of five, and he failed the first trial.

Trial Two. The producers wanted to film a total of five trials, which we agreed to as long as it was made clear that the formal test was over. In the second trial Gary, #2, reported a noticeable improvement. Dr. Yuen had selected Don, #5, so he failed again and was now 0/2.

But, as we prepared for the third trial, Randi and I noticed some activity surrounding Don, #5, who, when he discovered that he was the one who was worked on in *Trial Two*, suddenly felt better! He was walking around proclaiming that his limp was gone. Even though when Dr. Fleishman tested him for *Trial Two* his pain was as intense as it was in *Trial One* and in the pre-exam, now he was claiming he felt better. After the sessions I interviewed him briefly: "Were you surprised that you suddenly felt better?" He said: "No, not at all, because I do a lot of energy work and I believe in this type of healing." That would seem to explain what happened when he discovered he had been Dr. Yuen's subject—a classic case of placebo effect.

In the third and fourth trials Dr. Yuen also missed, leaving him at 0/4. Finally, in the fifth trial, he got a hit, which was expected to occur by chance (five subjects over five trials gives a 20 percent probability of success). He selected Mary, #1, and she reported that her lower back felt "appreciably better." It should be noted that Mary is a former ballet dancer who is still lean and muscular. With each trial, in order to test the level of her pain (to rate it on a 1–10 scale), she stretched (and quite limberly at that). I noticed that each time she seemed more flexible than the time before. My wife is a dancer and she informs me that such an improvement in lower back pain would be expected just from the stretching. So even Dr. Yuen's 1/5 performance is questionable.

Randi and I recognize, of course, that this protocol was less quantifiable than we would have liked because of the subjectivity of reporting pain, and the fact that it would have been better to have five new patients with each round so there could be no improvement effect from moving about, as in the case with Mary. Again, this was a demonstration, not a controlled test. Nevertheless, even in these less than ideal conditions, Dr. Yuen did no better than random chance. As Randi and I explained to the producers: the null hypothesis (that Dr. Yuen would do no better than chance) could not be rejected. Our knowledge filters weeded out the claims from the facts to reach a simple and solid conclusion that Dr. Yuen, under controlled conditions, could not do what he claims he could do.

NATURE CANNOT BE FOOLED

In October 1997, NASA launched the nuclear-powered spacecraft Cassini to explore the planet Saturn and its moons. Concerns were expressed publicly by some groups that the on-board plutonium could pose a public safety threat should the rocket crash on take off, or should the spacecraft later burn up in the earth's atmosphere when it makes a close fly-by. Since I live less than a mile from JPL, the builder of the spacecraft, this was of special interest to me as this is where protesters gather to air their concerns. As far as I could tell from talking to various experts, the probability of a premature rocket explosion or a spacecraft reentry into the earth's atmosphere was miniscule; and if it did crash the risks of nuclear contamination were so low that I would be much better off worrying about more prosaic dangers, such as automobile accidents or dietary cholesterol. Yet the day before the launch there were protesters marching in front of the JPL entrance gate with antinuclear placards. Why did these folks not trust the assurances of the JPL scientists? One obvious reason could be because our government has lied to us about matters nuclear before. Remember the downwinders in the western United States during the Cold War (revealed decades later)? Would you trust a government that, without public knowledge or permission, detonated nuclear weapons in order to monitor the effects of the downwind fallout over American cities?

Nuclear energy in general provides a paradox of trust between the government and the public. In his book, *Imagined Worlds*, Freeman Dyson observes "So long as it is allowed to fail, nuclear energy can do no great harm."[2] How's that again? Nuclear energy is so greatly feared in this country, Dyson explains, that zero tolerance of failure has become the unattainable standard, forcing authorities to present it as cleaner, safer, cheaper, and more profitable than anyone could possibly deliver. This has resulted in what can only be called deceitful practices of doctoring accounts to show profits, and rules and guidelines to be written with political and ideological standards in mind, not scientific and technological. By comparison, Dyson demonstrates that since the 1920s there have been roughly 100,000 different varieties of airplanes and jets, out of which only about 100 have survived. Applying a Darwinian model, Dyson concludes: "Because of the rigorous selection, the few surviving airplanes are astonishingly reliable, economical, and safe."[3] The paradox is that we cannot afford the risk of 100,000 nuclear designs to get 100 reliable, economical, and safe nuclear reactors. But we want them, so we force our authorities to present an illusion, and we accept the illusion as truth.

Even an enterprise as seemingly pure in the spirit of youthful adventure and wide-eyed exploration as the space program, turns out to be fraught with bureaucratic inefficiencies and governmental deceit from the very beginning: Kennedy's famous declaration to Congress on May 25, 1961, that the United States "should commit itself to achieving the goal, before this decade is out, of landing a man on the moon and returning him safely to the earth," was primarily a political maneuver aimed at finding something, anything, at which to beat the Soviets. His advisors thought we might have a good shot at the moon, and thus began the space race that ended in the wildly successful (and enormously expensive) Apollo program. Bureaucratic agencies do not die easily, however, and when it looked like Nixon was going to abandon manned space exploration altogether, NASA responded by taking steps that would severely compromise its integrity, as well as that of the government.

NASA designed a space shuttle that would be all things to all agencies, but that has ended up being few things to few agencies. The Department of Defense went along with NASA's plan to use the shuttle as a surveillance device in a national emergency (in the event of a pending war, send up the shuttle to see what the enemy is up to). They even built a shuttle pad at Vandenberg Air Force Base in California for just such a purpose. No one actually believed the shuttle would be employed in national defense, and the pad has gone unused, but it is the sizzle that sells the steak. NASA also argued that the shuttle would be the most economical way to launch satellites, an absurd idea belied by the simple calculation that it costs five times as much to put a satellite into orbit on the shuttle as it would using the relatively inefficient Saturn booster it replaced. NASA promised 572 shuttle missions by 1991; 35 actually flew.

The eroding trust between the public and NASA liquefied into quicksand on January 28, 1986, when the space shuttle *Challenger* exploded in midlaunch, triggering a commission investigation during which Caltech physicist Richard

Figure 2. Richard Feynman Demonstrates the Effects of Temperature on the Space Shuttle O-ring. Feynman concluded in his report to NASA: "For a successful technology, reality must take precedence over public relations, for Nature cannot be fooled."

Feynman brilliantly exposed the O-ring cold-weather problem. With a flair for the dramatic (and guided by NASA insider General Douglas Kutyna), Feynman dipped an O-ring sample in a glass of ice water at a press conference to reveal its fatal flaw—in the cold (like the morning of the shuttle launch) the O-ring loses its resilience so it cannot fill the gap in the expanding rocket booster joints. Holding up the deformed O-ring for all to see, and with characteristic understatement, Feynman told the committee: "I believe that has some significance for our problem."[4] But by now NASA was in full self-deception mode. The commission was nothing more than a public relations rubber stamp. With countless unanswered questions about the disaster, they recommended that "NASA continue to receive the support of the administration and the nation. The agency constitutes a national resource and plays a critical role in space exploration and development. It also provides a symbol of national pride and technological leadership. The Commission applauds NASA's spectacular achievements of the past and anticipates impressive achievements to come."[5]

Feynman would not be a party to this political chimera. He was hired to solve a scientific puzzle, not a public relations one. His concluding statement to NASA, which lies buried by the agency in an appendix, should serve as a motto when examining the pronouncements of authorities: "Let us make recommendations to ensure that NASA officials deal in a world of reality. . . . NASA owes it to the citizens from whom it asks support to be frank, honest, and informative, so that these citizens can make the wisest decisions for the use of their limited resources. For a successful technology, reality must take precedence over public relations, for Nature cannot be fooled."[6]

Unfortunately humans can be fooled, and the line between reality and illusion is often a blurry one.

2

THEORIES OF EVERYTHING
Nonsense in the Name of Science

IN 1950, MARTIN GARDNER published an article in the *Antioch Review* entitled "The Hermit Scientist," about what we would today call pseudoscientists.[1] It was Gardner's first-ever publication of a skeptical nature, and it launched not only a lifetime of critical analysis of fringe claims, but in 1952 (at the urging of his literary agent John T. Elliott) the article was expanded into a book-length treatment of the subject under the title *In the Name of Science*, with the descriptive subtitle "An entertaining survey of the high priests and cultists of science, past and present." Published by Putnam, the book sold so poorly that it was quickly remaindered and lay dormant until 1957, when it was republished by Dover and has come down to us as *Fads and Fallacies in the Name of Science*, still in print and arguably *the* skeptic classic of the past half-century.[2] (Gardner realized his book had made it when he turned on the radio "at 3 a.m. one morning, when I was giving a bottle of milk to my newborn son, and being startled to hear a voice say, 'Mr. Gardner is a liar.' It was John Campbell, Jr., editor of *Astounding Science Fiction*, expressing his anger over the book's chapter on dianetics.")[3]

When we call something a "classic" we mean that it has staying power, relevance that transcends generations—a work that is not just for one age but all ages. Centuries after their composition we listen with fresh ears to Beethoven, Bach, and Mozart; we stand in line for hours to see Leonardo's Mona Lisa smiling across the epochs; and despite universal education and a population of potential writers orders of magnitude larger than in his day, no one has ever approached Shakespeare in the breadth and depth of his oeuvre.

There is, however, something discomforting about calling a skeptic treatise a classic. In our business, progress means the attenuation of irrationality, the dismantling of nonsense, and the reversal of pseudoscience. If we are doing our job (one presumes and hopes), a book ticking off the absurdities of one generation should be out of date by the next, a mere historical curiosity about the silly things they believed back then. (Charles Mackay's description of the nineteenth-century craze for tulips—tulipomania—comes to mind, although even parts of this classic from a previous century, *Extraordinary Popular Delusions and the Madness of Crowds*, still have relevance today.) So it is troubling to dig through Martin Gardner's classic and discover that most of the fiddle-faddle of 1950 survives in 2000, and in some cases has far surpassed what troubled this young writer then as he embarked on a journey that led him through the looking glass of science turned upside down.

As we shall see in this brief tour, much of what was popular in 1950 remains in vogue today, with a few far outstripping anything Gardner could have imagined then. Thankfully some have faded from view, although even in these few cases they have mostly mutated into similar claims where only the names have been changed to protect the credulous. As a tribute to Martin Gardner on the 50th anniversary of his skeptical fountainhead we will look back, and forward, at theories of everything then, and now. As Gardner wrote then: "Cranks vary widely in both knowledge and intelligence. Some produce crudely written pamphlets, usually published by the author himself, with long titles, and pictures of the author on the cover."[4] The ideas may have changed but the cranks haven't, as we discovered at *Skeptic* magazine since its founding in 1992, as articles and book manuscripts regularly come in that purport to explain, well, *everything* with a single overarching idea.

THEORIES OF EVERYTHING—1950

What caught the attention of a youthful Martin Gardner half a century ago? The "hermit scientist," working alone and usually ignored by mainstream scientists. "Such neglect, of course, only strengthens the convictions of the self-declared genius," Gardner concluded in his original 1950 paper. "Thus it is that probably no scientist of importance will present the bewildered public with detailed proofs that the earth did not twice stop whirling in Old Testament times, or that neuroses bear no relation to the experiences of an embryo in the

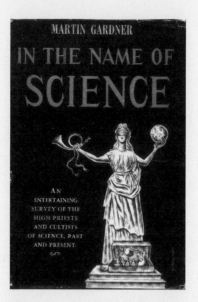

Figure 3. Martin Gardner's *In the Name of Science*—the "bible" of the modern skeptical movement.

mother's womb" (referring to L. Ron Hubbard's dianetics theory that negative engrams are imprinted in the fetus's brain while in the womb).[5]

Gardner was, however, wrong by half in his prognostications: "The current flurry of discussion about Velikovsky and Hubbard will soon subside, and their books will begin to gather dust on library shelves." While Velikovskians are a quaint few surviving in the interstices of fringe culture, L. Ron Hubbard has been canonized by the Church of Scientology and deified as the founding saint of a world religion.[6]

In the first chapter of *In the Name of Science*, Gardner picks up where he left off, noting that "tens of thousands of mentally ill people throughout the country entered 'dianetic reveries' in which they moved back along their 'time track' and tried to recall unpleasant experiences they had when they were embryos."[7] Half a century later Scientology has converted those reveries into a worldwide cult of personality surrounding L. Ron Hubbard, that targets celebrities for membership and generates hundreds of millions of dollars in tax-free revenue as an IRS-approved "religion."

Today UFOs are big business, but in 1950 Gardner could not have known that the nascent flying saucer craze would turn into an alien industry, but it was off to a good start: "Since flying saucers were first reported in 1947, countless individuals have been convinced that the earth is under observation by visitors from another planet." Absence of evidence then was no more a barrier to belief than it is today, and believers proffered the same conspiratorial explanations for the dearth of proof: "I have heard many readers of the saucer books upbraid the government in no uncertain terms for its stubborn refusal to release the 'truth' about the elusive platters. The administration's 'hush-hush policy' is angrily cited as proof that our military and political leaders have lost all faith in the wisdom of the American people."[8]

From his perspective in 1950 Gardner was even then bemoaning the fact that some beliefs never seem to go out of vogue, as he recalled H. L. Mencken's quip from the 1920s "that if you heave an egg out of a Pullman car window anywhere in the United States you are likely to hit a fundamentalist." Gardner cautions that when presumably religious superstition should be on the wane how easy it is "to forget that thousands of high school teachers of biology, in many of our southern states, are still afraid to teach the theory of evolution for fear of losing their jobs."[9] Today, bleeding Kansas enjoins the fight as the creationist virus spreads northward.

Thankfully there has been some progress. Now largely antiquated are Gardner's chapters on believers in a flat earth, a hollow earth, worlds in collision (Velikovsky), Atlantis and Lemuria, Alfred William Lawson's strange "Law of Penetrability and Zig-Zag-and-Swirl movement" (that makes Newton's gravitational theories "but a primer lesson, and the lessons of Copernicus and Galileo are but infinitesimal grains of knowledge"), numerous anti-Einstein wannabe scientists, Roger Babson's "Gravity Research Foundation" (whose stated goal was to find a "gravity screen" capable of blocking the effects of gravity like "a sheet of steel cuts off a light beam"), Lysenkoism, Wilhelm Reich's orgonomy, and Alfred Korzybski's general semantics.

But, disturbingly, a good two-thirds of the book is still current, including Gardner's discussions of homeopathy, naturopathy, osteopathy, iridiagnosis (today called iridology—"reading" the iris of the eye to determine bodily malfunctions), food faddists, impaired vision cures, and other forms of medical quackery (now known as alternative medicine, or complementary medicine), Edgar Cayce, the Great Pyramid's alleged mystical powers (believers today are known among skeptics as pyramidiots), Charles Fort (aka the Forteans, "dedicated to the frustration of science, and the haven of lost causes" says Gardner), handwriting analysis (aka graphoanalysis), ESP and PK, reincarnation (Bridey Murphy is out but others are in), dowsing rods, eccentric sexual theories, theories of racial differences between human groups, and, as already discussed, flying saucers, creationism, and dianetics.

And the motives of the hermit scientists have not changed either. Gardner recounts the day that Groucho Marx interviewed Louisiana State Senator Dudley J. LeBlanc about the "miracle" cure-all vitamin and mineral tonic called Hadacol that the senator had invented. Groucho asked him what it was good for. LeBlanc answered with uncharacteristic honesty: "It was good for five and a half million for me last year."[10]

What I find especially valuable about *In the Name of Science* are Gardner's insights into the difference between science and pseudoscience — that sometimes fuzzy boundary between sense and nonsense, the normal and the paranormal. Martin begins by showing that the fuzziness is due to the fact the we are dealing with a continuum scale, not a binary choice. On the one extreme we have ideas that are most certainly false, "such as the dianetic view that a one-day-old embryo can make sound recordings of its mother's conversation." In the borderlands middle "are theories advanced as working hypotheses, but highly debatable because of the lack of sufficient data," and for this Martin selects a most propitious example: "the theory that the universe is expanding." That theory would now fall onto the spectrum at the other extreme end of "theories almost certainly true, such as the belief that the earth is round or that men and beasts are distant cousins."[11]

Subtle thinker that he is, however, Gardner recognizes a second continuum in the claimants themselves, in the form of "scientific competence," "ranging from obviously admirable scientists, to men of equally obvious incompetence." There are, he notes, "men whose theories are on the borderline of sanity, men competent in one field and not others, men competent at one period of life and not at others," not to mention "a type of self-styled scientist who can legitimately be called a crank." How can we tell if someone is a crank? "If a man persists in advancing views that are contradicted by all available evidence, and which offer no reasonable grounds for serious consideration, he will rightfully be dubbed a crank by his colleagues."[12]

There is here, however, a more serious tangential problem in science in general and within each of us in particular, in the form of striking the right balance between orthodoxy and heresy, between tradition and change, between keeping your mind open enough to be able to accept radical new ideas, but not so open that your brains fall out. Most ideas that come down the pike are fiddle-faddle nonsense that can be safely ignored. But because every once in awhile one of them may turn out to be revolutionary, we cannot simply dismiss all eccentricities out of hand. Gardner offers us some advice on how to spot a crank.

1. "First and most important of these traits is that cranks work in almost total isolation from their colleagues."[13] Cranks typically do not understand how the scientific process works—that they need to try out their ideas on colleagues, attend conferences, and publish their hypotheses in peer-reviewed journals before announcing to the world their startling discovery. Of course, when you explain this to them they say that they are misunderstood, that their ideas are too radical for the conservative establishment to accept, and that science is not ready for their revolution. Yet as Gardner noted half a century ago (and it still rings true today):

> Nothing could be further from the truth. Scientific journals today are filled with bizarre theories. Often the quickest road to fame is to overturn a firmly-held belief. Einstein's work on relativity is the outstanding example. Although it met with considerable opposition at first, it was on the whole an intelligent opposition. . . . In a surprisingly short time, his relativity theories won almost universal acceptance, and one of the greatest revolutions in the history of science quietly took place.[14]

But, the crank snorts, what about Galileo, tried for heresy for supporting heliocentrism, or Giordano Bruno, burned at the stake for proposing an infinity of worlds, or Ignasz Semmelweiss, outcast from the medical profession for his heretical idea of sterilization before surgery? Might the crank be another Galileo, Bruno, or Semmelweiss? Maybe, but unlikely. For every unjustly persecuted scientist whose ideas have become canonized in the annals of science history, there are a thousand, perhaps ten thousand, eccentrics with goofball notions too quirky to have even been recorded by his contemporaries and thus lost in the dustbin of history.

2. "A second characteristic of the pseudo-scientist, which greatly strengthens his isolation, is a tendency toward paranoia." This paranoia manifests itself in several ways in the eccentric: (1) "He considers himself a genius." (2) "He regards his colleagues, without exception, as ignorant blockheads." (3) "He believes himself unjustly persecuted and discriminated against. The recognized societies refuse to let him lecture. The journals reject his papers and either ignore his books or assign them to 'enemies' for review. It is all part of a dastardly plot. It never occurs to the crank that this opposition may be due to error in his work." (4) "He has strong compulsions to focus his attacks on the greatest scientists and the best-established theories. When Newton was the outstanding name in physics, eccentric works in that science were violently anti-Newton.

Today, with Einstein the father-symbol of authority, a crank theory of physics is likely to attack Einstein." (5) "He often has a tendency to write in a complex jargon, in many cases making use of terms and phrases he himself has coined." And, of course, because he has been locked out of the establishment channels of communication, "He speaks before organizations he himself has founded, contributes to journals he himself may edit, and—until recently—publishes books only when he or his followers can raise sufficient funds to have them printed privately."[15]

Keep these criteria at the forefront when we shortly look at some of the theories of everything that have poured into *Skeptic*'s offices over the past decade. Should we be surprised that fifty years later crankdom flourishes? No, Gardner concluded in closing his introduction: "If the present trend continues, we can expect a wide variety of these men, with theories yet unimaginable, to put in their appearance in the years immediately ahead. They will write impressive books, give inspiring lectures, organize exciting cults. They may achieve a following of one—or one million. In any case, it will be well for ourselves and for society if we are on our guard against them."[16] And so we are.

THEORIES OF EVERYTHING—2000

On Venice Boulevard in west Los Angeles, next to a bus stop and Postal Instant Press, sits what has to be one of the strangest repository of weird things in the world—The Museum of Jurassic Technology. One exhibit of relevance for our discussion is entitled *No One May Ever Have the Same Knowledge Again,* a series of letters written to the astronomers at the Mount Wilson Observatory (high atop the San Gabriel mountains above Pasadena) between 1915 and 1935. The title comes from the first letter in the collection (published by the museum in a booklet of the same title) dated July 7, 1915, from one Alice May Williams of Auckland, New Zealand: "I want to tell you I am not after money & I am not a fraud. I believe I have some knowledge which you gentlemen should have. If I die my knowledge may die with me, & no one may ever have the same knowledge again."[17]

Williams's knowledge was derived "in half-sleep trance," and includes such breathless discoveries as "The Planet Mars is inhabited by human spirits like us can talk eat & drink wear clothes, but have great power. They are something people of this earth have never seen. They are kept to do work overhead. They also work our wireless gramphones, machinery, Moving Pictures Talking Pictures and all that sort of thing [*sic*]." Another letter from one May Bernard Wiltse to the astronomer George Hale dated December 3, 1932, announced that

"In 1916 I went to Washington, D.C. and transmuted silver into gold for the United States government and I have their reports. BUT IT WAS HUSHED up for reasons I cannot explain."[18] Of course.

Surprisingly, given our name, at *Skeptic* we also routinely receive such letters, along with essays and articles for publication consideration, some of which I store in a file called "Theories of Everything." These are mostly attempts at constructing all-encompassing explanatory theories. They usually deal with the physical sciences, alleging to prove that Newton, Einstein, and Hawking are wrong and that this person's ten-page, single-spaced typed essay without references contains the secret to the cosmos. As this book was going to press, for example, I received this interesting letter (original grammar and spelling) from a man whose webpage address is (I'm not making this up) theory-of-everything.com:

> Hi! I am writing to you in hope you would be interested in my discovery to make some sense out of this I'll start with: Einstein spent his last years trying to discover a theory which unites all laws of physics (forces of nature) in 1 theory = theory of everything, as you know some think that a near answer to that is the string theory. I believe I discovered the right (correct) theory: the theory of logic = the theory which unites (makes, proves) all laws of physics (forces of nature) = the theory of everything my webpage is at http://theory-of-everything.com so if you see that you understand and you're interested in working in a lab with me then let's do it.
>
> I've sent it to many scientists but nobody interested so far in it many of them said: boy, don't be stupid, nobody can discover such a theory, theory of everything, it's too crazy! it's impossible! it would be like knowing the mind and law of GOD! if you're not interested please don't write, spare me your critical view. so god to me is Logic therefore the law of God to me is the law of Logic.

The webpage provides a single page describing his theory of everything, which is about as lucid as his letter. My files abound with such grandiose ideas, each of whom is unaware of all the similar attempts being made by other amateurs. For example, in March 1997 I received by E-mail an "Introduction to the Unified Field Theory" written by Barry R. Brownlee, who explained that "completion of the Unified field theory is proof Einstein's 'key to the universe' does exist." This problem has long eluded physicists, but no more. Barry Brownlee reveals for the first time The Unified Field Theory (spelling and grammar errors in original): "I have discovered the previously unknown invisible particle that fully explains light and all forms of energy. I call it the froton particle. I have tested every way imaginable, and even tested it as the missing and vital information needed to complete the unified field theory. Thus far it has never failed to be right. I have called for debate from all corners of the world and have done all at my own expense. Einstein didn't know about the froton particle—but you will. I cant email a drum roll, and dont want to pontificate. The knowledge of this nature inspires a reverence that I choose not to ignor. Inappropriately or not, if only out of respect for Einstein and physicists as far back as Aristotle—Ive chosen the words to introduce the unified field theory. 'Behold the froton particle . . . the key to the universe.' "

Another manuscript entitled *Infinite Dynamics*, presents "the opposing force of structure as the singular force of reality" that "establishes the preeminence of philosophy," goes "beyond Einstein," and represents "an historic, classical, artistic, scientific, philosophical paradigm shift"! Another called *Photonics* promises "Einstein's unified-field theory now complete, all forces finally unified," and as a bonus, "fundamental cause of gravity found." A paper called "The Angel Fragmentation Scenario" provides answers to such questions as "What is spirit," "Where did all of the matter and energy that form our material universe come from?," and "If God can instantly create angels, why would He bother to take 15 billion years to create humans?"

One UFO observer wrote to ask if I wanted to see his book that "is the first consistent theory of anti-gravity in the history of mankind," revealing the secrets to "time warp interstellar travel, wormhole engineering and Dipole Antigravity Drive, gravitational free energy, and alien visitation." An Australian woman sent me proof that "the speed of light has not been measured." Her book manuscript, *Time Ends—The New Reality*, proves that "light's velocity is so fast it is literally everywhere/everywhen at once. This would include being in our past and future simultaneously."

Another correspondent submitted his "Question that lurks behind all other questions," explaining that his *ONE LAW* series offers a "theory of everything" (his description). "Failing to incorporate the natural cycles of life and death into their design, governments created by Man have no universal sanction to exist or 'natural' authority to govern." Another fellow calling himself "the world's most controversial author" (you will not have heard of him so he will remain anonymous) has penned such memorable books as *The Most Powerful Idea Ever Discovered, Explosive Revolutionary Ideas,* and *What the Establishment Doesn't Want You to Know About Government and Politics.*

Theories of Everything are by definition grandiose, but it would be hard to top one I received from a Canadian gentleman who submitted a book manuscript simply entitled *Good News*, with the following descriptive reading line: "Electronics can now amplify your senses, ability, and power; billions of times. It shows the *one* true nature of all things; that eliminates disagreements, arguments, and wars. It unites all people in a world federation; that will become all-knowing, all-powerful, and eternal. People will live everywhere in an eternal universe." The author explains in a cover letter that sixty years ago he discovered "how all particles and objects are made, their *absolute characteristics, and the eternal Natural Laws* that govern them. And, how they make all the objects and Life in the Universe. They explain everything in the Universe, and everything everyone has ever seen, heard, felt, tasted, and smelled." Now *that's* a Theory of Everything!

Mathematics is ripe for the picking by cranks, and, in fact, whole books have been written on the subject. Underwood Dudley's *Mathematical Cranks* and *Numerology*, both published by the Mathematical Association of America, are wonderful compendiums of hundreds of examples of numerological nonsense. We have gotten our fair share of those as well. For example, we received a query letter from a gentleman in Brooklyn, New York, about his book now being written (and soon to be published) called *The Greatest Mathematician of All Time*. Who might you guess? Euclid? Gauss? Newton? Gödel? Nope. The greatest mathematician of all time is the author of this book!

> I'm expressing my own point of view about math, a perhaps radical point of view, one which I believe differs in significant ways from that of the authority. I question what in math is interesting or has value and meaning and what are we trying to accomplish, what direction are we headed in. I'm also interested in the psychology, the motivations, and regard that as a part of math, itself. I regard the book as a math book and in it I maintain that, although I have only a Master's in math, I deserve a Ph.D., largely based on the merits of the book, which I look upon as my doctoral thesis. I give an argument as to why I should have a Ph.D., and why, in fact, I may even be a reasonable candidate for the greatest mathematician of all time, from some point of view.

That "point of view," of course, is obviously relative.

Theoretically, it must be said, it is possible that this fellow really is the greatest mathematician of all time (although I'm not sure who produces such rankings — perhaps the Mathematical Association of America?), but if probability theory has any validity the likelihood of this statement being true is about the same as it is for the Loch Ness monster, Big Foot, and other chimeras.

Mr. T. L. Delphinus's 1999 mathematical treatise, *God Answers the New Age, Skeptics and Post-Christian Society, Mathematically,* shares the same likelihood of veracity. This 146-page, self-published booklet is filled with numerological blather of which one example will suffice to give a flavor for the book. "The number SEVEN is God's number of completion and/or perfection." Seven appears 287 times in the Old Testament ($287 \div 41 = 7$), the word "seventh" occurs 98 times ($98 \div 14 = 7$), the word "seven-fold" appears seven times, the word "seventy" 56 times ($56 \div 8 = 7$), there are seven days of creation, seven deadly sins, God rested on the seventh day, and so forth. Nothing new here. Numerologists have been "discovering" these relationships for centuries. But Delphinus reveals that he has found a secret in medium James Van Praagh's book *Talking to Heaven.* On page seven Van Praagh writes "My prayer was answered when I was eight years old." By "going across seven letters from the first letter of the sentence," Delphinus found a code hidden in Van Praagh's line—the Hebrew name of Jesus:

```
12  345671
My prayer was answered when I was eight years old.
      E

12  345671 234 567
My prayer was answered when I was eight years old.
      E     S

12  345671 234 567 2345  67
My prayer was answered when I was eight years old.
      E     S       H

12  345671 234 567 2345  6712 3 456 7
My prayer was answered when I was eight years old.
      E     S       H       E

12  345671 234 567 2345  6712 3 456 7
My prayer was answered when I was eight years old.
      E     S       H       E     A

12  345671 234 567 2345  6712 3 456 7123456712345
 67
My prayer was answered when I was eight years old.
 Y    E     S       H       E     A
```

HE'S ALIVE!!!!

The author's point is that God is revealing Himself through James Van Praagh, therefore "I ask you, all of you new agers and post-Christians, won't you make the decision to throw away your new age spiritual books and come back to the one true living God, the God of the Bible?" Of course, Delphinus realizes that he has misspelled Jesus's Hebrew name (should be Yeshua, not Yeshea), but he rationalizes it by arguing that God "even left you saving grace by allowing each and every one of you to demonstrate before Him that you have FAITH the size of a 'mustard seed.' That faith resides in the fact that there is one letter that appears to be in error in His signature." Q.E.D. for G.O.D.

Air disasters always lend themselves to conspiratorial thinking, and if numerological connections can be made it drives the conspiratorialists into a frenzy. Richard Hoagland, the man who singlehandedly put the "face on Mars" into pop culture, posted on his Web page (www.enterprisemission.com) factoids about the crash of Egypt Air Flight 990, including:

- $990 \div 3 = 330$.
- There were 33 passengers on the previous leg of its trip.
- There were 33 Egyptian military officers on board.
- The plane was at 33,000 feet when it began its fatal dive.
- It disappeared exactly 33 minutes after takeoff.
- It had logged just over 33,000 flight hours.
- Comet Encke was dead on the horizon and visible from 33,000 feet.
- The period of the comet is 3.3 years and its perihelion to the sun is .33 of the earth's distance, or 33 million miles!
- To the ancient Egyptians, stellar objects on the horizon were thought to be in transition to the underworld, the abode of Osiris, God of death and resurrection. To them, the horizon symbolically represented the brief period between life and death.
- At virtually the same moment, from the view of the Pyramids of Giza (the belt stars of Orion) Orion's middle belt star is dead on the horizon (representing the spiritual evolution at the moment of death) due West (death).

Hopefully someone informed the National Transportation and Safety Board of this vital information.

The Great Pyramid, of course, is a long-time favorite among those on the borderlands of science, and we have received our fair share of speculative notions. One manuscript entitled *The Darkness Gradually Lightens*, for example, deals "with the original function of the Great Pyramid as revealed by study of the structural engineering and by the commentary contained within the cryptic language of the *Book of the Dead* and *Pyramid Texts*." The original function, it turns out, is that "the upper chamber of the Great Pyramid is a blast-containment structure, with an insulated heat-shield ceiling and a multi-portcullis fire-door. It follows that the stone 'coffer' contained an energy source, and that it was reloaded externally and introduced via the reusable portcullis. The energy explosion was released externally via two narrow vent shafts, forming beams recorded in the ancient texts as 'a flame before the wind to the end of the sky and to the end of the earth . . . his beams flood the world with light.' " The figure below of the King's Chamber is said by the author to represent this blast furnace.

Along similar lines, Thomas O. Mills kindly sent me an autographed and numbered (#420) copy of *The Truth*, in which he set out "to prove the Hopi

Figure 4. The King's Chamber in the Great Pyramid, allegedly a blast furnace.

Myth of Creation." What's the connection to the Great Pyramid? After a re-counting of the Hopi myths of a separate creation, Mills speculates that since it appears from mitocondrial DNA that Neanderthals are a separate species from modern humans, "perhaps, as the Hopi have taught for hundreds of millennia, they were from a previous world and could not talk or meet the Creators ex-pectations." Since these Hopi myths indicate that they "traveled east over the ocean for a long period of time," Mills suggests we look west "past Polynesia, West Samoa, Tonga, New Guinea, Indonesia and then come to the Gulf of Aden and the Red Sea. Egypt and the Pyramids. Could the Pyramids be what the Hopis were talking about? Ancient Egyptians considered the area south of Gaza as the actual location of the 'First Time' or their 'Garden of Eden.' Could this be the location where mankind was created by the Creator and Spider-woman?" Mills's figure below represents the connection between the Hopi and the Egyptians: "The Hopi myth starts by saying that the Creator instructed his nephew to set up his (the Creator's) plan on Earth or one for the Creator, one for the Nephew and seven for the Worlds to come." If you view the pyramid complex looking to the east from above you see the alignment in FIGURE 5.

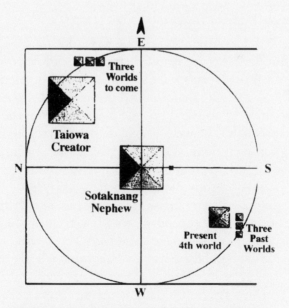

Figure 5. Alignment of the three pyramids at Giza allegedly matches that of the belt in the con-stellation of Orion.

Why pyramids? Balance is a key concept in Hopi mythology, so Mills wonders: "Could it be that the Pyramids were built to keep the Earth rotating in perfect balance? My children placed small star stickers on one blade of the ceiling fan in their room and could not figure out why the fan started wobbling and making noise at low speed. Once I removed the stickers the fan returned to normal. I have read that a person standing on the Earth is actually traveling at 1000 mph as we travel through space. So speed and weight have some effect on the balance of a moving object, like a top, centrifugal force. If the object slows down the force is reduced and the object wobbles and falls. Does the size of the top really matter? Could it be the size of the Planet and just in balance by the slightest of margins? [sic]" The pyramids were the Hopi Indians' way of keeping earth in the balance. Al Gore take note.

SWIMMING IN THE EDDIES WITH THE FLOTSAM AND JETSAM

These "hermit scientists," as Martin Gardner originally called them, are, among their variegated characteristics, outsiders. It is not that outsiders cannot make important contributions to science. They can and have. But in order to think outside the box one first must know what is inside the box (it's called graduate school), one must be able to convince those in the box that the box needs reinventing (it's called peer review), and, of course, one must be right (it's called research). Far from scientists being unaccepting of radical new ideas, any scientist worth his salt would love to witness or be part of a scientific revolution. But science is conservative. It cannot afford not to be. It makes rigid demands on its participants in order to weed out the bad ideas from the good.

Despite this conservatism scientific revolutions do happen, and not all that infrequently. But for every lone genius working away in solitude that shifted the paradigm, shattered the pedestal, or smashed the status quo, ten thousand quacks working away in solitude didn't understand the paradigm, couldn't find the pedestal, or whiffed when swinging at the status quo. While I find it odd that people send such articles and books to a magazine well-known for being skeptical of even ordinary extraordinary claims, there is value in reading them because they provide data on how belief systems work. Clearly the problem here is not one of education or intelligence (many of the authors of these treatises have M.D.s and Ph.D.s tagged onto their names on their letterhead). The problem is there seems to be no screening process, no filter to weed out fantasy from reality. Anyone with a modicum of intelligence, an active imagination, a handful of popular books on science and religion, and some free time, can turn these out at will. And the new world order of electronic publishing and instant Internet distribution is making it more and more difficult to discriminate between reality and fantasy. The knowledge filters of the past—peer-reviewed journals, conference paper reviews, journalistic integrity—are being bypassed by a more direct and unfiltered link to the world at large. Harmless you say? Ask Timothy McVeigh where he got his information on the evil doings of the government, as well as how to build bombs.

In exploring these fringe ideas we are veering out of the main channel and into the eddies, swimming with the flotsam and jetsam in search of a good idea. There are not many to be found but it is worth the trip if, for no other reason, we gain a deeper understanding of the workings of the mind of the pattern-seeking, storytelling species that calls itself *Homo sapiens*—wise man.

3
ONLY GOD
CAN DO THAT?
Cloning Tests the Moral Borderlands of Science

WHERE SCIENTISTS TAKE VOLUMES to express complex ideas, poets can often say more in a few couplets, such as T. S. Eliot's summary of the human career:

> Birth, and copulation, and death.
> That's all the facts,
> when you come to brass tacks:
> Birth, copulation, and death.

Now even these most basic facts of life are being compromised with the prospect of what looks to be one of the most revolutionary sciences in history, human cloning and genetic engineering, threatening, say some critics, to reinvent birth, do away with copulation, and stave off death. Are the critics of cloning and genetic engineering that extreme in their fears and condemnations about tweaking the human genome? Some are. And they hail from the highest quarters. On February 24, 1997, on the heels of Ian Wilmut's dramatic announcement that he had achieved cloning of a sheep called Dolly[1] (so named because the original cell was isolated from a mammary gland, to which Dolly Parton responded "I'm honored. There's no such thing as baa-aa-aa-d publicity"), President Clinton sent a letter to Dr. Harold Shapiro, president of Princeton University and chair of the National Bioethics Advisory Commission, requesting that the collective "undertake a thorough review of the legal and ethical issues associated with the use of this technology, and report back to me within ninety days with recommendations on possible federal actions to prevent its abuse."[2] Note Clinton's suggested slant for the commission to counsel prohibition, not even pretending to be neutral on the subject.

Three months later Shapiro (who is, interestingly, an identical twin — a natural clone of sorts), in conjunction with the seventeen-member commission, submitted to Clinton a 120-page report entitled *Cloning Human Beings*.[3] Although the report is both thorough and thoughtful, summarizing every major position on cloning from a scientific, moral, and legal perspective, and despite the clearly stated acknowledgment that more research is needed, the advice to the president was unequivocal: "It seems clear to all of us, however, given the current stage of science in this area, that any attempt to clone human beings via somatic cell nuclear transfer techniques is uncertain in its prospects, is unacceptably dangerous to the fetus and, therefore, morally unacceptable." Even though Shapiro admitted to Clinton that "the Commission itself is unable to agree at this time on all the ethical issues that surround the issue of cloning human beings in this manner," he was confident that "moral consensus on this issue should be easily achieved."[4] Loosely translated: "We've got a few dissidents in the group, Mr. President, but don't worry, we'll straighten 'em out and help you put a stop to this thing."

How does one achieve moral consensus on a complex scientific subject — a new technique called *somatic cell nuclear transfer*, to be exact — about which so little is known? Easy: call it cloning, associate it with all that is evil in Western civilization (mentioning Hitler and the Nazis is always helpful), focus the public's attention on any conceivable misuse, and follow the slippery slope to genetic Ludditism. The National Bioethics Advisory Commission is not this extreme, but a nonprofit advocacy organization out of Cambridge, Massachusetts called the Council for Responsible Genetics and representing over 1,000 scientists, bioethicists, and religious leaders concerned about the social and ethical implications of genetic technologies, did just this in their position statement issued shortly after the Dolly story broke:

I. We call upon the nations of the world to prohibit the cloning of human beings, by incorporating such prohibitions into their national laws and statutes.

II. We call upon the United Nations to take the initial steps by constituting an International Tribunal to articulate the concerns arising in different

nations, cultures, religions and belief systems, with respect to the potential cloning of humans.

III. We call upon the Congress of the United States to pass legislation to:
  1) Prohibit the cloning of humans either through embryo splitting or
     nuclear transfer.
  2) To exclude animals and plants, their organs, tissues, cells or mole-
     cules from patenting, whether naturally occurring or cloned.

IV. We call upon the citizens of the world and their institutions, including
    the media, to promote a vigorous public debate regarding the cloning
    of animals, and in particular, which practices are acceptable and which
    are are not.[5]

In an appalling breach of logic and blatant use of fear-mongering rhetoric,
the council compared cloning with the worst evils in human history (the statement is so egregious that I provide the entire quote lest the reader think I am
setting up a straw man):

> In the course of human history our species has recognized many behaviors
> that are counter to the interests of the survival, development and flourishing
> of individuals within civilization. Among these are involuntary servitude, or
> slavery; torture, the use of poison gas, the use of biological weapons, and
> human experimentation without consent. Human societies are working on
> preventing other destructive practices such as child labor, environmental deg-
> radation, nuclear war and global warming.
>
> The cloning of sheep and monkeys opens up the specter of human clon-
> ing. The fundamental character of this activity is to transform humans into
> commodities, to devalue the relationship of humans to each other and to
> their culture. Just as the 13th Amendment outlawed slavery, and other laws
> prohibited torture, child labor, and other forms of human exploitation, the
> time has come to prohibit human cloning. We therefore call upon the United
> States, individual nations, and the United Nations to declare the cloning of
> humans beings an immoral and illegal activity.[6]

Not so loosely translated: cloning is the moral equivalent of slavery, torture,
chemical and biological warfare, child labor, and all manner of exploitation.
Apparently this attitude reflects that of the American population at large. A
Time/CNN poll of 1,005 adults, conducted the week of the Dolly media feeding
frenzy (reported in the March 10 issue), found that 93 percent of Americans
oppose the cloning of humans, and 65 percent oppose the cloning of animals.

Most interestingly, only 7 percent said they would want to clone themselves, thereby gainsaying the fear that human cloning will be rampant once the technology is mastered.[7]

I do not want to oversimplify the ethical issues involved in this debate, and I do agree with the National Bioethics Advisory Commission that in an educated democracy open discussion of controversial subjects is vital to the smooth functioning of the citizenry (I even participated in one such debate in the Op-Ed pages of the *Los Angeles Times*, with Patrick Dixon taking the con position).[8] I recognize that extreme positions like that of the Council for Responsible Genetics can be found among human cloning zealots (Richard Seed being a fine example of the reckless abandonment of science and rationality in what appears to me nothing more than a personal quest for immortality, or at least infamy).[9] To that end I take to heart the sage advice of Proverbs 10:19: "In the multitude of words there wanteth not sin; but he that refraineth his lips is wise." I have nothing particularly new to add to the reasoned arguments and responses offered by both sides in this debate, already covered elsewhere in the flood of books and articles released over the past two years.[10] But there are two misunderstandings about science held by critics of cloning, one particular and one general, that I find especially interesting for what they reveal about our deepest fears: the *Identical Personhood Myth* and the *Playing God Myth*.

THE IDENTICAL PERSONHOOD MYTH

An especially odd conjunction in the cloning debates is the assumption made by almost everyone from John Q. Public to Dr. Gene O. Type, that cloning will produce an identical *person*—that genes make the man, and environment counts for nothing. The irony is that this classic fallacy of genetic determinism is being proffered as an argument against cloning by those who traditionally embrace strong environmental determinism. Shouldn't their argument be the exact opposite, i.e., "clone all you like, you'll never produce another you because environment matters more than heredity"? Countless quotes fill the pages of articles I clipped the past several years, but one will do, this from the relentless king of genetic Luddites, Jeremy Rifkin, an environmental determinist if there ever was one: "It's a horrendous crime to make a Xerox of someone. You're putting a human into a genetic straitjacket. For the first time, we've taken the principles of industrial design—quality control, predictability—and applied them to a human being."[11]

Behavior geneticists and evolutionary psychologists—among the strongest proponents of genetic influentialism—demonstrate in their research precisely how environments interact with heredity to shape behavior and personality. This *interactionism* starts when genes code for biochemical reactions, which regulate physiological changes, which govern biological systems, which impact neurological actions, which induce psychological states, which cause behaviors; these behaviors, in turn, interact with the environment, which change the behaviors, which influence psychological states, which alter neurological actions, which transform biological systems, which modify physiological changes, which transfigure biochemical reactions. And all of this happens in a complex interactive feedback loop between genes and environment throughout development and into adulthood. Two examples make my point that identical personhood is a myth.

(1) In her 1999 book on twins, *Entwined Lives*, behavioral geneticist and evolutionary psychologist Dr. Nancy Segal reveals through mountains of research that genes influence our behavior and personality in innumerable ways that can no longer be ignored. Comparing identical twins reared apart with identical twins reared together, fraternal (nonidentical) twins reared together, siblings reared together, and pseudotwins (genetically different adopted children) reared together, identical twins reared apart are more alike on almost all measures than the comparison groups (including a number of striking similarities between identicals reared apart—from the sublime, such as Harold Shapiro and his twin both growing up to become university presidents, to the ridiculous, such as a preference for a rare Swedish toothpaste called *Vademecum*). But what struck me in reading Segal's book was just how different twins can be. Even for such characteristics as height and weight, which have a heritability of over 90 percent, I was struck by how different many of these identical twins turned out to be (photographs generously appear throughout Segal's book). And when we consider traits that show heritabilities in the 50 percent range, it is clear how much environment counts.[12]

Maybe it is a personal preference on my part for seeing the glass as half full (I am by nature a rather sanguine person), but when I read that 50 percent of the variance among individuals in such attributions as personality, religious preferences, and sociopolitical attitudes is accounted for by genetics, I immediately thought, "Hot damn, that leaves half for me to tweak in the direction I want." Segal and her colleagues who study twins have revealed that heredity counts a great deal more than it was recently fashionable to believe, and their science is solid (unlike that of the pseudoscientific eugenicists earlier this century). But Segal also explained that these twin studies reveal something else: "Genetic influence does not mean that behaviors are fixed, but the ease, immediacy and magnitude of behavioral change vary from trait to trait, and from person to person."[13] It is from that variance that unique personhood, even between identical twins, emerges.

(2) In the film version of Ira Levin's 1976 novel, *The Boys From Brazil*, Gregory Peck plays the evil Dr. Josef Mengele, who has come out of hiding in the jungles of South America to orchestrate a plan for a master race that begins with the cloning of der Führer himself (using blood and tissue samples that survived his death). Laurence Olivier's Nazi hunter Lieberman, when he understands the full intent of Mengele's scheme, inquires of a scientist colleague if cloning another Hitler (or seven or eight) is really possible, declaring "It's monstrous!" The scientist explains to him that it could be used for good as well, but that in any case genes are only half the story: "Wouldn't you want to live in a world of Mozarts and Picassos? Of course, it's only a dream. Not only would you have to reproduce the genetic code of the donor, but the environ-

Figure 6. The ultimate fear of cloning is represented in a scene in *The Boys from Brazil* (1978), with Gregory Peck as Dr. Josef Mengele, who cloned numerous Adolf Hitlers.

mental background as well."[14] Mengele's problem throughout the film, in fact, is duplicating Hitler's quirky and complex life history.

This turns out to be a clearly impossible task because historical events are highly *contingent*, by which I mean *a conjuncture of events occurring without design*. Contingencies, in turn, interact with *necessities*, or *constraining circumstances compelling a certain course of action*.[15] Rewind the timeline of Hitler's life and play it again and the chances of an Austrian corporal's life ending as Germany's Führer in a bombed-out Berlin bunker with over fifty million dead in his wake are virtually nil. In the course of a life, or even just a childhood, the number and complexities of these conjunctures are so beyond computation that there is no possible way to duplicate a personal life, let alone the cultural history of the period. And the past is governed by both contingencies and necessities—the sometimes small, apparently insignificant, and usually unexpected events of life, in conjunction with the large and powerful laws of nature and trends of history. The past, including each and every one of our personal histories, is constructed by both contingencies and necessities, therefore I combine the two into one term to express this interdependence—*contingent-necessity*—taken to mean: *a conjuncture of events compelling a certain course of action by constraining prior conditions*. Karl Marx said it best in *The Eighteenth Brumaire* with these now classic lines: "Men make their own history, but they do not make it just as they please; they do not make it under circumstances chosen by themselves, but under circumstances directly found, given and transmitted from the past."[16]

Cloning Hitler's environment would require not only emulating the circumstances transmitted from the Europe of the 1910s and 1920s, including fin-de-siécle Vienna, Wilhelmine Munich, the Great War, Hitler's gassing in the trenches causing temporary blindness, his subsequent recovery in the hospital in which he formulated many of his ideas about the Jews, the political turmoil of 1918–1919 that influenced his political thinking, his years as a starving artist and political neophyte in beer halls and pastry shops, and the nascent formation of the Nazi party with all the inside bickering and back-stabbing before he could come to power. Where, except in a Hollywood script, could such a sequence be duplicated in enough detail to reproduce the same person?

To examine just one crucial event in Hitler's rise to power, Ron Rosenbaum shows "just how precarious, contingent, that end result was—the extent to which Hindenburg's fateful decision to call upon Hitler was the product of unpredictable factors of chance and personality, rumor and accident, rather than the inevitable product of historical abstractions."[17] Or go back a little further to Hitler's capricious childhood as the first surviving child of his mother. She pampered and favored him over two older stepchildren from her husband's earlier marriage. But since they were considerably older than Hitler, he soon found himself the eldest son with three younger siblings. So what?

Applying Frank Sulloway's birth order model to this scenario, we see that personality, including a militant personality, is strongly shaped by family dynamics: "In terms of birth order, conservative ideologies will typically attract firstborns. Among conservatives, firstborns will gravitate toward those ideologies, such as fascism, that are tough-minded."[18] Hitler's twisted road to becoming a firstborn, conservative, tough-minded, fascist ideologue is contingently complex. Can we possibly duplicate it? It would seem unlikely. Or to take it back one more generation, there is the confusing genealogy of Alois Hitler with his mother, Maria Schicklgruber, and the mysterious father who might even have been Jewish. So it would also require cloning Hitler's doting mother, his stern father, which would require cloning their parents and the environments of their cultures, and so on. Add all of this together and you've got a unique person, no more but no less than any of us. Go ahead and clone a thousand Hitlers—I'll give you Vegas odds that not one of them will develop into a German Führer who launches another monstrous political and military regime.

THE PLAYING GOD MYTH

Even this issue of genetic and environmental personhood, contentious as it may be, along with all the other ethical concerns, is secondary or tertiary to the deeper fear that strikes people's hearts when it comes to the scientific manipulation of the human genome. This dread was stated forthrightly by President Clinton in his initial press conference: "There is much about cloning that we still do not know. But this much we do know: Any discovery that touches upon human creation is not simply a matter of scientific inquiry, it is a matter of morality and spirituality as well."[19] In fact even before the National Bioethics Advisory Commission submitted its report to the White House, President Clinton had instituted a ban on federal funding related to research on the cloning of humans, and asked that the private sector follow suit. Later that year, when the maverick Chicago physicist, Dr. Richard Seed, announced his intention to become the first human to be cloned, Clinton held another press conference urging Congress to ban human cloning altogether (not just research funds) before Seed could seed himself: "Personally, I believe that human cloning raises deep concerns, given our cherished concepts of faith and humanity."[20]

What's spirituality and faith got to do with it? A lot, according to the *Time/*

CNN poll, which revealed that 74 percent of Americans answered "yes" to the question "Is it against God's will to clone human beings?"[21] How can anyone presume to know God's will? Through the Bible, of course, but even here it is not clear what is proper or moral, thus leading religious leaders to disagree on how to respond to cloning and genetic engineering. Sifting through a sizable body of religious literature on the subject, the strongest objection seems to be that cloning, in the words of the Catholic theologian Albert Moraczewski (echoing the Pope's 1987 *Donum Vitae* that condemned it) would "jeopardize the personal and unique identity of the clone (or clones) as well as the person whose genome was thus duplicated." Twins do not count, Moraczewski explained, since they are not the "source or maker of the other," meaning only God can do that.[22] The Christian Life Commission of the Southern Baptist Convention, on March 6, 1997, passed a resolution that called on Congress to "make human cloning unlawful" and for "all nations of the world to make efforts to prevent the cloning of any human being." Finally (just to show ecumenical diversity), Fred Rosner, a Jewish bioethicist, wrote that cloning can be considered as "encroaching on the Creator's domain."[23]

So that is what it really comes down to: the fear that science is unduly infringing on religion's turf. Actually, this theme has been circulating for decades, from Howard and Rifkin's 1977 *Who Should Play God?: The Artificial Creation of Life and What it Means for the Future of the Human Race,*[24] to Ted Peters's 1997 *Playing God?: Genetic Determinism and Human Freedom,*[25] to a flurry of Godly warnings following the Dolly brouhaha. The message is clear: science can only go so far. Kenneth Woodward, opining in *Newsweek,* suggested as much: "Perhaps the message of Dolly is that society should reconsider its casual ethical slide toward assuming mastery over human life. Do we really want to play God?"[26] A Conrad editorial cartoon in the *Los Angeles Times* captured the country's mood: a modification of Michelangelo's creation of man fresco on the ceiling of the Sistine Chapel shows two cloned men touching index fingers with a caption that reads "The scientists who play God."[27]

What's God got to do with it? Plenty, in our culture. But most people inappropriately intertwine science and religion—enterprises with goals and methods that could not be more disparate—and as a consequence are either overly offended by religion's perceived encroachment into science, or are unnecessarily threatened by science's alleged intrusion into religion. Consider the climactic scene of Robert Wise's 1951 science fiction film classic, *The Day the Earth Stood Still.* The space alien Klaatu (played by Michael Rennie, who goes by the earthly name "Mr. Carpenter" in this Jesus allegory) is killed by a fear-mongering gov-

Figure 7. The limits of science are presented in the film, *The Day the Earth Stood Still* (1951), where viewers were told by the film industry's Breen censorship Board that "only God" can raise the dead.

ernment agency, then resurrected by his robot charge Gort. Astonished by the power of this alien technology, Patricia Neal's Mary Magdalene-like character (after delivering what has become one of the most memorable lines in science fiction history—"Gort, Klaatu barada nikto") inquires whether control over life and death is in the cards for the science of the future. Klaatu assures her that such powers belong only to the "Almighty Spirit," and that his life extension is good only "for a limited period" the duration of which "no one can tell." Telling indeed. In Edmund North's original script Gort resurrects Klaatu without limitation. But the film industry's Breen Board (a censorship committee established for self-regulation) told the producers: "Only God can do that."

This Promethean theme of limiting knowledge is a common one not only in science fiction, but in science fact. For every mythic Icarus who flew too close to the sun there are real life scientists who got their wings clipped for daring to push their frontiers too far. Birth control? Only God can do that. Artificial insemination? Only God can do that. Life extension? Only God can do that. Euthanasia? Only God can do that.

We should not be surprised, then, that when a British government advisory commission, bucking Clinton's lead, encouraged the legalization of research into cloning human tissues and organs for therapeutic uses, they were met with fierce opposition from both religious and secular groups. Cloning? Only God can do that.

What precisely did this Human Genetics Advisory Commission recommend? From the wails of doom and gloom we heard, one would think they had suggested a scheme of harvesting body parts from cloned adults, á la Robin Cook's *Coma*. On the contrary. The recommendations could not have been more judiciously worded: ". . . we believe that it would not be right at this stage to rule out limited research using such techniques, which could be of great benefit to seriously ill people."[28]

To technophobes who resist any venture into forbidden knowledge (while simultaneously partaking of every medical breakthrough that benefits them per-

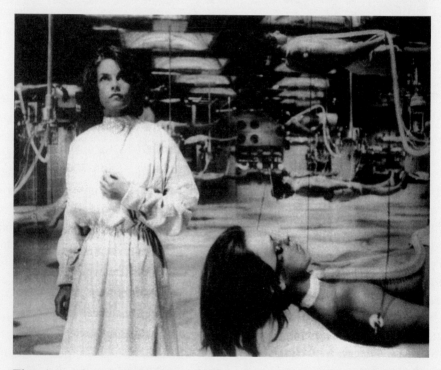

Figure 8. The evils and limits of science are portrayed most dramatically in the film *Coma* (1978), where Genevieve Bujold discovers an organ factory run by greedy medical doctors.

sonally), such cautious forays into the future are the slippery slope into the scientistic hell where vultures will peck at us for eternity. But let's step back for a moment. What do we have to fear? The mass hysteria and moral panic surrounding cloning is nothing more than the historically common rejection of new technologies, coupled to the additional angst produced when medical advances fly too close to religion's sun. In the 1940s, when artificial insemination was first introduced, critics called it adultery. "Only God can do that" say the religious Luddites. "Only Nature can do that" cry the secular Luddites.

In fact, nature is already cloning humans. They're called identical twins. Why aren't moralists crying for legislation against twinning? Because it happens naturally, and according to the Law of Ludditism, "Only God/Nature can do that."

Nonsense! Most of us are alive because of medical technologies and social hygiene practices that have doubled the average life span in this century. What's Godly or natural about heart-lung transplants, triple bypass surgeries, vaccinations, or radiation treatment? What's Godly or natural about birth control, in vitro fertilization, embryo transfer, and other fully-sanctioned birth enhancement technologies? Absolutely nothing. Yet we cheerfully accept these advances because we have benefited from them and, more important, grown accustomed to them.

Why not lift the ban on all research into cloning—including humans—and see what happens? Let's run the social experiment and analyze the data. The null hypothesis says that nothing evil will befall humanity. The cloning cons say the experimental results will reject the null hypothesis. The cloning pros say they will not. The only way to find out is to run the experiment. In the borderlands between science and pseudoscience, the best method to determine which fuzzy category a claim belongs is to test it. Why not do that here? Most of the horror-laden scenarios proposed by moralists are already addressed by law—a clone, like a twin, is a human being, and you cannot harvest the tissues or organs of a twin. A clone, like a twin, is a person no less than any other human. Even with an identical genome, a contingently unique history guarantees a unique personhood. In any case, cloning is going to happen whether it is banned or not, so why not err on the side of freedom and allow scientists to freely explore the possibilities—not to play God, but to do science.

In 1818 Mary Shelley warned in her novel, *Frankenstein, or the Modern Prometheus*, that "supremely frightful would be the effect of any human endeavour to mock the stupendous mechanism of the Creator of the world."[29] The censors took her words to heart in the final cut of James Whale's 1931 film version starring Boris Karloff. In the riveting laboratory scene when the monster is brought to life, Dr. Frankenstein roars "It's alive. It's alive. In the name of God. . . ." At that moment his lips keep moving but his voice is cut off. The censors deleted the rest of the sentence—the forbidden words that have frightened cultures from ancient Greece to modern America: ". . . now I know what it feels like to be God."

Scientists don't want to be God. They just want to solve scientific problems. Only scientists can do that. Let them do it.

Advertisement for *Frankenstein* (1931). *Courtesy of Scott MacQueen.*

Figure 9. The quintessential icon of science gone mad, James Whale's 1931 *Frankenstein* has Boris Karloff's monster represent all that we must fear if we don't keep scientists in check.

4
BLOOD, SWEAT, AND FEARS
Racial Differences and What They Really Mean

IN *AN ESSAY ON MAN*, the nineteenth-century English poet and essayist Alexander Pope elucidated the pitfalls of speculating on ultimate causes derived from immediate events:

> In vain the sage, with retrospective eye,
> Would from th' apparent what conclude the why,
> Infer the motive from the deed, and show
> That what we chanced was what we meant to do.

Pope's wise words were in the back of my mind as I began writing this essay on a miserably cold and rainy March 5, 2000 Sunday morning, simultaneously watching the elite runners in the Los Angeles Marathon—just a handful among the 23,000 weekend warriors who braved the elements—cross the finish line. Although I have run the L.A. Marathon before, and even once completed a marathon after first swimming 2.4 miles in the open ocean and riding a bike 112 miles in the Hawaiian Ironman triathlon, I would not have given the results a second glance were it not for a book I had just read that called my attention to a certain characteristic common in the top five finishers' profiles. They were: (1) Benson Mutisya Mbithi, 2:11:55, (2) Mark Yatich, 2:16:43, (3) Peter Ndirangu Nairobi, 2:17:42, (4) Simon Bor, 2:20:12, and (5) Christopher Cheboiboch, 2:20:41.[1]

It was not the times of the top five finishers that stood out in this year's race, since they were well below world and course records (understandably, considering the conditions). What was startling was their country of origin. All were from Kenya. Coincidence? Hardly. Meaningful? To some, yes; to others, no; to science, maybe. That is the subject of the book I had just read, Jon Entine's controversial *Taboo: Why Black Athletes Dominate Sports and Why We're Afraid to Talk About It.*[2]

I will not dissemble and pretend that I was not aware of the controversy surrounding the claim that blacks are better athletes than whites due to special breeding and being closer to the brood creation of humanity in Africa. I've been an athlete and sports fan all my life and recall the vitriolic reaction to Jimmy "the Greek" Synder's 1988 off-the-cuff remarks at a restaurant about black slaves being bred for superior physicality (on Martin Luther King Day, no less, with a camera crew present): "The black is a better athlete because he's been bred to be that way. During slave trading, the slave owner would breed his big woman so that he would have a big black kid, see. That's where it all started." Blacks, Snyder explained, could "jump higher and run faster" because of their "high thighs and big size."[3]

I even saw live the now-infamous 1987 ABC *Nightline* show (occasioned by a celebration of Jackie Robinson's shattering of the color barrier in baseball) when Ted Koppel asked Los Angeles Dodger baseball executive Al Campanis why there were no blacks in upper management. Campanis said that blacks "may not have some of the necessities" for such positions. "Do you really believe that?" Koppel rejoined. "Well, I don't say all of them," Campanis demurred, "but they certainly are short in some areas. How many quarterbacks do you have, how many pitchers do you have that are black?" After continuing with his folk lesson in sports physiology, Campanis noted why blacks do not compete in elite swimming: "because they don't have buoyancy."[4] Whites are floaters, blacks are sinkers.

Campanis's attempts to explain himself opened the gates into the largely unspoken but pervasive attitudes held by many whites about blacks, even whites who would not consider themselves racist. "I have never said that blacks aren't intelligent, but they may not have the desire to be in the front office," Campanis continued. "I know that they have wanted to manage, and many of them have managed. But they are outstanding athletes, very God-gifted and they're very wonderful people. They are gifted with great musculature and various other things. They are fleet of foot, and this is why there are a number of black ballplayers in the major leagues."[5] Blacks are fast around the bases, slow around the boardroom.

As University of Texas Professor John Hoberman explained in his 1998 book *Darwin's Athletes*,[6] even many blacks embrace part of the thesis. Dallas Cowboys all-star player Calvin Hill, a Yale graduate, opined: "On the plantation, a strong black man was mated with a strong black woman. [Blacks] were simply bred for physical qualities."[7] San Francisco '49ers wide receiver Bernie Casey explained: "Think of what the African slaves were forced to endure in this country merely to survive. Black athletes are their descendants."[8] Even the liberal champion of cultural determinism, Jesse Jackson, in a 1977 CBS *60 Minutes* segment on his P.U.S.H. program for black school kids, made a case for heredity over environment when he stated (in response to sociologists' environmental explanations for black's poorer school performances) that "If we [blacks] can run faster, jump higher, and shoot a basketball straighter [than whites] on those same inadequate diets . . ." then there is no excuse.[9] It is time, Jackson argued, for blacks to start living up to their potentials.

With such comments from both blacks and whites it is understandable why some blacks, such as the noted U.C. Berkeley sports sociologist Harry Edwards, respond so strongly, and usually wrongly, in the opposite extreme of environmental determinism. On a March 8, 2000 radio show I hosted with Entine, Edwards, and Hoberman, Edwards actually made the argument that the only reason blacks dominate NBA basketball, despite more than equal opportunity for whites to make it to the top, was that at this period of time the "black style" of basketball happens to be popular instead of the "white style" popular in the 1950s, and that neither "style" was in any way superior. My cohost Larry Mantle and I, both enthusiastic Laker fans, gave each other a knowing glance of acknowledgment that this was, of course, utter nonsense.[10]

Somewhere between Edwards's extreme environmental determinism and Campanis's radical biological determinism lies the truth about the cause and meaning of black-white differences in sports. But the Campanis episode was enlightening because these were not the remarks of a rabid bigot spewing racial epithets; rather, Campanis had spent decades in close proximity and in tight friendship with some of the greatest black ballplayers of the twentieth century, so his comments were emblematic of the common attitudes shared by many, perhaps most, lay people and sports enthusiasts who know just enough to speculate in a social Darwinian mode about how and why blacks dominate in some fields but not others, and what these differences tell us about the human condition.

What do these differences mean? The answer depends on what it is you want to know. I shall address this subject neither to embrace the theory nor to debunk it; rather, the question itself raises a number of other questions and problems in this field of research that makes most grand and sweeping conclusions problematic at best. Race differences research is an example of a borderlands science—much of the particular research itself is good science, but much of the speculative theorizing about causes and explanations is pure pseudoscience, so I place the overall field in the borderlands region. Much research is still to be conducted, and less ideological steering of the data needs to be infused.

FROM THE PARTICULAR TO THE GENERAL: DO BLACK ATHLETES
DOMINATE SPORTS?

If you are a basketball, football, or track-and-field fan, the black-white differences are obvious and real. You'd have to be blind not to see the gaping abyss any given day of the week on any one of the numerous 24-hour-a-day sports channels. Kenyans dominate marathon running, but you'll likely never see one line up for the 100-meter dash. On the other hand, blacks whose origins can be traced to West Africa own the 100-meter dash, but will not likely soon be taking home the $35,000 automobile awarded to the L.A. Marathon winner. And it could be a long while before we see a white man on the winner's platform at either distance. As Entine carefully documents, at the moment "every men's world record at every commonly-run track distance belongs to a runner of African descent," and the domination of particular distances are determined, it would seem, by the ancestral origin of the athlete, with West Africans reigning over distances from 100 meters to 400 meters, and East and North Africans prevailing in races from 800 meters to the marathon.[11]

But my first quibble with the debate is how quickly it shifts from Kenyans winning marathons or West Africans owning the 100-meter dash to, as stated in Entine's subtitle, "why black athletes dominate sports." I understand a publisher's desire to economize cover verbiage and maximize marketability (the inside of *Taboo* is, appropriately, filled with qualifiers, caveats, and nuances), but the simple fact is that black athletes do not dominate sports. They do not dominate speed skating, figure skating, ice hockey, gymnastics, swimming, diving, archery, downhill skiing, cross-country skiing, biathlons, triathlons, Ping Pong, tennis, golf, wrestling, rugby, rowing, canoeing, strong-man competitions, auto racing, motorcycle racing, and on and on.

In my own sport of cycling, in which I competed at elite ultra-marathon distances (200 miles to 3,000 miles) for ten years,[12] there are almost no blacks to be found in the pack. Where are all those West African sprinters at velodrome track races? Where are all those Kenyans in long-distance road races or ultra-marathon events? They are almost nowhere to be found. In fact, in over a century of professional bicycle racing there has been only one undisputed black champion—Marshall W. "Major" Taylor. And Taylor's reign was a century ago! He started racing in 1896 and within three years he became only the second black athlete to win a world championship in any sport, and this was at a time when bicycle racing was as big as baseball and boxing. Since there were few automobiles and no airplanes, cyclists were the fastest humans on earth and were rewarded accordingly with lucrative winnings and more than fifteen minutes of fame. Major Taylor was the first black athlete in any sport to be a member of an integrated team, the first to land a commercial sponsor, and the first to hold world records, including the prestigious mile record. He competed internationally and is still revered in France as one of the greatest sprint cyclists of all time. The fact that outside cycling circles he is completely unknown in America tells us something about the influence of culture on sports.[13]

By the theory proferred by Entine and others, there is no reason blacks should not be prominent in cycling since the physical requirements are so similar to

Marshall W. "Major" Taylor

Figure 10. Marshall W. "Major" Taylor, the greatest black cyclist ever, dominated the sport from 1899–1910. Where have all the great black cyclists gone? Into sports open to them. Not many would put up with the racism Taylor faced, or handle it with such nobility. "Notwithstanding the bitterness and cruel practices of the white bicycle riders, their friends and sympathizers, against me, I hold no animosity towards any man. Life is too short for a man to hold bitterness in his heart. As the late Booker T. Washington, the great Negro educator, so beautifully expressed it, 'I shall allow no man to narrow my soul and drag me down, by making me hate him.' "
—Courtesy of Andrew Ritchie, *Major Taylor: The Extraordinary Career of a Champion Bicycle Racer* (1988, Johns Hopkins University Press).

running. There are no longer racial barriers, as witnessed by the wide range of colors and nationalities that fill out the pelotons throughout Europe and the Americas. The reason they don't is obvious, says Dr. Ed Burke, a sports physiologist at the University of Colorado in Boulder: "No money, no publicity, no grass roots program. Why would gifted American athletes, with so many lucrative opportunities in other sports, choose cycling?"[14] In Europe working class fathers introduce their sons to the sport at an early age where they can be nursed through junior cycling programs until they turn professional and permanently bootstrap themselves into the middle classes. But there are not that many blacks in Europe, and in America no such social structure exists. Bottom line: in cycling, culture trumps biology.

(After Major Taylor, many cite the black sprinter Nelson Vails, since he took the silver medal on the track in the 1984 Olympics. But this is problematic because the East Germans boycotted that Olympics, and they were dominating the sport in those years, having thoroughly trounced both Vails and the 1984 gold medalist, Mark Gorski, in the world championships the year before. After Vails, Scott Berryman was a national sprint champion, and 19-year-old Gideon Massie recently won the Jr. Worlds on the track and is an Olympic hopeful for 2004. The few other isolated cases—Shaums March in downhill mountain biking and Josh Weir on the road—only further calls our attention to the dearth of blacks in cycling.)

Would blacks dominate cycling *caterus parabus*? The problem is that all other things are never equal so it is impossible to say until the natural experiment is actually run. There is no reason why they shouldn't, by the arguments put forth by Entine, since track cycling is not so different than sprinting, and road cycling is not dissimilar from marathon running in terms of the physical demands on the athlete. But we simply do not know, and thus, it would be unwise to speculate. For that matter, the *caterus parabus* assumption never holds true in the messy real world, so this whole question of race and sports is fraught with complications making it exceptionally difficult to say with scientific certainty what these differences really mean.

THE HINDSIGHT BIAS: DID EVOLUTION SHAPE BLACK BODIES BEST FOR RUNNING?

Tiger Woods may very well be the greatest golfer of all time. Although he is not "pure" black, he is considered to be by most, especially and including the black community. Thus, he very well could inspire other blacks to go into the sport. What if this were to happen on such a scale that blacks came to dominate golf like they have football and basketball? Would the explanation for this dominance be role modeling coupled to cultural momentum, or would we hear about how blacks are naturally gifted as golfers because of their superior ability to swing a club and judge moving objects at a distance due to the fact that they are closer to the African homeland of humanity, the Environment of Evolutionary Adaptation (or EEA, as evolutionary psychologists call the Pleistocene period of human evolution)?

In cognitive psychology there is a fallacy of thought known as the *hindsight bias,* which states that however things turn out we tend to look back to justify that particular arrangement with a set of causal explanatory variables presumably applicable to all situations.[15] Looking back it is easy to construct plausible scenarios for how matters turned out; rearrange the outcome and we are equally skilled at finding reasons why it was inevitable it should be so.

Consider professional basketball. At the moment blacks dominate the sport and it is tempting to slip into the adaptationist mode of Darwinian speculation and suggest that the reason is because blacks are naturally superior at running, jumping, twisting, turning, hang time, and all the rest that goes into the modern game. Then it is only a step removed from suggesting, as does Entine and U.C. at Berkeley anthropologist Vince Sarich,[16] that the reason for these natural abilities is that since humans evolved in Africa where they became bipedal, populations that migrated to other areas of the globe diluted those pure abilities through adaptations to other environments—e.g., colder climates led to shorter, stockier torsos (Bergmann's Rule) and smaller arms and legs (Allen's Rule)—thereby attenuating the ability to run and jump. African blacks, however, are closer to the EEA and thus their abilities are genetically undiluted and therefore purer. (For basketball, however, I would point out the remarkable range of skin tone one sees on the court. I grant that races may exist as fuzzy sets where the boundaries are blurred but the interiors represent a type we might at least provisionally agree represents a group labeled "black" or "white." But the step from racial group differences on a basketball court to racial evolutionary differences in the Pleistocene is a significant one, and it is here where the *hindsight bias* kicks in.)

But let's go back in time and consider the age when Jews dominated basketball and see what sorts of arguments were made for their "natural" abilities in this sport. In the 1920s, 1930s, and 1940s basketball was an east coast, inner-city, blue-collar immigrant game largely dominated by the oppressed ethnic group of that age, the Jews. Like blacks decades later, the Jews went into professions and sports open to them. As Entine so wonderfully tracks the history in *Taboo,* according to Harry Sitwack, star player of the South Philadelphia Hebrew Association (SPHA), "The Jews never got much into football or baseball. They were too crowded [with other players] then. Every Jewish boy was playing basketball. Every phone pole had a peach basket on it. And every one of those Jewish kids dreamed of playing for the SPHA's."[17] The reason why is obvious, right? Cultural trends and socioeconomic opportunities set within an autocatalytic (self-generating) feedback loop led more and more Jews to go into the game until they came to dominate it. Wrong. As Entine shows, according to the wisdom of the scientific experts of the day the Jews were naturally superior:

> Writers opined that Jews were genetically and culturally built to stand up under the strain and stamina of the hoop game. It was suggested that they had an advantage because short men have better balance and more foot speed. They were also thought to have sharper eyes, which of course cut against the other stereotype that they suffered from myopia and had to wear glasses. And it was said they were clever. "The reason, I suspect, that basketball appeals to the hebrew with his Oriental background," wrote Paul Gallico, sports editor of the *New York Daily News* and one of the premier sports writers of the 1930s, "is that the game places a premium on an alert, scheming mind, flashy trickiness, artful dodging and general smart aleckness."[18]

Figure 11. Jewish dominance of basketball in the 1930s was portrayed in this poster emphasizing the "fact" that the game requires "an alert, scheming mind, flashy trickiness, artful dodging and general smart aleckness," traits to be found in Jews.

By the late 1940s Jews moved into other professions and sports and, Entine notes, "the torch of urban athleticism was passed on to the newest immigrants, mostly blacks who had migrated north from dying southern plantations. . . . It would not be long before the stereotype of the 'scheming . . . trickiness' of the Jews was replaced by that of the 'natural athleticism' of Negroes."[19] If Jews were dominating basketball today instead of blacks, what explanatory models, in hindsight, would we be constructing? If, in thirty years, Asians come to control the game would we offer some equally plausible "natural" reason for their governance?

Does this mean that blacks are not really better than whites in basketball? No. I would be shocked if it turned out that what we are witnessing is nothing more than a culturally dominant "black style" of play. But because of the *hindsight bias* I cannot be certain that we are not being fooled.

THE CONFIRMATION BIAS: WHY ARE WE SO INTERESTED IN BLACK-WHITE DIFFERENCES?

Why, it seems reasonable to ask, are we so interested in black-white differences in sports? Why not Asian-Caucasian differences? Why has no one written a book entitled *Why Asians Dominate Ping Pong and Why We're Afraid to Talk About It?* The reason is obvious: because no one cares that Asians are the masters at ping pong. This is America, and what Americans care about are black-white differences. By way of analogy, no first-century B.C. Egyptian would have wondered if Cleopatra was black, but twentieth-century Americans have entertained that very question.[20]

The *confirmation bias* holds that we have a tendency to seek confirmatory data that support our already-held beliefs, and ignore disconfirmatory evidence that might counter those beliefs.[21] We all do this. Liberals read the paper and see greedy Republicans trying to rig the system so that the rich can become richer. Conservatives read the same paper and see bleeding-heart liberals robbing the rich of their hard-earned dollars to support welfare queens on crack. Context is everything and the *confirmation bias* makes it very difficult for any of us to take an objective perspective on our own beliefs.

Yes, there are black-white difference in sports, and there may even be good physical reasons for these differences. But, as noted above, the vast majority of sports are not dominated by blacks. Why don't we hear about them? Because they don't interest us, or they do not support our preconceived notions about the importance of black-white race questions. Out of the literally hundreds of popular sports played in the world today, blacks dominate only three: basketball, football, and track-and-field. That's it. That's what all the fuss is about. Why do we focus on those three? Because we live in America where race questions, and these three sports, are prominent.

I am not arguing that it is scientifically untenable or morally wrong to focus on these differences, but I am curious why those particular differences are of such interest to some people. Is it nothing more than some people like chocolate pudding and others tapioca? I doubt it. I suspect the *confirmation bias* directs our attention to differences most likely to support already-held beliefs about race differences.

Consider the example presented by Vince Sarich in his article for the special issue of *Skeptic* on race differences in sports. Sarich presents the selection for high oil content in corn kernels as an example of how "the underlying genetic variability has not been running out." Vince cites this in support of his argument that in a brief period of time genomes can be selected to make dramatic shifts in phenotypic characters, such as "the tripling in size of the human brain over the last two million years." He goes on to note that in only 4,000 years Polynesian peoples migrated from a small population (or populations) living on the eastern end of New Guinea, to islands all over the Pacific, from New Zealand in the south, Hawaii in the north, and Easter Island in the east. "Doing so meant crossing thousands of miles of ocean that could be chillingly cold at night, and in large outriggers where upper body strength would have been at a premium, thus apparently selecting primarily for larger body size (Bergmann's Rule), and by extension, proportionately even larger upper bodies."[22]

But is corn the best genetic comparison to make with humans? Let's use a mammal, a running mammal—thoroughbred racehorses. Here we find rather disconfirming evidence that the underlying genetic variability of thoroughbreds long ago ran out despite the vigilant efforts of highly motivated horse breeders with millions of dollars at stake for a horse who could knock off a second or two. The Kentucky Derby is the most prestigious of all thoroughbred races and has been run since 1875 when, by the way, thirteen of the fifteen jockeys were blacks. In fact, black jockeys dominated the Derby for the first thirty years, winning half of all races. The first race was 1.5 miles and was won in 2:37. In 1896 the distance was lowered to its present length of 1.25 miles and was won by Ben Brush in a time of 2:07. As evident in the table below (given in five-year increments), since 1950 the horses are just not getting any faster.[23]

1900	Lt. Gibson	2:06
1905	Agile	2:10
1910	Donau	2:06
1915	Regret	2:05
1920	Paul Jones	2:09
1925	Flying Ebony	2:07
1930	Gallant Fox	2:07
1935	Omaha	2:07
1940	Gallahadion	2:05
1945	Hoop Jr.	2:07
1950	Middle Ground	2:01
1955	Swaps	2:01
1960	Ventian Way	2:02
1965	Lucky Debonair	2:01
1970	Dust Commander	2:03
1975	Foolish Pleasure	2:02
1980	Genuine Risk	2:02
1985	Spend a Buck	2:00
1990	Unbridled	2:02
1995	Thunder Gulch	2:02

The greatest thoroughbred racehorse of all time, Secretariat, is the only horse to break the two minute barrier at 1:59.2. If this is not an example of an upper ceiling of genetic variability, I don't know what is.

Also, how do we know that Polynesians were selected for bigger upper bodies by either Bergmann's Rule for cold climates, or some other selection factor for stronger rowers? We don't. We haven't a clue as to how Polynesians got bigger upper bodies. This is a just-so story pure and simple. Sarich might be right. Who knows? I don't. He doesn't, either. Maybe those Polynesians got into an evolutionary arms race (pardon the pun but the example fits) where there were selection pressures having nothing to do with rowing or cold temperatures on New Guinea that led to bigger upper bodies, and these were the men and women most likely to make it to Hawaii. Or, maybe there were sexual selection pressures whereby for whatever cultural reason women preferred men with big arms and chests, thereby directing the gene pool toward bigger upper bodies, that were later well-adopted for ultra-marathon rowing. Or, maybe large torsos were accidentally selected (exapted) when some other trait was preferred or needed by Polynesians, and that trait was genetically linked (pleiotropy) to upper body size.[24] Or maybe Polynesian women needed stronger arms and upper bodies because, due to cultural preferences, they had to both carry their babies and forage for food, and this led to all Polynesians having larger upper bodies. Or, or, or.

BLOOD OR SWEAT?: THE NATURE-NURTURE DEBATE IN SPORTS

In the middle of the 1985 3,000-mile nonstop transcontinental bicycle Race Across America, I was pedaling my way across Arkansas when the ABC *Wide World of Sports* camera crew pulled up alongside to inquire how I felt about my third place position—way ahead of the main pack, but too far behind to catch the leaders. I answered: "I should have picked better parents."

I got the quote from the renowned sports physiologist Per-Olof Astrand at a 1967 exercise symposium: "I am convinced that anyone interested in winning Olympic gold medals must select his or her parents very carefully."[25] At the time I regretted repeating it because I meant no disrespect for my always-supportive parents. But it was an accurate self-assessment for I had done everything I could do to win the race, including training over 500 miles a week in the months before, observing a strict diet, employing weight training, utilizing massage therapists and trainers, and more. My body fat was 4.5 percent, and at age thirty-one I was as strong and fast as I had ever been or would be. Nevertheless, it was apparent I was not going to win the race. Why? Because despite maximizing my nurture, the upper ceiling of my physical nature had been reached and was still below that of the two riders ahead of me.

This vignette is emblematic of the larger discussion in sports physiology on the relative roles of heredity and environment. In 1971, the exercise physiologist V. Klissouras, for example, reported that 81–86 percent of the variance in aerobic capacity, as measured by VO^2 uptake, is accounted for by genetics. In 1973, he confirmed his earlier findings in another study that showed that only 20–30 percent of the variance in aerobic capacity can be accounted for by the environment—i.e., training.[26]

Randy Ice, the sports physiologist who has been testing Race Across America cyclists for the past 18 years, estimates that 60–70 percent of the variability between cyclists in aerobic capacity is genetically determined.[27] Others estimate similar percentages for anaerobic threshold, workload capacity, fast twitch/slow twitch muscle fiber ratio, maximum heart rate, and many other physiological parameters that determine athletic performance. In other words, the difference between Pee Wee Herman and Eddy Merckx (the greatest cyclist of all time) is largely due to biology.

Now, let's be clear that no one—not Vince Sarich and Jon Entine on one end nor, hopefully, Harry Edwards on the other—is arguing that athletic ability is determined entirely by either genetics or environment. Obviously it is a mixture of the two. The controversy arises over what the ratio is, the evidence for that ratio, and the possible evolutionary origins of the difference. What surprised me in reading Entine's book and Sarich's article—both making the best case for evolutionary origins of biologically-based racial group differences in athletic ability—was the dearth of hard evidence and the need to make inferences and sizable leaps of logic.

Although Entine's book is promoted as if it were a polemic for the hereditary position, he confesses that even in his best case example of Kenyan marathon runners, we cannot say for certain if they are "great long distance runners because of a genetic advantage or because their high-altitude lifestyle serves as a lifelong training program." It's a chicken-and-egg dilemma, Entine admits: "Did the altitude reconfigure the lungs of Kenyan endurance runners or was a genetic predisposition induced by the altitude? Is that nature or nurture . . . or both?"[28]

It is both. But proving a particular percentage of each is tricky business. "Most theories, including those in genetics, rely on circumstantial evidence tested against common sense, known science, and the course of history," Entine explains. "That scientists may yet not be able to identify the chromosomes that contribute to specific athletic skills does not mean that genes don't play a defining role. . . ."[29] Clearly that is so. But the real debate is not *if*; it is how, and how much. It is here where the science is weak and our biases strong.

In his article Vince Sarich also makes a comparison of measured differences of racial morphological distances within our species, as well as those of other primate species, to argue that lots of morphological features vary widely both within and between species, so we should not be surprised to find equally dramatic differences in athletic abilities within the human species.[30] This is flawed logic. Assuming the data are accurate that human racial groups vary in measured cranial and facial measurements as much as or even more than chimpanzees and gorillas vary both within their own species and even between their species, so what? What has this got to do with athletic ability? What do cranial and facial feature variations between human groups and primate species have to do with black-white athletic differences in humans? Is it possible that some traits vary more than other traits? Of course. It is specious reasoning to jump from one set of characteristics (facial features) to another (running). Also, running may be a far more complicated set of variables for genetic coding than cranial and facial features. Running ability depends on a host of variables—fast twitch/slow twitch muscle fiber ratio, VO^2 uptake capacity (how efficiently oxygen is exchanged from the lungs into the blood), lung capacity, maximum heart rate, anaerobic threshold figures (that determine the level one can sustain work output), measures of strength versus endurance, etc.[31] Do we know that all of this is coded by a similar size cohort of genes that codes for cranial and facial features? I put this question to Savich, who responded as follows:

> I don't know that I, or any biologist I know, would care to be on the record arguing that running is far more complex (that is, more genetic loci are involved) than cranial and facial structure, or the reverse. We don't have a clue either way. I have the sense that you'd like to have structural and physiological proxies for running ability; that is, characteristics whose variation is correlated with variation in running ability. So would I. Now obviously such characteristics must exist, and just as obviously we don't know very much about them. We are a very long way from predicting running ability from either genotypic or phenotypic data, and asking for such is to, in effect, end the discussion by denying that there is, at this time, anything much to discuss "scientifically."
>
> Rather than throwing up our hands or sitting on our hands, what I've argued in my piece is that we go back to where we started—look at the levels of interpopulational ("racial") variation present in such features as we do have data on, and see how variation in running ability compares in that context. It's pretty much what Darwin did when looking at the variation among domestic breeds of animals and inferring from that how change could take place over geological time.[32]

Fair enough. I don't know and Sarich doesn't either. No one knows. And agreed, this doesn't mean we should throw in the towel. No doubt some black-white differences in some sports are heavily influenced by genetics and might possibly even have an evolutionary basis of origin. But *proving* that supposition is another matter entirely. As it is, to be fair, for the extreme environmental position. Harry Edwards, for example, argued on my radio show that Kenyans are tenacious trainers, rising at 5:00 A.M. every morning to run mountains at high altitude. But that's just the *hindsight* and *confirmation biases* at work again, where we examine the winner of a race to see what ingredients went into the winning formula. It ignores all the other hardworking jocks who also got up every morning at 5:00 A.M. (oh, don't I remember it so painfully well?) but didn't take the gold. Or the other winners who slept in until 8:00 A.M. and went for a leisurely jog. Training alone won't get you to the finish line first. Neither will genetics. To be a champion you need both.

MASTER OF MY FATE

We are all products of an evolutionary history of biological descent. Modifying Astrand's phrase, our parents have been very carefully chosen for us — by natural selection. Yet we are what we are because of our biology *in interaction with the environment*. Theoretically we can separate them through twin studies and behavioral genetics. But in practice they cannot be separated. Even the daunted statistical percentages used in describing the relative influence of heredity and environment are descriptive for large populations, not individuals. Even the most complete knowledge of a person will not allow us to predict the future of this individual, because the laws for making such predictions are built around populations.

The key element here is the *range of possibilities*. Behavior geneticists call it the *genetic reaction range*, or the biological parameters within which environmental conditions may take effect. We all have a biological limit, for example, on how

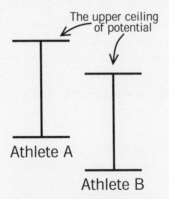

Figure 12. Athlete *A* may be biologically superior to athlete *B*, but such environmental variables as coaching, diet, training, and the will to win can lead *B* to beat *A* every time. We are free to select the optimal environmental conditions that will allow us to rise to the height of our biological potentials.

fast we can ride a 40k time trial or run a 10k. There is a range from lowest to highest that establishes the parameters of our performance. In the diagram below, athlete *A* has a higher genetic reaction range than athlete *B*. But there is overlap of the ranges, and this is the key to where such environmental factors as nutrition, training, coaching, and desire take effect. *A* may be more "gifted" than *B*, but this does not mean he will always or even ever beat *B*. If *B* performs at his best and *A* is only at 50 percent of his potential, then the genetic advantage is negated. *Inheritability of talent does not mean inevitability of success, and vice versa.*

Why do some black athletes dominate some sports? For the same reason that some white athletes dominate some other sports, and some Asian athletes dominate still other sports—a combination of biological factors and cultural influences. We do not know for sure how to tease apart these variables, but we've got some reasonably good indications. What do the differences really mean? My answer is a consilience of both positions: *We are free to select the optimal environmental conditions that will allow us to rise to the height of our biological potentials.*

In this sense athletic success is measured not just against *other's* performances, but against the upper ceiling of our own ability. To succeed is to have done one's absolute best as measured against the high mark of one's personal range of possibilities. To win is not just to have crossed the finish line first, but to cross the finish line in the fastest time possible within the allowable genetic reaction range. The poet William Ernest Henley expressed this concept well in his stirring *Invictus*:

> Out of the night that covers me,
> Black as the pit from pole to pole,
> I thank whatever gods may be
> For my unconquerable soul.
> It matters not how straight the gate,
> How charged with punishments the scroll,
> I am the master of my fate:
> I am the captain of my soul.

5
THE PARADOX OF THE PARADIGM
Punctuated Equilibrium and the Nature of Revolutionary Science

> Down went the owners—greedy men whom hope of gain allured:
> > Oh, dry the starting tear, for they were heavily insured.
> > —W. S. Gilbert, *The 'Bab' Ballads*, "Etiquette"

STEPHEN JAY GOULD CAN FIND meaning and metaphor in the most unusual of literary places, so perhaps we can consider this consoling advice of his favorite operatic authors in the light of ambitious proprietors of scientific ideas who have apparently been rejected, as later exonerated by the insurance of the truth. But how can we know today who will be villified or venerated tomorrow? As paranormalists are fond of saying (after citing such notable blunders as Lord Kelvin's paper "proving" that heavier-than-air craft could not fly), "they laughed at the Wright Brothers." The standard rejoinder, made by skeptics for both levity and effect, is: "They also laughed at the Marx Brothers."

The point is that specific historical references to wrongly rejected theories is not a general principle that applies to all cases of intellectual rebuff. Every instance of dismissal has its peculiar set of historical contingencies that led to that outcome. Historical abnegation does not automatically equal future vindication. For every Columbus, Copernicus, and Galileo who turned out to be right, there are a thousand Velikovskys (*Worlds in Collision*), von Danikens (ancient astronauts), and Newmans (perpetual motion machines) who turned out to be wrong.

This is why scientists and skeptics bristle when they hear descriptions such as "revolutionary," "earth-shattering," and "paradigm shift" freely thrown about by any and all would-be (and wanna-be) revolutionaries. To reverse the analysis, however, just because some quacks and flimflam artists (and genuinely honest thinkers) making claims of a new paradigm are wrong, does not mean that *all* challenging new ideas will go the way of colliding planets, ancient astronauts, and perpetual motion machines. We must examine each claim on its own.

In 1992 *Skeptic* magazine marked the 150th anniversary of Charles Darwin's first essay on natural selection, and the 20th anniversary of Niles Eldredge's and Stephen Jay Gould's first paper on punctuated equilibrium, by considering their status as paradigms. Few would challenge the idea that Darwin's theory of evolution by natural selection triggered a paradigm shift, but many are skeptical that punctuated equilibrium deserves equal status as a new paradigm. Since Darwinism is alive and well as we begin the twenty-first century it seems paradoxical to even consider the question. Darwinism displaced creationism, but itself has not been displaced, so no other paradigm shift can have occurred.

This is what I call the *paradox of the paradigm*. It is a false dichotomy created, in part, by our assumption that only one paradigm may rule a scientific field at any one time, and that paradigms can only "shift" from one to another, instead of building upon one another (and cohabitating within the same field). What I wish to argue is that there exists simultaneously an overarching Darwinian paradigm and a subsidiary punctuated equilibrium paradigm, both constituting paradigm shifts (with the former significantly broader in scope and the latter more narrowly focused), and that they presently and peacefully coexist and share overlapping methods and models. The paradigm paradox disappears when we define with semantic precision science, paradigm, and paradigm shift, and eschew the either-or fallacy of a false alternative choice by seeing punctuated equilibrium as a paradigm set within a larger Darwinian paradigm.

THE SCIENCE OF PARADIGMS

Science is a specific way of thinking and acting common to most members of a scientific group, as a tool to understand information about the past or present. More formally, I define science as *a set of cognitive and behavioral methods to describe and interpret observed or inferred phenomenon, past or present, aimed at building a testable body of knowledge open to rejection or confirmation.* Cognitive methods include hunches, guesses, ideas, hypotheses, theories, and paradigms; behavioral methods include background research, data collection and organization, colleague collaboration and communication, experiments, correlation of findings, statistical analyses, manuscript preparation, conference presentations, and paper and book publications.

There are two major methodologies in the sciences—experimental and historical. Experimental scientists (e.g., physicists, geneticists, experimental psychologists) constitute what most people think of when they think of scientists in the laboratory with their particle accelerators, fruit flies, and rats. But historical scientists (e.g., cosmologists, paleontologists, archaeologists) are no less rigorous in their cognitive and behavioral methods to describe and interpret past phenomena, and they share the same goal as experimental scientists of building a testable body of knowledge open to rejection or confirmation. Unfortunately a hierarchical order exists in the academy, as well as in the general public, in two orthogonal directions: (1) experimental sciences higher than historical sciences, (2) physical sciences higher than biological sciences higher than social sciences. Within both of these there exists a corresponding ranking from hard science to soft (with experimental physicists on top and social scientists and historians on the bottom), further discoloring our perceptions of how science is done. The sooner we can overcome what is known colloquially as "physics envy," the deeper will be our understanding of the nature of the scientific enterprise.

One common element within both the experimental and historical sciences, as well as within the physical, biological, and social sciences, is that they all operate within defined paradigms, as originally described by Thomas Kuhn in 1962 as a way of thinking that defines the "normal science" of an age, founded on "past scientific achievements . . . that some particular scientific community acknowledges for a time as supplying the foundation for its further practice."[1] Kuhn's concept of the paradigm has achieved nearly cult status in both elite and populist circles (even motivation speakers—as populist as they come—speak of shifting paradigms). But he has been challenged time and again for his multiple usages of the term without semantic clarification.[2] His 1977 expanded meaning of "all shared group commitments, all components of what I now wish to call the disciplinary matrix," still fails to give the reader a sense of just what Kuhn means by paradigm.[3]

Because of this lack of clarity, and based on the definition of science above, I define a paradigm as *framework(s) shared by most members of a scientific community, to describe and interpret observed or inferred phenomena, past or present, aimed at building a testable body of knowledge open to rejection or confirmation.* The singular/plural option and the modifier "shared by most" is included to allow for competing paradigms to coexist, compete with, and sometimes displace old paradigms, and to show that a paradigm(s) may exist even if all scientists working in the field do not accept it/them. Philosopher Michael Ruse, in fact, identified four usages of "paradigm" in his attempt to answer the question "Is the theory of punctuated equilibria a new paradigm?"[4] These include:

(1) *Sociological*, focusing on "a group of people who come together, feeling themselves as having a shared outlook (whether they do really, or not), and to an extent separating themselves off from other scientists."

(2) *Psychological*, where individuals within the paradigm literally and figuratively see the world differently from those outside the paradigm. An analogy can be made to people viewing the reversible figures in perceptual experiments, for example, the old woman/young woman shifting figure where the perception of one precludes the perception of the other.

(3) *Epistemological*, where "one's ways of doing science are bound up with the paradigm" because the research techniques, problems, and solutions are determined by the hypotheses, models, theories, and laws.

(4) *Ontological*, where in the deepest sense "what there is depends crucially on what paradigm you hold. For Priestley, there literally was no such thing as oxygen. . . . In the case of Lavoisier, he not only believed in oxygen: oxygen existed."

In my definition of paradigm the shared cognitive framework for interpreting observed or inferred phenomena can be used in the sociological, psychological, and epistemological sense. To make it wholly ontological, however, risks drawing the conclusion that one paradigm is as good as any other paradigm because there is no outside source for corroboration. Tea-leaf reading and economic forecasting, sheep's livers and meteorological maps, astrology and astronomy, all equally determine reality if one fully accepts the ontological construct of a paradigm. But paradigms are not equal in their ability to understand, predict, or control nature. As difficult as it is for economists and meteorologists to understand, predict, and control the actions of the economy and the weather, they are still better at it than tea-leaf readers and sheep's liver diviners.

The other component of science that makes it different from all other paradigms and allows us to resolve the paradigm paradox is that it has a self-correcting feature that operates, after a fashion, like natural selection functions in nature. Science, like nature, preserves the gains and eradicates the mistakes. When paradigms shift (e.g., during scientific revolutions) scientists do not necessarily abandon the entire paradigm any more than a new species is begun from scratch. Rather, what remains useful in the paradigm is retained, as new features are added and new interpretations given, just as in homologous features of organisms the basic structures remain the same while new changes are constructed around it. Thus, I define a *paradigm shift as a new cognitive framework, shared by a minority in the early stages and a majority in the later, that significantly changes the description and interpretation of observed or inferred phenomena, past or present, aimed at improving the testable body of knowledge open to rejection or confirmation.*

As Einstein observed about his own new paradigm of relativity (which added to Newtonian physics but did not displace it):

> Creating a new theory is not like destroying an old barn and erecting a skyscraper in its place. It is rather like climbing a mountain, gaining new and wider views, discovering unexpected connections between our starting point and its rich environment. But the point from which we started out still exists and can be seen, although it appears smaller and forms a tiny part of our broad view gained by the mastery of the obstacles on our adventurous way up.[5]

The shift from one paradigm to another may be a mark of improvement in the understanding of causality, the prediction of future events, or the alteration of the environment. It is, in fact, the attempt to refine and improve the paradigm that may ultimately lead to either its demise or to the sharing of the field with another paradigm, as anomalous data unaccounted for by the old paradigm (as well as old data accounted for but capable of reinterpretation) fit into the new paradigm in a more complete way.

Science allows for both cumulative growth and paradigmatic change. This is *scientific progress,* which I define as *the cumulative growth of a system of knowledge over time, in which useful features are retained and non-useful features are abandoned, based on the rejection or confirmation of testable knowledge.*

THE PUNCTUATED EQUILIBRIUM PARADIGM

A deeper question to ask about paradigms is what causes them to shift and who is most likely to be involved in the shift? Kuhn answers the question this way: "Almost always the men who achieve these fundamental inventions of a new paradigm have either been very young or very new to the field whose paradigm they change."[6] Kuhn was reflecting Max Planck's famous quip: "An important scientific innovation rarely makes its way by gradually winning over and converting its opponents. What does happen is that its opponents gradually die out and that the growing generation is familiarized with the idea from the beginning."[7] In his 1996 book, *Born to Rebel*, social scientist Frank Sulloway presented experimental and historical evidence for the relationship between age and receptivity to radical ideas, with openness related to youthfulness (see chapter 6 for a complete discussion).[8]

It was in 1972 that two young newcomers to the field of paleontology and evolutionary biology, Niles Eldredge and Stephen Jay Gould, presented the theory of punctuated equilibrium. What Eldredge and Gould proposed is a model of nonlinear change—long periods of equilibrium punctuated by, in geological terms, "sudden" change. This appears to contrast sharply with the Darwinian gradualistic model of linear change—slow and steady (and so minute it cannot be observed) transformation that given enough time can produce significant change. Thus its challenge to the Darwinian paradigm might be considered by some to be a paradigm shift. Michael Ruse called punctuated equilibria a paradigm "as far as the sociological aspect is concerned," but he expressly denies it paradigm status at the psychological, epistemological, and ontological levels.[9] We shall see.

The development of the theory of punctuated equilibrium was stimulated by Tom Schopf, who in 1971 organized a symposium integrating evolutionary biology with paleontology. The goal was to apply theories of modern biological change to the history of life. Eldredge had already done this with a 1971 paper in the prestigious journal *Evolution*, under the title "The Allopatric Model and Phylogeny in Paleozoic Invertebrates."[10] Schopf then directed Gould and Eldredge to collaborate on a paper applying theories of speciation to the fossil record, and this resulted in a paper published in 1972 in the volume *Models in Paleobiology* (with Schopf as the editor). This paper was entitled "Punctuated Equilibria: An Alternative to Phyletic Gradualism."[11] Gould explained that he coined the term but "the ideas came mostly from Niles, with yours truly acting as a sounding board and eventual scribe."[12] In brief, they argued that Darwin's linear model of change could not account for the apparent lack of transitional species in the fossil record. Darwin himself was acutely aware of this and stated so up front in the *Origin of Species*: "Why then is not every geological formation and every stratum full of such intermediate links? Geology assuredly does not reveal any such finely graduated organic chain; and this, perhaps, is the gravest objection which can be urged against my theory."[13]

Ever since the *Origin* the missing transitional forms have vexed paleontologists and evolutionary biologists. Collectively both groups have tended to ignore the problem, usually dismissing it as an artifact of a spotty fossil record. (This is actually a reasonable argument considering the exceptionally low probability of any dead animal escaping the jaws and stomachs of scavengers and detritus feeders, reaching the stage of fossilization, and then somehow finding its way back to the surface through geological forces and contingent events to be discovered millions of years later. It's a wonder we have as many fossils as we do.) Eldredge and Gould, however, see the gaps in the fossil record not as missing evidence of gradualism but as extant evidence of punctuation. Stability of species is so enduring that they leave plenty of fossils (comparatively speaking) in the strata while in their stable state. The change from one species to another, however, happens relatively quickly (on a geological time scale) in "a small subpopulation of the ancestral form," and occurs "in an isolated area at the periphery of the range," thus leaving behind few fossils. Therefore, the authors conclude, "breaks in the fossil record are real; they express the way in which evolution occurs, not the fragments of an imperfect record."[14]

Punctuated equilibrium is primarily the application of Ernst Mayr's theory of allopatric speciation to the history of life. Mayr's theory states that living species most commonly give rise to a new species when a small group breaks away (the "founder" population) and becomes geographically (and thus reproductively) isolated from the ancestral group. This new founder group (the "peripheral isolate"), as long as it remains small and detached, may experience relatively rapid change (large populations tend to sustain genetic homogeneity). The speciational change happens so rapidly that few fossils are left to record it. But once changed into a new species they will retain their phenotype for a considerable time, living in relatively large populations and leaving behind many well preserved fossils. (See FIGURE 13.) Millions of years later this process results in a fossil record that records mostly the equilibrium. The punctuation is there in the blanks.

Eldredge and Gould claim in this first paper that "the idea of punctuated equilibria is just as much a preconceived picture as that of phyletic gradualism," and that their "interpretations are as colored by our preconceptions as are the claims of the champions of phyletic gradualism." There is, however, a sense of paradigmatic progress when they note that "the picture of punctuated equilibria is more in accord with the process of speciation as understood by modern evolutionists."[15] It is not just that the gaps in the fossil record can now be ignored, but that they are real data. Thus, the gradualistic "tree of life" depicted by Darwin in the *Origin*, appears to be in conflict with the punctuated model of Eldredge and Gould. If punctuated equilibrium is a paradigm, this would appear

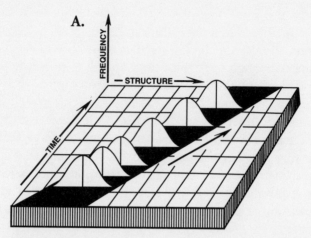

Figure 13. Competing or complementary paradigms? A. Above, the gradualistic model of shifting means of species characteristics through time (from Moore, et al., 1952). B. The punctuated equilibrium model, below, with static species abruptly giving rise to new species through geological time (from Eldredge and Gould, 1972).

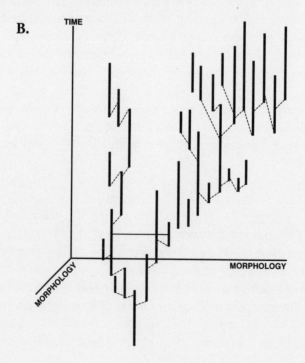

to be a paradigm shift, and thus we would be forced to accept the problem of the paradigm paradox and choose between the two competing models of evolutionary change.

The reaction to the theory, in Gould's words, "provoked a major brouhaha, still continuing, but now in much more productive directions."[16] Initially, says Gould, paleontologists missed the connection with allopatric speciation because "they had not studied evolutionary theory . . . or had not considered its translation to geological time." Evolutionary biologists "also failed to grasp the implication, primarily because they did not think at geological scales."[17] Though more in acceptance now, the theory at first received a thorough round of bashing for both good and bad reasons, the latter of which, Gould observes, include the "misunderstanding of basic content"; association with creationists who misrepresented the theory as spelling the demise of Darwin and all evolutionary theory; and, "this is harder to say but cannot be ignored, a few colleagues allowed personal jealousy to cloud their judgment."[18] Of course, the critical pounding could also be because Eldredge and Gould are wrong. But I think something else is going on here. The veracity of punctuated equilibrium aside, the paradigm paradox has forced observers to judge punctuated equilibrium as either completely right or totally wrong, when it can clearly be judged in fuzzy shades of correctness or wrongness, depending on the specific cases under question. In fact, Ruse notes that Eldredge and Gould "have polarized evolutionists in such a way that punctuated equilibria theory has defining paradigm properties at the social level."[19] Why must paradigms polarize? Because of this unresolved paradox.

Of course, we cannot judge a book by its author. As Gould confesses, "the worst possible person to ask about the genesis of a theory is the generator himself."[20] The ideal person to ask is a second generation student of the first generators, which I found in Occidental College world-class paleontologist Donald Prothero, who was a college freshman in 1973 when his paleontology class was assigned the new Raup and Stanley textbook, *Principles of Paleontology*, focusing on theoretical issues of fossil interpretation. Is punctuated equilibrium a paradigm, and was there a paradigm shift? Applying my definition of each, we can restate the question in several parts.

1. *Was punctuated equilibrium a new cognitive framework?* Yes and no. Yes, says Don Prothero, who writes that before punctuated equilibrium, "Virtually all the paleontology textbooks of the time were simply compendia of fossils. The meetings of the Paleontological Society at the Geological Society of America convention were dominated by descriptive papers." After the introduction

of the theory, new theoretical journals sprang up, old journals changed their emphasis from description to theory, and paleontological conferences were "packed with mind-boggling theoretical papers."[21]

No says Ernst Mayr, who makes it clear that *he* "was the first author to develop a detailed model of the connection between speciation, evolutionary rates, and macroevolution" and thus he finds it curious "that the theory was completely ignored by paleontologists until brought to light by Eldredge and Gould."[22] Mayr recalls that "In 1954 I was already fully aware of the macroevolutionary consequences of my theory," quoting himself as saying that "rapidly evolving peripherally isolated populations may be the place of origin of many evolutionary novelties. Their isolation and comparatively small size may explain phenomena of rapid evolution and lack of documentation in the fossil record, hitherto puzzling to the palaeontologist."[23] In a 1999 interview with Mayr (still going strong at the remarkable age of 95), he clarified for me the proper priority for the paradigm of punctuated equilibrium:

> I published that theory in a 1954 paper and I clearly related it to paleontology. Darwin argued that the fossil record is very incomplete because some species fossilize better than others. But what I derived from my research in the South Sea islands is that you get these isolated little populations for which it is much easier to make a genetic restructuring because it is small so it takes rather few steps to become a new species. Being a small local population that changes very rapidly I noted that you are never going to find them in the fossil record. My essential point was that gradual populational shifts in founder populations appear in the fossil record as gaps.[24]

I then pointed out to Mayr that Eldredge and Gould did credit him, citing his 1963 book *Animal Species and Evolution* several times. To this Mayr responded: "Gould was for three years my course assistant at Harvard where I presented this theory again and again, so he thoroughly knew it, so did Eldredge. In fact, Eldredge in his 1971 paper credited me with it. But that was lost over time."[25]

Was it lost over time? All professionals I have spoken to about punctuated equilibrium recognize this fact, as they do Niles Eldredge's solo paper published in *Evolution* in 1971. As Prothero concludes, however, it was the joint Eldredge and Gould paper published in 1972 that "has been the focus of all the controversy." Even Mayr admits: "Whether one accepts this theory, rejects, it, or greatly modifies it, there can be no doubt that it had a major impact on paleontology and evolutionary biology."[26]

What this historical development provides is further evidence for the social and psychological nature of paradigms. There are many reasons for the eighteen-year delay between Mayr's 1954 paper and Eldredge's and Gould's 1972 paper, having to do with the recent completion of the modern synthesis in evolutionary biology and, along sociological lines, who was proferring the theory. In a pure and unsullied scientific enterprise it should not matter who makes the discovery, when, and how it is presented. But science is not the objective process we would like it to be, and these factors do make a difference.

2. *Was punctuated equilibrium shared by a minority in the early stages and by a majority in the later?* Again, we must answer yes and no. Yes, says Prothero, and the "young Turks" who cut their paleontological teeth on the theory "are now middle-aged" and their influence "dominates the profession."[27] No, say Daniel Dennett, Richard Dawkins, and Michael Ruse, philosopher, zoologist, and philosopher respectively.[28] Dennett calls Gould "the boy who cried wolf," a "failed revolutionary," and "Refuter of Orthodox Darwinism."[29] Dawkins calls punctuated equilibrium a "tempest in a teapot," "bad poetic science," and says that Gould unfairly downplays the differences between rapid gradualism and macro-mutational saltation that "depend upon totally different mechanisms and they have radically different implications for Darwinian controversies."[30]

Dawkins is right on this last count, but as I read the original Eldredge and Gould 1972 paper, they are not arguing for punctuated equilibrium as anything more than a description of rapid gradualism reflected in the fossil record as gaps. A quarter century later, of course, much more has been made for punctuated equilibrium, occasionally by the authors but more often by the public. (My favorite example comes from an *X-Files* episode where the skeptical scientist Scully attempts to explain to her believing partner Mulder that the rational explanation for a suddenly mutated cancer-eating man is none other than punctuated equilibrium!) Michael Ruse believes that one reason for the confusion on this point is that punctuated equilibrium has gone through three phases, from a modest new description of the fossil record in the 1970s, to a radical new theory about evolutionary change in the 1980s, back to a more reserved tier of a multitiered hierarchical model of evolutionary change that incorporates both gradualism and punctuation.[31] (I should also point out that none of the most vocal critics of the theory—Dennett, Dawkins, and Ruse—are paleontologists. If the theory has limited application we should not be surprised if it is not openly utilized by those outside its boundaries.)

Ruse attempted a quantitative analysis of Gould's writings through the *Science Citation Index*, concluding that "virtually nobody (including evolutionists) outside of the paleontological community builds on Gould's theory of punctuated equilibria."[32] Ruse's critical interpretation, however, does not follow from the data. He begins by tallying up the number of citations of Gould's major works on punctuated equilibrium, including the original 1972 paper, the 1977 paper, "Punctuated Equilibria: The Tempo and Mode of Evolution Reconsidered," the 1980 paper, "Is a New and General Theory of Evolution Emerging?" and the 1982 paper, "The Meaning of Punctuated Equilibrium and Its Role in Validating a Hierarchical Approach to Macroevolution" (the first two coauthored with Eldredge). The grand total number of citations between 1972 and 1994 is 1,311, which Ruse admits is "respectable." But respectable (or not) compared to what? Ruse compares these four papers to the citation figures of four books by Edward O. Wilson: *The Theory of Island Biogeography, The Insect Societies, Sociobiology*, and *On Human Nature*. From this comparison Ruse concludes that "punctuated equilibria theory seems not to be in the same category as MacArthur and Wilson's island biography or Wilson's sociobiology." Ruse then totals the citations to everything Gould has written in two key scientific journals: *Paleobiology* and *Evolution*. In *Paleobiology* between 1975 and 1994 "35 percent refer to something by Gould, but only 13 percent refer to punctuated equilibria and a mere 4 percent respond favorably." In *Evolution* in the same time frame "9.8 percent refer to something by Gould, but only 2.1 percent to punctuated equilibria and a mere 0.4 percent respond favorably." Ruse then concludes: "The average working evolutionist is no better off with Gould than without him."[33]

What can we make of this analysis in our consideration of punctuated equilibrium as a paradigm? First, I applaud Ruse for his attempt to quantify a subjective evaluation, something almost unheard of in the historical profession. But has he made a fair comparison? Has he controlled for intervening variables that could account for the differences? No. Has he established a baseline from which to compare punctuated equilibrium to other revolutions in science? No. Comparing citation rates of scientific papers with scientific books is unsound because, with few exceptions, books are almost always of greater influence and impact than papers. And to compare a narrowly restricted theory like punctuated equilibrium to the much broader biogeography, and especially to the maximally encompassing sociobiology, is untenable. Punctuated equilibrium applies only to the fossil record and is mainly of interest to paleontologists. Biogeography applies to not only the fossil record, but to modern species and speciational processes, and is of interest to zoologists, botanists, ecologists, environmentalists, and field biologists. And sociobiology applies to all social animals from ants to humans and is of interest to anyone concerned with animal or human behavior, which is to say, almost everyone working in both the biological and social sciences, not to mention the general public's fascination with all things genetic. Also, according to Prothero, the journal *Evolution* is hardly read by paleontologists at all since its editorial slant is heavily weighted toward molecular biology, genetics and population genetics, and other biological subjects that have little or nothing to do with either punctuated equilibrium or the general working subjects of professional paleontologists. Finally, what does a citation rate of 13 percent (in *Paleobiology*) and 2.1 percent (in *Evolution*) mean? Compared to what? Perhaps other theories in *Paleobiology* merit only 6 percent, or maybe 25 percent. Without a comparison there is no way to know if Gould's figures are robust or weak. And shouldn't Eldredge's citation figures be included in this analysis since he was, after all, the first author of the original paper? Why is Eldredge largely left out of this discussion? Could it be that Gould's name is bigger, and bigger targets are easier to hit, especially from a distance?

3. *Did punctuated equilibrium significantly change the description and interpretation of observed or inferred phenomena?* This is the most important component of the sociological definition of a paradigm, but at this point in its history the answer could only be a provisional one. Prothero certainly thinks so, and most of his paleontological colleagues would agree. To me, Ruse's tally of 13 percent of all papers in *Paleobiology* referencing punctuated equilibrium sounds more than respectable; it seems quite high, considering the number of papers one sees in that journal that would have no reason to discuss punctuated equilibrium at all. But, again, without a formal survey of working paleontologists, and a quantitative comparison to other paradigms or revolutions, a comparison baseline, and preset operational definitions of judging criteria, there is no way to know if 13 percent is high or low.

4. *As a new paradigm, did punctuated equilibrium improve the testable body of knowledge that was open to rejection or confirmation?* That is, setting aside its cognitive components, historical acceptance or rejection, and changed perceptions, is it a superior model of nature? Again we are forced to offer a maximally equivocating "it depends." Prothero's extensive search through the empirical literature leads him to conclude that "among microscopic protistans, gradualism does seem to prevail," but "among more complex organisms . . . the opposite consensus had developed."[34] In hundreds of studies, including his own examination of "all the mammals with a reasonably complete record from the Eocene-Oligocene beds of the Big Badlands of South Dakota and related areas in Wy-

oming and Nebraska," Prothero concludes that "all of the Badlands mammals were static through millions of years, or speciated abruptly."[35] (See FIGURE 14). My own informal survey of paleontologists and evolutionary biologists at numerous conferences leads me to conclude that punctuated equilibrium applies to some fossil lineages, but not others. It is an accurate description of some specific evolutionary processes, but it is not universal.

It must be said again that most of the attacks on the punctuated equilibrium model have come from outside paleontological circles. Communities of knowledge share a set of common interests and methods that are most applicable to them and what they do, and less so to other communities. These other knowl-

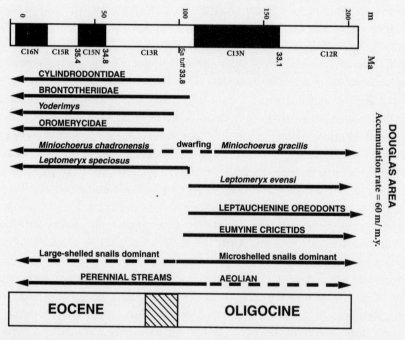

Figure 14. Evolutionary patterns at the Eocene-Oligocene transition (33–34 million years ago) recorded in strata near Douglas, Wyoming. On the top is the magnetic polarity time scale. In the middle are the ranges of species and families through the climatic shift. Most show prolonged stasis, followed by rapid speciation or extinction. All other mammals (not shown) exhibit no change through the interval. On the bottom are the climatic indicators that independently show that a major cooling event occurred at this time, even though most mammals did not track this climatic change.

edge explorers, of course, can, and do, borrow from other fields, but the models they swipe for temporary adoption will likely not have the universal appeal they may have to their originators. Thus it is that we see punctuated equilibrium—a model that describes the fossil record—most useful to those who specialize in studying the fossil record.

Still, it is useful to listen to critics outside the field for they may bring fresh insight to a problem. A case in point is Brown University cell biologist Kenneth Miller who, in his splendid book *Finding Darwin's God*, wonders if all of the brouhaha isn't over a fuzzy and fluid definition of the time scale under question.[36] Perhaps, he suggests, punctuated equilibrium and gradualism are the same models operating at different time scales. Let's return to the original source of the metaphor of the tree of life with its many branches of speciation—Charles Darwin's *Origin of Species* (FIGURE 15). What did Darwin say about the gradual versus punctuated nature of the tree of life? Miller calls out this quote from *The Origin of Species:* "But I must here remark that I do not suppose that the process ever goes on so regularly as is represented in the diagram, though in itself made somewhat irregular, nor that it goes on continuously; it is far more probable that each form remains for long periods unaltered, and then again undergoes modification."[37] It sounds like Darwin is saying that species remain stable over long periods of time, then undergo rapid speciational change. That's what Miller concludes: "A visitor to the land of evolution-speak might be forgiven for coming to the conclusion that the controversy over gradualism versus punctuated equilibrium was a wee bit contrived. And so it was."[38]

Was it? Not so fast. Miller pulls this quote out of the 6th edition of the *Origin*. In the first edition the sentence ends at "though in itself made somewhat irregular." The rest of the sentence, including the all important "each form remains for long periods unaltered, and then again undergoes modification," was added later. Why? Because, although he had discovered the mode of evolution—natural selection—he had not yet determined the tempo. And *that* is the difference between Darwin's tree of life and the iconography of punctuated equilibrium. Darwin does not tell us how rapid this modification process is, or, more importantly *what* the process is. He couldn't have, because in his time it was still unclear how populations shifted morphologically and behaviorally into new species (regardless of whether they did it rapidly or slowly), from one large population to another large population (peripatrically), or from a large population to a small population (allopatrically through the founder effect) that then develops back into a large population as a new species. In fact, this remains a point of great interest not only to paleontologists but to zoologists, botanists,

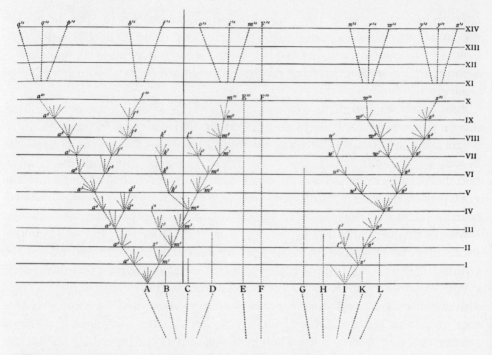

Figure 15. Darwin's tree of life from the *Origin of Species*.

biogeographers, and ecologists. What Darwin did identify in the first edition of the *Origin* was that there is a great range of evolutionary tempo:

> Species of different genera and classes have not changed at the same rate, or in the same degree. . . . The productions of the land seem to change at a quicker rate than those of the sea. . . . I believe in no fixed law of development, causing all the inhabitants of a country to change abruptly, or simultaneously, or to an equal degree. The process of modification must be extremely slow. The variability of each species is quite independent of that of all others.[39]

Darwin never explained why such variation in the speed of evolutionary change should exist. What Mayr, Eldredge, and Gould presented as something new and beyond what Darwin had said in the *Origin* was a mechanism—allopatric speciation—applied to the fossil record. That is what makes punctuated equilibrium a new paradigm—building on but not displacing Darwin and Darwinian gradualism.

To the extent that punctuated equilibrium constitutes a new paradigm (at least in paleontological circles), why has it received so much attention? Kenneth Miller frets about it because he doesn't want to give creationists any more targets to shoot at, and punctuated equilibrium has been a favorite theory of creationists looking for any angle they can find to tear down the Darwinian citadel. The creationists aside, however, it seems reasonable to ask why Ernst Mayr's 1954 paper didn't trigger a paradigm shift? The short answer is that he was the wrong triggerman. As a fifty-year-old biologist Ernst was not the "young Turk" needed to lead a paleontological revolution. The longer answer is found in the man who did champion punctuated equilibrium—Stephen Jay Gould—arguably the most prominent expositor of evolution of the past thirty years (and dubbed America's "Evolutionist Laureate"). Whether it was Mayr's idea or Eldredge's, it was, by his own admission, not Gould's idea, and yet it is his name most noticeably attached to it. As much as we may harbor a distaste for the social nature of science, the fact is *who* is doing the saying sometimes matters as much as what is being said. Even his critics admit that no one says it more often and with greater eloquence than Gould. Though he frequently calls himself a trades-man, and denies the polymathic modifier, one suspects that the gentleman doth protest too much. His monthly essays in *Natural History* range across the intellectual spectrum, and while they do indeed usually link up at some thematic middle, it is at the edges that Gould's reputation has grown well beyond the boundaries of his science, leading to a simultaneous overabundance of both credit and critique.

Carl Sagan certainly experienced this Janus-faced problem, and in chapter 10 I compare his accomplishments to those of other eminent scientists, one of whom is Gould. Gould matches Sagan in every category, including a National Magazine Award for his column "This View of Life," a National Book Award for *The Panda's Thumb,* a National Book Critics Circle Award for *The Mismeasure of Man,* the Phi Beta Kappa Book Award for *Hen's Teeth and Horse's Toes,* and a Pulitzer Prize Finalist for *Wonderful Life,* for which Gould characteristically commented "close but, as they say, no cigar."[40] At the time of this writing he has pulled down no less than forty-four honorary doctorates, published 593 scientific articles (including forty-five in *Science* and *Nature*), and written twenty books (only three of which were coauthored). Sixty-six major fellowships, medals, and awards bear witness to the depth and scope of his accomplishments in both the sciences and humanities: Fellow of AAAS, MacArthur Foundation "genius" Fellowship, Scientist of the Year from *Discover* magazine, Humanist Laureate from the Academy of Humanism, the Silver Medal from the Zoological Society of London, the "Skeptic of the Year" award from the Skeptics Society, the Edinburgh Medal from the City of Edinburgh, and the Britannica Award and Gold Medal for dissemination of public knowledge, among others. With such awards come enough invitations to fill a calendar and earn prodigious frequent flyer miles. For those he cannot accommodate, a form letter is sent, written in vintage Gouldian style—cerebral but to the point:

> I can only beg your indulgence and ask you to understand an asymmetry that operates cruelly (since it produces tension and incomprehension) but that leads to an ineluctable (however regrettable) result. The asymmetry: you want an hour or two, perhaps a day, of my time—not much compared to what you think I might provide (exaggerated, I suspect, but I won't struggle to disillusion you). From that point of view, I should comply—not to do so could only be callousness or unkindness on my part. But now try to understand my side of the asymmetry: I receive on average (I promise that I am not exaggerating) two invitations to travel and lecture per day, about 25 unsolicited manuscripts per month asking for comments, 20 or so requests for letters of recommendation per month, about 15 books with requests for jacket blurbs. . . . I am one frail human being with heavy family responsibilities, in uncertain health and with a burning desire (never diminished) to write and research my own material. Thus, I simply cannot do what you ask. I can only beg your understanding and extend to you my sincere thanks for thinking of me. . . .

(I must confess to being the recipient of the letter for a request made in youthful ignorance of the way the world works.)

Gould's numbers also match those of Edward O. Wilson, Jared Diamond, and Ernst Mayr (some lower, some higher—see figures in chapter 10). My point is that with such accolades and public recognition comes an inevitable truncating and short-shrifting of the complexities of the scientific process and the subtleties of allocating credit or critique where it is due and in appropriate measure. It is simply much easier to say or write "Gould's theory of punctuated equilibrium" than it is "the theory of punctuated equilibrium, first proposed in 1954 by Ernst Mayr, published again in 1971 by Niles Eldredge, and solidified in 1972 by Niles Eldredge and Stephen Jay Gould. . . ."

What the future holds for punctuated equilibrium remains to be seen, but we have learned from this story that theories and paradigms are social in nature and the marketing of an idea is at least as important as its creation (though both in the end, one hopes, give way to evidence). Gould did not achieve his status as one of the best known and most respected writers and scientists today by coauthoring a paper on punctuated equilibrium. That was part of it, but his reputation has promoted the theory far more than the reverse. If there was a paradigm shift, the reason it was triggered in 1972 instead of 1954 or even 1971 is, primarily, because it was Gould who pulled the trigger.

Why, we might ask, was it Gould who led this paradigm shift? One way to get to an answer to this question is to examine his personality. As we shall see in the next section of the book, personality traits influence receptivity or resistance to revolutionary ideas. University of California, Berkeley social scientist Frank Sulloway has developed a model describing this relationship between personality and orthodoxy/heresy (described in detail in the next chapter), and to further test his hypothesis we had eight of Gould's colleagues take the Five Factor Personality Inventory, also known as the "Big 5," representing the five most dominant traits that best explain personality. They are *Conscientiousness, Agreeableness, Openness to Experience, Extroversion*, and *Neuroticism*. To test these, we had subjects complete a survey of 40 adjective pairs on a 1–9 scale, such as these:

> *I see Stephen Jay Gould as someone who is*:
> —Stubborn/headstrong 1 2 3 4 5 6 7 8 9 Acquiescent/compliant
> —Untraditional 1 2 3 4 5 6 7 8 9 Traditional
> —Leisurely 1 2 3 4 5 6 7 8 9 Energetic/fast paced
> —Rarely depressed/sad 1 2 3 4 5 6 7 8 9 Often depressed/sad
> —Deliberate 1 2 3 4 5 6 7 8 9 Hasty/impulsive
> —Modest 1 2 3 4 5 6 7 8 9 Arrogant

Sulloway and I ran a correlation on all 40 adjective pair for all eight raters, and came up with a .92 interrater reliability index, an exceptionally high figure that gives us confidence that we have a good handle on the personality of this scientific revolutionary. The results are presented in Table 1 as percentile scores, comparing Gould to the more than 100,000 people already in Sulloway's database.

Gould scores exceptionally high in *openness to experence*, which is a key personality trait in the development of a revolutionary. But not all radical ideas are equal, so it helps to be high in *conscientiousness* to help one weed through the bogus revolutionary ideas. Gould is also exceptionally high in *conscientiousness*,

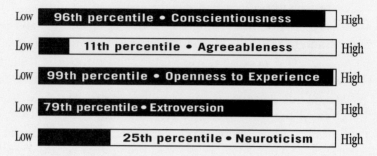

GOULD'S PERSONALITY
Ratings on the "Big 5" Personality Traits

Low   **96th percentile • Conscientiousness**   High

Low   **11th percentile • Agreeableness**   High

Low   **99th percentile • Openness to Experience**   High

Low   **79th percentile • Extroversion**   High

Low   **25th percentile • Neuroticism**   High

aiding him in finding that essential tension Thomas Kuhn speaks of between orthodoxy and heresy—finding the balance between being open-minded enough to recognize quality new ideas, but not too open that even nutty ideas are treated with equal respect. That balance is reinforced by Gould's low score on agreeableness: he's a tough-minded intellectual who does not suffer fools gladly, and that is a good trait to have when one is in the rough and tumble world of ideas where most are wrong, a few are acceptable, and only a handful really stick as legitimate revolutions. Gould has the personality profile that lends itself to leading and supporting scientific revolutions, without also being taken in by what might turn out to be a failed revolution. It would appear that punctuated equilibrium is one such successful revolution, and these data help us further understand why it was Gould who led it.

My assessment of punctuated equilibrium is that it is an improved theory for explaining the fossil record and thus meets the criteria for being progressive science. That is, in the cumulative growth of Darwinian gradualism, useful features were retained and nonuseful feature abandoned, based on the rejection or confirmation of testable knowledge. But Darwinian gradualism has not been displaced by punctuated equilibrium, just modified to include the latter as a more accurate description and interpretation of some sequences in the fossil record. Whatever it is, punctuated equilibrium is not a new mechanism of evolutionary change. It is a new description, and seeing it as such helps us resolve the paradigm paradox. Its status as new paradigm is restricted to this level—a nontrivial contribution to the science, but not on the level of the Darwinian paradigm.

THE DARWINIAN PARADIGM

Unlike most figures of intense historical study who become less chimerical, the more we get to know Charles Darwin, the higher our regard seems to be. There exists today a sizable and ever-growing Darwin industry that is one of the largest bodies of historical literature in the science history profession. The Correspondence Project of the Darwin Archives at Cambridge University Library, that promises to continue unabated for another couple of decades, is testimony to the intense interest the "sage of Down" has sparked in the historical community.[41] The editor of that project, Frederick Burkhardt, informs me that beginning with 1859 there will be a single volume of correspondence for every year of Darwin's life through 1882. The entire set should weigh in sometime in the twenty-first century at over thirty volumes!

In 1985 David Kohn edited *The Darwinian Heritage*, an exhaustive 1,138-page handbook on the state of the Darwin industry a century following the naturalist's death. An 80-page bibliography lists nearly 3,200 entries pertaining to Darwin, and Kohn appropriately opens the work with an introduction entitled "A High Regard for Darwin," in which he states: "what characterizes the present community is a belief in the importance of Darwin. This high regard for Darwin is its central tenet."[42]

There are three historical trends that help account for this Darwinian renaissance: (1) changes in the history of science profession in the early 1960s that began to contextualize scientists within their cultures; (2) changes in the biological profession, especially the successful application of Darwinism after the modern synthesis; and (3) changes in the general culture that placed a greater emphasis on education in the physical and biological sciences, including a reintroduction of Darwin and evolutionary theory in response to creationist attacks in particular, and to the Sputnik scare in general. Throughout, Darwin has emerged a hero of greater proportions than in his own lifetime or the century following his death in 1882. This hero-myth image has fed the Darwin industry and continues to support a Darwinian paradigm.

1. The Evolution of Darwinian Historiography

The historical contextualization of Darwin by professional historians of science has lifted Darwin to a high plane while simultaneously placing him in his proper culture. This historiographical development, however, has taken three decades to reach maturity and occurred in two phases triggered by two centennial celebrations in 1959 (for the publication of the *Origin of Species*) and 1982 (for Darwin's death).

The first phase began slowly, with considerable misunderstanding of Darwin, who was typically seen as a first-rate scientist but only second-rate thinker.[43] Jacques Barzun, for example, calls Darwin "A great assembler of facts and a poor joiner of ideas, a man who does not belong with the great thinkers."[44] Gertrude Himmelfarb writes that Darwin was "limited intellectually and insensitively culturally."[45] This denigration of Darwin to one who merely happened along at the right place and time triggered a number of contrasting responses that lofted Darwin to a plane of which his work would have a hard time reaching. Michael Ghiselin's well-received book, *The Triumph of the Darwinian Method,* for example, is as triumphant as it is overstated, as he opens with this sharp antithesis to Himmelfarb and Barzun: "In 1859 there began what ultimately may prove to be the greatest revolution in the history of thought."[46] Ghiselin's Whiggish interpretation of Darwin leads him to set up such either-or questions as, "Was Darwin just a naturalist and a good observer, or was he a theoretician of the first rank?," forcing the reader to conclude the latter since it is clear he was more than just the former.

In this understandable reaction there are two forms of historical bias that have affected the development of the study of Darwin: one is the "Whiggish" history of arranging constructed characterizations in linear sequence toward those currently in favor; the other is using current models of understanding to judge the effectiveness of past ideas. In Ghiselin's case, imposing twentieth-century philosophical models on a nineteenth-century problem leads to the pigeonholing and reorganization of a thinker's work into modern categories of meaning. This is precisely what he does when he concludes that Darwin was using the modern hypothetico-deductive method, which he again contrasts in a limiting two-alternative choice for the reader: "This declaration of confidence in the power of the modern hypothetico-deductive scientific method in revealing the past contrasts sharply with the view of 'inductionists,' who believe that the facts must be accepted 'as they are'."[47] Once again, we are forced to conclude that either Darwin was a pure inductionist, which we know is impossible (since no one does science in a cultural vacuum), or he was "ahead of his time" as a hypothetico-deductive thinker. Ghiselin opts for the latter, of course, and because of this concludes that "the entire corpus of Darwinian writings constitutes a unitary system of interconnected ideas."[48] Ghiselin consummates the *Triumph* with an even balder statement than his opening one: "Whether one likes it or not, the age belongs to Darwin."[49]

Darwin himself confused the historical record on this point. In the *Autobiography* he claims to have "worked on true Baconian principles, and without any theory collected facts on a wholesale scale." It is clear from his earliest notebooks in the 1830s on transmutation, man, mind, and morals, not to mention his even earlier theorizing on coral reefs while aboard the *Beagle*, that Darwin worked on anything but true Baconian principles. He freely and intelligently speculated and theorized about a large number of important subjects that would only much later succumb to empirical verification or rejection. In a letter to Henry Fawcett on September 18, 1861, in response to a debate about whether geologists should just collect and classify rocks, or theorize about their origin and development, Darwin made this statement that more honestly reflects his scientific work: "About thirty years ago there was much talk that geologists ought only to observe and not theorize, and I well remember someone saying that at this rate a man might as well go into a gravel-pit and count the pebbles and describe the colours. How odd it is that anyone should not see that all observation must be for or against some view if it is to be of any service!"[50]

As the Darwin industry matured in the 1980s, especially in works by professional historians of science schooled in the art of contextualization, it became obvious that the intellectual and ideological development of Charles Darwin was far more complex (and less linearly consistent) than these early historical works revealed.[51] To be fair, though, the history of science as a field has undergone considerable change since 1959. We have learned to strike a balance between anachronistic and diachronistic history—the past from today's perspective versus that of the historical character. We have also found a functional medium between internalist and the externalist history—those who see science as a steady and ineluctable march toward truth and those who see it as yet another cultural tradition, little different than myth and religion.

Historian of science Robert Richards hits the mark between internalist and externalist positions when he notes that ". . . ideas become historically linked only by passing through minds trapped in flesh, human minds which respond to logical implications, evidentiary support, and the legacy of scientific problems and concepts, though also to high emotion, religious feeling, class attitudes, and even perhaps Oedipal anxieties."[52] Similarly, the Dutch historian of science, R. Hooykaas finds the scale of anachronism and diachronism evenly weighted when the historian can sit balanced between the two:

> In order to judge fairly, the historian has to approach the thinking, observing and experimenting of the forebears with a sympathetic understanding: he

must possess a power of imagination sufficiently great to "forget" what became known after the period he is studying. At the same time, he must be able to confront earlier views with the actual ones, in order to be understood by the modern reader and in order to make history something really alive, of a more than purely antiquarian interest.[53]

With Darwin we have finally found that balance between intellectual development and cultural context, between what Darwin hath wrought and what his culture begat.

2. The Evolution of Darwinian Biology

In the biological sciences the importance of much of Darwin's work, particularly on natural and sexual selection, has increased significantly since the completion of the modern synthesis in the middle of this century, a full hundred years after the publication of the *Origin*. By the 1950s Darwin and Darwinism emerged far stronger than ever before. We have, as it were, "rediscovered" Darwin because we need him today.

In the Darwin industry this usefulness helps explain the success of Ghiselin's book. But as the field matured, such internalistic interpretations were countered by more historically accurate works on the impact Darwin really had on his contemporaries. In fact, it would be difficult to find a more dramatically opposing view to Ghiselin's (without returning to the Barzun/Himmelfarb brand) than that argued in Peter Bowler's *The Non-Darwinian Revolution*. For example, where Ghiselin claims, *"The Origin of Species . . .* effected an immediate and cataclysmic shift in outlook, casting into doubt ideas that had seemed basic to man's conception of the entire universe,"[54] Bowler responds: "the parts of Darwin's theory now recognized as important by biologists had comparatively little impact on late nineteenth-century thought."[55] For Bowler there was a *revolution* in the rejection of creationism and essentialism, but it was *non-Darwinian* because it was a *developmental*, not an evolutionary revolution:

> I suggest that it is unreasonable to believe that a theory that failed to impress the scientists of the time could have brought about a major cultural revolution. Once this point is accepted, one is led to suspect that the traditional interpretation of the Darwinian Revolution is a myth based on a distorted image of Darwin's effect both on science and on the emergence of modern thought.[56]

It is these biases that have influenced the historical work of such biologists as Ghiselin and, especially, Ernst Mayr whose *Growth of Biological Thought*[57] and *Toward a New Philosophy of Biology*[58] are, despite claims to the contrary, unabashedly anachronistic and internalistic. For example, Mayr claims that "we have Darwinism, but not Newtonianism, Planckism, Einsteinism, or Heisenbergism," and that because of "this exceptional status . . . it would be difficult to refute the claim that the Darwinian revolution was the greatest of all intellectual revolutions in the history of mankind."[59] It would be hard to top this claim, but Mayr does in his analysis of the man himself. Darwin was "brilliant," "bold," and so keen an "observer," "philosophical theoretician," and "experimentalist" that "the world has so far seen such a combination only once, and this accounts for Darwin's unique greatness."[60] And, unbelievably, Mayr makes this statement that leaves the historian's jaw agape: "Most students of the history of ideas believe that the Darwinian revolution was the most fundamental of all intellectual revolutions in the history of mankind. While such revolutions as those brought about by Copernicus, Newton, Lavoisier, or Einstein affected only one particular branch of science, the Darwinian revolution affected every thinking man."[61]

I am amazed that a man of such depth and culture as Ernst Mayr, whose *Growth of Biological Thought* I enthusiastically devoured in my earliest studies of the history of science, could make such a historically naive claim. The dean of science historians today, I. B. Cohen, for example, makes an important distinction between scientific and ideological revolutions, demonstrating that the revolutions created by Copernicus, Newton, Lavoisier, and Einstein all extended well beyond the confines of their particular branch of science.[62] Likewise, historian Margaret Jacob has not only used the word *Newtonianism*, but provides an adequate description of its effect throughout European culture in the century following Newton, far beyond the boundaries of the physical sciences (for example, extending even into the social realm of economic and sociological theories).[63]

It is important to note, however, that we need not take anything away from Darwin in the defense of other scientists and revolutions, and the clarification of their relative effects. Our regard for Darwin has climbed steadily with the ascent of both evolutionary biology and the history of that science, independent and irrespective of the other giants upon whose shoulders we stand, leading us to a clearer focus of the larger picture of nature and the nature of science.

3. The Evolution of Darwinian Culture

In the 1920s a perceived degeneration of the moral fiber of America was increasingly linked to Darwin's theory of evolution. "Ramming poison down the throats of our children is nothing compared with damning their soul with the teaching of evolution," proclaimed T. T. Martin in 1923, a supporter of the fundamentalist orator William Jennings Bryan.[64] Fundamentalists gathered in force to squelch the moral degeneration by nipping the problem at its source — the teaching of Darwinian evolutionary theory in public schools. Florida passed an antievolution law in 1923, followed two years later in Tennessee with the Butler Act, making it "unlawful for any teacher in any of the Universities, Normals and all other public schools of the state . . . to teach any theory that denies the story of the Divine Creation of man as taught in the Bible, and to teach instead that man has descended from a lower order of animals."[65]

A challenge to the bill by the ACLU resulted in the infamous Scopes trial that produced a supposed "moral" victory for Mssrs. Scopes, Darrow, and Mencken, who focused the attention of the country on the matter of whether knowledge and scientific "truth" could be regulated by law. The effects, however, were just the opposite of popular myth. The controversy generated by the trial made textbook publishers and state boards of education nervous about dealing with Darwin and evolution in any manner, and the two were quickly expunged from texts. In a before-and-after comparison of textbooks, Judith Grabiner and Peter Miller concluded: "Believing that they had won in the forum of public opinion, the evolutionists of the late 1920s in fact lost on their original battleground — teaching of evolution in the high schools — as judged by the content of the average high school biology textbooks [that] declined after the Scopes trial."[66] A quick check of the table of contents or index of any book from this period will generally find "evolution" and "Darwin" missing.

Historical trends of this nature do not often change overnight, but this one did on October 4, 1957, the day the Soviet Union launched Sputnik I, which in turn launched the cold war space and science race. The post-Sputnik scare jolted American science educational leaders into reintroducing Darwin and evolution back into the public school curriculum. By 1961 the Biological Science Curriculum Study of the National Science Foundation produced an outline for the teaching of biology that would include the man and his theory.

In short, an idol of biology was needed and Darwin fit the bill. Creationist attacks on Darwin in their quest for "equal time" in the 1970s and 1980s only enhanced the naturalist's reputation, as scientists circled the wagons in support.[67] The ongoing debate about the proper relationship of science and religion focuses on two fronts: cosmology and the origins and evolution of the universe, and evolutionary biology and the origins and evolution of life (especially humans). Thus, in addition to his resurrection by historians of science and evolutionary biologists, Darwin has resurfaced as *the* cultural icon of the second front in this great debate. But where Copernicus, Galileo, Newton, and Einstein—revolutionary leaders of the first front—hardly warrant a mention by creationists and other opponents of science, Darwin stands alone as the symbol of all that is wrong with the modern scientific worldview from their perspective.

DARWIN'S SECULAR SAINTHOOD

Throughout this interplay of history, biology, and culture, a central figure stands out in Darwin, and a primary ideology in Darwinism. In a review of Adrian Desmond's and James Moore's biography, *Darwin*, historian of science James Rogers observed that "Darwin is well on his way to attaining secular sainthood for a second time."[68] Indeed, no other paradigm in the biological sciences comes close in impact or importance. There is a psychological component in the identification of a paradigm—whether one is for or against it—with a single individual. Historians of science have created a Darwin industry. Biologists are Darwinians or Neo-Darwinians. Creationists are non-Darwinians or anti-Darwinians. Critics ask "Was Darwin Wrong?" or "Did Darwin Get It Right?" Even Alfred Wallace said he was more Darwinian than Darwin himself. In its economy of thought the mind classifies, categorizes, and pigeonholes people and ideas. The process is simplified with one person—a single name to be modified with the appropriate prefix or suffix.

But the phenomenon is much more than just neural networking. There is another factor in the construction of hero-myths that is part of our desire to achieve greater recognition. Whether this is done vicariously through heroes and hero-myths, or these figures are used as role models for actual achievement, the effect is the same. "We have not even to risk the adventure alone, for the heroes of all time have gone before us," explained the mythologist, Joseph Campbell. "We have only to follow the thread of the hero path. . . ." A hero is a product of talent and timing. The extent of the influence depends on courage and ambition. "A hero," says Campbell, "is someone who has given his or her life to something bigger than oneself."[69]

In his historic quest for *The Hero With a Thousand Faces*, Campbell has found a common theme in all hero myths in a sequential adventure of *separation—initiation—return*, where the hero "ventures forth from the world of common day into a region of supernatural wonder: fabulous forces are there encountered and a decisive victory is won: the hero comes back from this mysterious adventure with the power to bestow boons on his fellow man."[70] Campbell reinforces his model with examples from mythic literature, such as the story of Prometheus, who stole fire from the gods and then descended to deliver it to man; as well as the story of Jason, who sailed through the Clashing Rocks and endured other hazards to wrest the Golden Fleece from the guardian dragon and return home to take his rightful place on the throne.

Darwin's career and life fits this hero-myth schemata, though the vagaries of real life make the progress within each stage, and the transition between stages, much less distinct than classic myth-stories. Darwin's *separation* stage, for example, began with the death of his mother when he was eight, continued when he went to the Edinburgh University at age sixteen to study medicine, was reinforced at Cambridge in his studies of natural history with Professor John Henslow, and was finalized on his circumnavigational five-year voyage in which "any idea of being a minister or a doctor or anything other than a scientist died a natural death on the *Beagle*." The trip was, in his own words "by far the most important event in my life."[71]

The *Beagle* voyage began Darwin's separation from creationism and his *initiation* into an entirely new paradigm of evolutionary thought. "When I see these Islands in sight of each other," Darwin observed at the Galapagos, "& possessed of but a scanty stock of animals, tenanted by these birds [mockingbirds], but slightly differing in structure & filling the same place in Nature, I must suspect they are only varieties. If there is the slightest foundation for these remarks the zoology of Archipelagoes — will be well worth examining; for such facts would undermine the stability of Species."[72] The completion of his initiation came in the late 1830s and early 1840s when it "appeared to me that by following the example of Lyell in Geology, and by collecting all facts which bore in any way on the variation of animals and plants under domestication and nature, some light might perhaps be thrown on the whole subject."[73] The methodology led him over the course of the next decade to realize that "we can allow satellites, planets, suns, universe, nay whole systems of universe to be governed by laws, but the smallest insect, we wish to be created at once by special act."[74] This, he thought, was absurd. Laws of nature applied everywhere or not at all. For twenty years he did just that, applying laws and principles to every nook and cranny of the natural world of plants and animals until he had the evidence he needed to support his bold theory on the origin of species.

Darwin's *return* stage of bestowing boons to mankind, however, was delayed two decades for several fascinating reasons brought to light by Desmond and Moore. The obvious underlying cultural reason is that evolution was damned as blasphemous by both church and state rulers and Darwin had no stomach for raising controversy for its own sake. More important was his need for peer acceptance. The publication in 1844 of *The Vestiges of Creation* by an anonymous author (Robert Chambers), and its subsequent severe lambasting by the scientific community as too speculative and unscientific, drove Darwin to delay publication until he had compiled enough data to make the most demanding empiricist blurry-eyed. Further, his friend and colleague Joseph Hooker, in a critique of the French botanist Frederic Gerard's book *On Species*, said that no scientist should "examine the question of species who has not minutely described many."[75] Darwin took his friend's indirect criticism seriously and spent the next eight years on a thorough examination of barnacles. This delaying process was elevated to a principle. Late in his life Darwin offered his own son George, who was preparing a critique of prayer, this admonition: "It is an old doctrine of mine that it is of foremost importance for a young author to publish only what is very good & new; so that the public may have faith in him, & read what he writes. . . . remember that an enemy might ask who is this man, & what is his age & what have been his special studies, that he shd. give to the world his opinions on the deepest subjects? This sneer might easily be avoided . . . but my advice is to pause, pause, pause."[76]

Darwin's *return* stage begins with the publication of the *Origin of Species* in 1859 and continues through his death in 1882. The final paragraph in the *Origin* demonstrates just how broad in scope was Darwin's thinking. Since this sentence remained virtually unchanged from its original autographed inscription in his 1842 "Sketch" of his theory, we know that Darwin had long before given himself to something bigger than oneself: "There is grandeur in this view of life, with its several powers, having been originally breathed into a few forms or into one; and that, whilst this planet has gone cycling on according to the fixed law of gravity, from so simple a beginning endless forms most beautiful and most wonderful have been, and are being, evolved."[77]

A paradigm shift occurs when people begin to look at the world or a particular problem in a whole new light, a metaphor that Darwin used again and again in the *Origin* ("Much light will be thrown . . ."). The shift was sudden and dramatic. For or against the theory it was Darwin and Darwinism that triggered the change. Although Thomas Huxley ("Darwin's bulldog") proclaimed that the *Origin* was "the most potent instrument for the extension of the realm of knowledge which has come into man's hands since Newton's Principia," Darwin's close friend Lyell held back his support for the theory a full nine years and then hinted at a modified version with providential design behind the whole scheme. Where Ernst Haeckel ("the German Darwin") promoted evolution in Germany, and Asa Gray ("the American Huxley") supported Darwin's theory in the United States, the astronomer John Herschel called natural selection the "law of higgledy-piggledy." The geologist and Anglican cleric Adam Sedgwick proclaimed that natural selection was a moral outrage and penned this ripping harangue in personal letter to Darwin:

> There is a moral or metaphysical part of nature as well as a physical. A man who denies this is deep in the mire of folly. You have ignored this link; and, if I do not mistake your meaning, you have done your best in one or two cases to break it. Were it possible (which thank God it is not) to break it, humanity, in my mind, would suffer a damage that might brutalize it, and sink the human race into a lower grade of degradation than any into which it has fallen since its written records tells us of its history.[78]

So immediate was the impact of Darwin that thirteen months after the publication of the *Origin* Henry Fawcett published "A Popular Exposition" in *Macmillan's Magazine* in which he observed:

> No scientific work that has been published within this century has excited so much general curiosity as the treatise of Mr. Darwin. It has for a time divided the scientific world with two great contending sections. A Darwinite and an anti-Darwinite are now the badges of opposed scientific parties.[79]

That sentence could have been written a hundred years later and it would still be relevant—a sign that the Darwinian paradigm was and is one of the most profound in all of human thought.

PART II
BORDERLANDS PEOPLE

Individuals who launch radical revolutions typically require strong determination, courage, and independence of mind. Unfortunately, their divergent ways of thinking have tended to condemn these bold thinkers to rejection, ridicule, and torment. Like Charles Darwin, who compared his belief in evolution to "confessing a murder," heterodox individuals have typically suffered for their revolutionary aspirations. Not every unorthodox thinker has succeeded, and not all of them have been right. But a surprising number of them have shared a deep and powerful bond. More often than not, they were born to rebel.

—Frank Sulloway, final paragraph, *Born to Rebel: Birth Order, Family Dynamics, and Creative Lives,* 1996

6
THE DAY THE EARTH MOVED
Copernicus's Heresy and Sulloway's Theory

New and revolutionary systems of science tend to be resisted rather than welcomed with open arms, because every successful scientist has a vested intellectual, social, and even financial interest in maintaining the status quo.

—I. B. Cohen, *Revolution in Science*, 1985

IN THE FIRST YEAR of the seventeenth century, the most prolific astronomical observer of the age—Tycho Brahe—joined forces with the greatest astronomical theoretician of the age—Johannes Kepler. Brahe, a roguish Danish nobleman who held the position of imperial mathematician in the court of the Holy Roman Emperor, invested enormous amounts of time and energy in compiling a remarkably complete database of astronomical observations. On the island of Hveen he constructed an observatory—Uraniborg—over which he ruled in royal splendor for twenty years. Along the walls of the observatory he constructed great quadrants for measuring celestial altitudes that brought the art of astronomical observation to the highest accuracy possible (all without the aid of a telescope). Yet, these data by themselves were inadequate to explain the workings of the cosmos. Kepler, a brilliant thinker who supported the new and radical Copernican heliocentric model of the universe, observed after a time spent with Brahe: "He lacks only the architect who would put all this to use."[1]

The flamboyant, raucous Brahe, who apparently spent almost as much time at the alehouse as he did the observatory, died prematurely, never to see the application of his life's work. "Let me not seem to have lived in vain," were his final, historically prescient words. "If God has sent us an observer like Tycho," Kepler responded, "it is in order that we should make use of him."[2]

The philosopher of science Karl Popper once observed that "Science is competitive—it is a collision of ideas with observations."[3] The Brahe-Kepler collaboration (awkward though it was) marks a transition in the evolution of thought that launched science on a track from which it has rarely wavered—the coalescence of theory and data. The collision of Kepler's ideas with Brahe's observations, followed by Galileo's experimental support and Newton's mathematical unification of this new "system of the world," fused the learned community's fragmented thoughts into a comprehensive model of a sun-centered solar system. This view, first described by Nicolaus Copernicus in his posthumously published *De Revolutionibus Orbium Coelestium (On the Revolutions of the Celestial Spheres)* in 1543, stimulated an intellectual revolution of epic proportions. But why, sixty years after publication, were Brahe and Kepler still struggling to construct a workable cosmology? Why did Brahe design a compromise system with planets orbiting the sun, and that whole system orbiting the centrally-located earth? (See FIGURE 19.) Why did Martin Luther and other churchmen, as well as political leaders, condemn Copernicus and his sun-centered model? In short, why did heliocentricism take a century and a half to reach relative acceptance?

Copernicus triggered one of the most monumental changes in the history of our perception of the world and ourselves. It did far more than affect the way we watched the sun rise and set everyday (even this descriptive language still lingers in an earth-centered system). Copernicus jolted humanity out of its egocentric complacency, out of a cognitive harmony that had resulted from its perception of itself as being at or near the center of the universe. Copernicus launched the decline of the medieval worldview, an encompassing picture that linked all aspects of the cosmos, the world, and the events of human life. Because of this, in part, the theory met considerable resistance and was not fully accepted for several generations. A psychological analysis of the normal opposition to new paradigms and heretical science, and a historical understanding of the medieval world picture, helps us grasp the reason for the delay in the acceptance of Copernicus. The Copernican revolution was sparked by Copernicus, but its flames were fanned by Kepler, Bruno, Galileo, and Newton, and would be more accurately described as occurring in the seventeenth century, not the sixteenth. This psychological and historical case study represents the normal reaction such revolutionary ideas receive in most fields of science throughout most of history. It is a classic case of how a borderlands, heretical science makes the transition to an accepted, normal science.

IDEOLOGICAL IMMUNITY AND HERETICAL-SCIENCE

Karl Popper classified paradigms of the magnitude and scope of Copernicus's as being beyond merely *scientific* revolutions. They are *ideological* revolutions. Popper described a scientific revolution as "a rational overthrow of an established scientific theory by a new one," whereas an ideological revolution includes "processes of 'social entrenchment' or perhaps 'social acceptance' of ideologies."[4] A scientific revolution changes the science, but not necessarily the culture. An ideological revolution (that is scientifically based) changes both the science and the culture in which it occurs.

It is these revolutions that experience the greatest resistance to acceptance, simply because they do have ideological implications. Social scientist Jay Snelson identified the cause of this resistance as the *ideological immune system*: "educated, intelligent, and successful adults rarely change their most fundamental presuppositions."[5] According to Snelson, the more knowledge an individual has accumulated, and the more well-founded his theories have become, the greater the confidence in his ideologies. The consequence of this, however, is that he builds up an "immunity" against new ideas that do not corroborate previous ones. Like a biological immune system that protects the body by warding off foreign bacteria and viruses, this system works against the acceptance of revolutionary ideas into the body of accepted knowledge.

Historians of science call this the "Planck Problem," after the physicist Max Planck who first identified it: "An important scientific innovation rarely makes its way by gradually winning over and converting its opponents. What does happen is that its opponents gradually die out and that the growing generation is familiarized with the idea from the beginning."[6] Social scientists are likewise familiar with the problem as expressed by the experimental psychologist E. G. Boring: "What would happen to science if its great men did not eventually die, no one can guess. What does happen is that a new man takes up the work of an older man without the constraint of inertia from his past, that he thinks, works and writes more simply and directly, and that thus from the old he creates something new that gradually itself accumulates inertia."[7] Evolutionary biologist Ernst Mayr has experienced the same phenomenon in his own field and would seem to agree with Snelson: "One can go so far as to claim that the resistance of a scientist to a new theory almost invariably is based on ideological reasons rather than on logical reasons or objections to the evidence on which the theory is based."[8] Mayr thinks ideological immunity is one of the reasons why those who might have "anticipated" Darwin and Wallace did not: "a considerable number of authors arrived at this conclusion before Darwin. And yet, the leading authorities in zoology, botany, and geology continued to reject evolution." Mayr notes that it was, indeed, the most educated, intelligent, and successful thinkers who "missed" the opportunity: "Since Lyell, Bentham, Hooker, Sedgwick, and Wollaston in England and their peers in France and Germany were highly intelligent and well-informed scientists, one cannot attribute their resistance to stupidity or ignorance."[9]

On one level, ideological immunity is purposefully built into the scientific enterprise as a way of maintaining the status quo long enough to test the validity of various new claims. The patriarch of science history, I. B. Cohen, explained what would happen to science if it were not conservative: "If every revolutionary new idea were welcomed with open arms, utter chaos would be the result. The rigid and brutal insistence on demonstration which is part of the resistance to change in science actually is a source of strength and stability. Many attempted or proposed revolutions simply do not pass the test."[10] History is replete with chronicles of the lone and martyred scientist working against his peers, and in the face of adversity from the authorities in and out of his own field of study. But can the revolutionary really expect the experts in the field to adopt unquestionably every new idea that comes along, without the necessary time allotted for verification? In the end, history rewards those who are at least provisionally right. Change does occur. The Ptolemaic geocentric universe was slowly displaced by Copernicus's heliocentric system. But why did it take so long, and what is the psychology of this process of resistance?

THE PSYCHOLOGY OF RESISTANCE

On a mild spring day in May 1995, as he settled into his office for another eye-straining session of data analysis and number crunching that would go long into the night, MIT social scientist Frank Sulloway opened a letter from the National Science Foundation. It was from the panel for the History and Philosophy of Science Program, a division of the NSF, and titled "Panel Recommendation on grant proposal SBR-9512062." It was a letter Sulloway had been anticipating for some time, as he drew no salary from the university and his bank account was precariously low. The brutally competitive world of grantsmanship is stressful enough when one has tenure. It is almost unbearable when one's livelihood depends on it.

Sulloway titled his proposal "Testing Theories of Scientific Change." What did he want to test? Max Planck's hypothesis about the relationship between age and intellectual receptivity. We can all cite anecdotes where Planck is right about change only coming after the old scientists pass on, but what about the exceptions? Were there older scientists who were innovative? No one had ever bothered to try to falsify the hypothesis. Specifically, what Sulloway proposed was using statistical methods to test innovation receptivity. The NSF panel turned him down flat. Testing historical hypotheses? Using science to analyze history? Was he joking? They rendered their judgment in no uncertain terms:

> One of the most pervasive issues discussed by the panelists was the approach the Principal Investigator was taking toward history. Many panelists thought that applying a heavy-duty statistical analysis to history is naive, inappropriate, and even peculiar. Is it really the case that generalizations in history should be tested with statistics, rather than be tested through a detailed examination of the sources? Some noted that it seemed as if the Principal Investigator was going back to 19th-century beliefs that history is a science which could uncover laws. Panelists were opposed to such a narrow view of history.[11]

Sulloway was shocked. How could a panel of scientists think that using statistics to test hypotheses is "naive, inappropriate, and even peculiar"? Then he remembered, these were not scientists. They were historians and philosophers. "Besides being an odd response to receive from the National *Science* Foundation where the principal criterion of grant evaluation is supposed to be 'scientific merit'," Sulloway recalled with some dismay, "this panel's criticisms confuse a method of research (hypothesis testing) with a theory of history. Testing is what makes an approach scientific, not the particular viewpoint that is endorsed." To emphasize the point, Sulloway pointed out that "Even the claim that history can be studied scientifically can only be evaluated properly through hypothesis testing."[12]

Sulloway pressed on anyway, tested this and many other historical hypothesis, and in 1990 published a paper on "Orthodoxy and Innovation in Science," and in 1996 came out with his pathbreaking book *Born to Rebel: Birth Order, Family Dynamics, and Creative Lives*. To test Planck's hypothesis, Sulloway conducted a multivariate correlational study examining the tendency toward rejection or receptivity of a new scientific theory based on such variables as "date of conversion to the new theory, age, sex, nationality, socioeconomic class, sibship size, degree of previous contact with the leaders of the new theory, religious and political attitudes, fields of scientific specialization, previous awards and honors, three independent measures of eminence, religious denomination, conflict with parents, travel, education attainment, physical handicaps, and parents' ages at birth." Using multiple regression models with these variables Sulloway discovered, quite surprisingly, that "birth order consistently emerges as the single best predictor of intellectual receptivity."[13] Although age is a factor in intellectual receptivity, of all these variables birth order was the most significant predictor of "attitudes toward innovation in science . . . in 2,784 participants in 28 highly diverse scientific controversies during the last four centuries."[14]

Consulting over a hundred historians of science, Sulloway had them "judge the stances taken by the participants in these debates," including the Copernican revolution and twenty-seven other scientific controversies, spanning in dates from 1543 to 1967 (see FIGURE 16 for a partial listing). Sulloway found that only 34 percent of firstborns supported the new ideas compared to 64 percent laterborns. Using a statistical test of significance this laterborn tendency for acceptance was significantly greater than the firstborns at the .0001 level, which means "the likelihood of this happening by chance is virtually nil" (specifically, it has a probability of one in ten thousand of being due to chance). Historically speaking this indicates that "laterborns have indeed generally introduced and supported other major conceptual transformations over the protests of their firstborn colleagues. Even when the principal leaders of the new theory occasionally turn out to be firstborns—as was the case with Newton, Einstein, and Lavoisier—the opponents as a whole are still predominantly firstborns, and the converts continue to be mostly laterborns."[15] Children without siblings, a "control group" of sorts, were sandwiched between firstborns and laterborns in their percentage of support for radical theories.

FIGURE 16 allows us to compare controversies to see that the pre-Galileo Copernican revolution ranks first in the order of magnitude of the birth-order correlation with the stance taken for or against it (with a correlation of .51), greater even than Einstein's controversial theory of relativity and the culture-shattering Darwinian evolution. It is also interesting to note that the post-Galileo Copernican revolution indicates that once the scientific support grew, coupled to the time to allow later generations to grow accustomed to the idea, forces of radicalism won out over those of conservatism. This conclusion is further reinforced by the data plotted in FIGURE 17 showing the changing sup-

Controversy	Years	Correlation Supporting	% First-Born Supporting	% Laterborns
Copernican revolution (pre-Galileo)	1543-1609	.51	22%	75%
Copernican revolution (post-Galileo)*	1610-1649	-.07	62%	54%
Relativity theory	1905-1927	.47	30%	76%
Phrenology	1799-1840	.42	39%	83%
Quantum hypothesis	1905-1911	.40	43%	82%
Darwinian revolution	1859-1870	.40	20%	61%
Hutton theory of of Earth's geology	1788-1829	.38	0%	35%
Harvey blood circulation theory	1628-1653	.37	57%	100%
Indeterminacy principle in physics	1918-1927	.34	36%	70%
Continental drift	1912-1967	.30	36%	68%
Semmelweis and puerperal fever theory	1842-1862	.24	50%	75%
Lister and antisepsis	1867-1880	.21	50%	73%
Newtonian revolution	1687-1750	.19	60%	79%

*Note shift in support. See text for explanation.

Figure 16. The Psychology of Resistance in the History of Science.
Birth-order effects in 28 scientific controversies from Copernicus (1543) to
Continental Drift (1967) turned out to be the strongest predictor of resistance
or acceptance of new and "heretical" ideas in the history of science. Controversies are listed in descending order of magnitude of the birth-order correlation (r) with stance taken in each controversy. The "Percent of Firstborns
Supporting" is relative to all firstborns in the sample, and the "Percent of
Laterborns Supporting" is relative to all laterborns in the sample. This allows
direct comparison between firstborn and laterborn support rates, since they
are corrected for the greater number of laterborns in the population. (Courtesy of Sulloway, 1996).

port for Copernicus's heretical science before and after Galileo. Before Galileo,
the religious and political implications of the sun-centered system were too powerful to be outweighed by the empirical evidence. After Galileo, time and the
telescope helped to attenuate the ideological objections.

Not all scientific theories are equally radical, of course, and in taking this into
consideration, Sulloway discovered a correlation between laterborns and the
degree of "liberal or radical leanings" of the controversy. He noted that laterborns "have also tended to prefer statistical or probabilistic views of the world

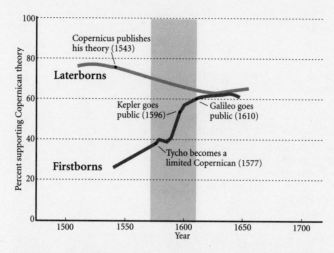

Figure 17. Changing Support of the Heretical-Science of Copernicus. The change of support coming for the Copernican revolution before and after Galileo's contribution of telescopic observations announced in 1610. Before Galileo, the ideological implications (e.g., religious and political) of the Copernican system were too strong to be overcome by the scientific evidence. After Galileo, time helped to attenuate ideological objections, in conjunction with compelling empirical evidence. (Courtesy of Sulloway, 1996.)

(Darwinian natural selection and quantum mechanics, for example) to a world-view premised on predictability and order." By contrast, Sulloway found that when firstborns did accept new theories, they were typically theories of the most conservative type, "theories that typically reaffirm the social, religious, and political status quo and that also emphasize hierarchy, order, and the possibility of complete scientific certainty."[16] In FIGURE 18 Sulloway plots the relative birth-order effects with the ideological context of the controversy—from conservative to radical—with Copernicanism primarily laterborn-led and quite radical.

The correlation between birth order and scientific open-mindedness is a powerful one, but *why* are laterborns more liberal and receptive to ideological change, while firstborns are more conservative and influenced by authority? What is the connection between birth order and personality? One possibility is that firstborns, being first, receive more attention from their parents than later-

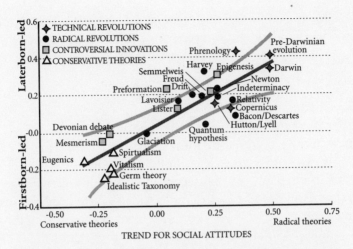

Figure 18. Ideological Immunity in the History of Science. The relative ranking of the conservative and radical qualities of 25 scientific controversies plotted against the effect birth order has in support of revolutionary ideas. As can be seen, there is a direct relationship between the size of birth-order effects and the ideological implications of controversies. The correlation between the two is a remarkable .77, significant at the .0001 level. Firstborn-led radical events or laterborn-led conservative events are virtually nonexistent. Firstborns tend to accept only the most conservative new theories; laterborns support scientific theories that possess radical or heretical tendencies. (Courtesy of Sulloway, 1996.)

borns, who tend to receive greater freedom and less indoctrination into the ideologies of and obedience to authorities. Firstborns generally have greater responsibilities, including the care and liability of their younger siblings. Laterborns are frequently a step removed from the parental authority, and thus less inclined to obey and adopt the beliefs of the higher authority.

Sulloway, of course, is not suggesting that birth order alone determines the ideological receptivity of radically new ideas. Far from it, in fact, as he notes that "birth order is hypothesized to be the occasion for psychologically formative influences operating within the family."[17] We might look at birth order to be a predisposing variable that sets the stage for numerous other variables, such as age, sex, social class, etc., to influence orthodoxy or innovation receptivity. In examining other variables and their interaction with birth order and ideological immunity, Sulloway found that "the general magnitude of birth-order effects is closely related to the ideological implications of scientific controversies."[18]

THE PSYCHOLOGY OF RESISTANCE TO COPERNICUS

The shift from laterborn to firstborn support for Copernicus after Galileo is an indication that the heretical nature of the theory was on the wane. This is due both to time—getting used to the idea—and to the influence of scientific observations in support of the theory. As Sulloway notes in support of Planck and Snelson, "time elapsed is a crucial factor in dissipating earlier birth-order effects."[19] The relative effects of birth order, ideological implications, scientific evidence, and time are complex in the Copernican revolution, so it behooves us to briefly examine the lives of the major players in the drama.

Nicolaus Copernicus (1473–1543) was laterborn (the youngest of four children) but was the most conservative of revolutionaries in Sulloway's pantheon of scientific heretics, with a predicted probability of .72 for leading the revolution. Called "The Timid Cannon" by one biographer (Koestler), at the age of 70, and on his deathbed, Copernicus finally agreed to allow his laterborn friend and colleague George Joachim Rheticus to publish *De Revolutionibus*. His "timidity" is not trivial, for Sulloway found an interaction effect whereby introversion (or "shyness") decreased the laterborn's radical tendencies. "Owing to the relationship between birth order and radical temperament," Sulloway explains, "it is not uncommon for shy laterborns, and especially shy lastborns, to become 'timid revolutionaries'."[20] Copernicus was one such introvert. His challenge to Ptolemy came slowly and deliberately after many years of study. He was well aware of other heliocentric theories, and in 1514 he privately circulated a brief manuscript entitled *De hypothesibus motuum coelestium a se constitutis commentariolus* ("A Commentary on the Theories of the Motions of Heavenly Objects from Their Arrangements"). Although the *Commentariolus* had the basic components of a moving earth that would be espoused in *De Revolutionibus*, it was never published. In 1533 he gave a series of lectures in Rome before an approving Pope Clement VII. Three years later a formal offer of publication was made, but still Copernicus hesitated. Finally, in 1540 Rheticus convinced him to publish, and on his deathbed it is believed he received the first copy. Though his actions seem more like that of a conservative firstborn, as we shall see later Copernicus knew his theory was radical and would not be readily received.

Tycho Brahe (1546–1601) was a second-born child who became, de facto, firstborn by the oddest of historical contingencies. Tycho's father Otto Brahe, governor of Helsingborg Castle, had a childless brother named Joergen to whom he had promised a son, to be adopted from Otto's family. The governor's wife, by a twist of chromosomal fate, had twin boys, seemingly able to provide both men with a suitable heir. But fate (and biology) works in mysterious ways. One of twins died and the survivor, Tycho, was abducted by his Uncle Joergen after Otto withdrew his promise. Tycho became the firstborn child to a wealthy country squire and vice-admiral, and before long inherited that wealth when Joergen died rescuing his king (Ferdinand II) who had fallen into a river. The series of events had the effect, in Sulloway's model, "of lowering Tycho's predicted probability of supporting new theories from 52 to 44 percent."[21]

There is no doubt that Tycho had the courage and intelligence to become the finest observational astronomer of his day, and to devise a clever system that, according to I. B. Cohen, is only disprovable by observing stellar parallax (the shifting of the background stars from one side of the earth's orbit to the other, which was not possible at the time).[22] Tycho was really not, however, a revolutionary, either scientifically or psychologically. Scientifically, Tycho rejected Copernicus's heliocentric system and instead proposed the one illustrated in FIGURE 19, which, though unique and effective, was incorrect. And if laterborns tend to prefer probabilistic views of the world in contrast to their firstborn colleagues who desire predictability and order in their cosmologies, then the functionally firstborn Tycho would seem to fit the bill perfectly. Arthur Koestler observes that a rather unspectacular partial eclipse of the sun was a great revelation for Tycho because of "the predictability of astronomical events—in total contrast to the unpredictability of a child's life among the temperamental Brahes." This effect, says Koestler, was the "opposite direction from both Copernicus's and Kepler's. It was not a speculative interest, but a passion for exact observation."[23]

The case of Johannes Kepler (1571–1630) is equally compelling. Although he was a firstborn, in Sulloway's model, Kepler had a 65 percent probability of supporting heretical ideas due primarily to the fact that he had a fairly radical childhood with considerable strife with his parents, whose marriage was an unhappy one. (Conflict with parents leads firstborns, who are strongly parent-identified, to become more like laterborns.) Kepler's father was away much of the time, fighting on the Catholic side in Holland, much to the dismay of his Protestant family. His mother, whom he described as "gossiping, and quarrelsome, of a bad disposition,"[24] was tried as a witch late in life, and Kepler had to come to her defense to have her life spared. As a boy he received early training

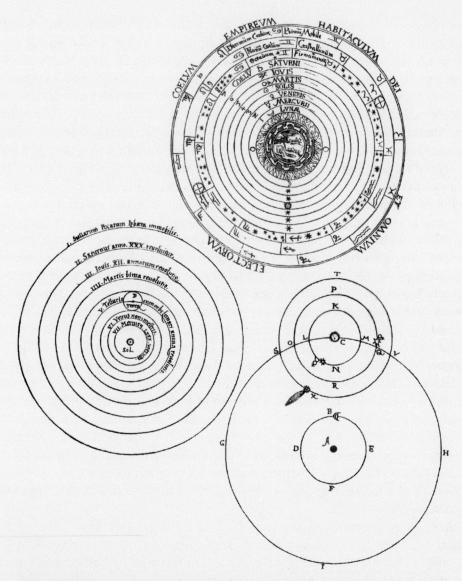

Figure 19. Three Contrasting Views of the Cosmos. Top: the medieval Aristotelian/Ptolemaic geocentric/geostatic world, from Petrus Apianus's *Cosmographia*. Middle: the Copernican heliocentric model from *De Revolutionibus Orbium Coelestium* (1543). Bottom: Tycho Brahe's compromise model in which the sun carried the planets along with it, as this solar system moved around the still centrally located earth.

as a revolutionary when he attended a Protestant seminary school that shaped young minds to battle Catholicism. At the age of sixteen Kepler began his university training and was fortunate to have a professor at Tubingen who believed in the Copernican system. His first great work, the *Mysterium Cosmographicum*, was read by Brahe, who was so impressed that he invited Kepler to join his research staff in 1600. When Brahe died the following year, Kepler inherited his position as imperial mathematician and promptly trashed the Tychonic system to work on his own, which eventually resulted in a refined and improved Copernican system that added the critical element of elliptical planetary orbits.

The story of Galileo (1564–1642) as a heretic-scientist is well known. Like Kepler he was a firstborn, but interactive effects of extroversion, conflict with his mother, and a progressive father, gave him a 60 percent probability of supporting heretical ideas. His father, for example, was himself a radical musical composer who had nothing but contempt for authority. "It appears to me," Vincento Galilei wrote, "that those who try to prove an assertion by relying simply on the weight of authority act very absurdly."[25] Many years later Galileo echoed his father: "Methinks that in the discussion of natural problems we ought not to begin at the authority of places of Scripture; but at sensible experiments and necessary demonstrations." Galileo not only accepted Copernicus, he did so in a manner that would raise the ire of the church. The more vocally he supported the Copernican system as a representation of reality (and not just another mathematical system), the more nervous the church became. On March 5, 1616, Cardinal Robert Bellarmine, concerned that Copernicanism might undermine Catholicity in its battle against Protestantism, decreed the theory to be "false and erroneous" and *De Revolutionibus* was put on the index of prohibited books. Restraining himself as long as he could, in 1632 Galileo finally burst out with the *Dialogue Concerning Two Chief World Systems*. The book was a masterful tool, set down in the style of a dialogue between two individuals, one a supporter of the Ptolemaic system, the other of the Copernican. The book's protagonist in support of Ptolemy was "Simplicio," whom Galileo presented as an irrational fool, and the "dialogue" is a systematic attack on the whole of Aristotelian physics and cosmology. The book was banned in August and Galileo called before the Inquisition in Rome the following February. Though he was forced to recant (for which he received a salutary penance "for the spiritual benefit of former heretics who had returned to the faith"), he is said to have mumbled under his breath, in true revolutionary style, *eppur si muove* — it moves anyway!

HOW HERETICAL WAS COPERNICUS?

Historians of science differ on the relative impact of the Copernican revolution on the European culture of the late Middle Ages and early modern period. A. C. Crombie argues for a conservative interpretation because "the Copernican revolution was no more than to assign the daily motion of the heavenly bodies to the rotation of the earth on its axis and their annual motions to the earth's revolution about the sun, and to work out, by the old devices of eccentrics and epicycles, the astronomical consequences of these postulates."[26] Thomas Kuhn, on the other hand, noted that not only did the Copernican system revise "the fundamental concepts of astronomy," it resulted in "many novel and immensely fruitful research programs, disclosing a host of previously unsuspected natural phenomena and revealing order in fields of experience that had been intractable to men governed by the ancient world view."[27] Deeper still in its impact, Kuhn called the Copernican system "a revolution in ideas, a transformation in man's conception of the universe and of his own relation to it."[28]

I. B. Cohen, however, challenged the existence of a "Copernican" revolution — scientific or ideological — when he observed that "Copernicus's writings and doctrines did not in their own day create any immediate radical change in the basic system of accepted astronomical theory, and only very slightly affected the practice of working astronomers."[29] Cohen traced the acceptance of Copernicanism "some half to three-quarters of a century after the publication of Copernicus's treatise,"and he bluntly concluded that the idea "that a Copernican revolution in science occurred goes counter to the evidence and is an invention of later historians."[30] Cohen's thorough review of astronomical literature from 1543 to 1600 "does not show any signs of a revolution," and that if there was a Copernican revolution, "it occurred in the seventeenth and not in the sixteenth century." In searching for the use of the word "revolution" in the science of that period (and related terms or concepts used in reference to Copernicus), Cohen concluded: "Many of the writers on science of the seventeenth century did not give much prominence to Copernicus, which is yet another indication that there had not been a Copernican revolution."[31]

The cause of this delay is due, in part, to the fact that the Copernican system was not superior astronomically to the system it proposed to replace — Ptolemy's geocentric model. If the earth moved, why did cannonballs travel the same distance when fired east or west? If the earth moved, why was Brahe, the best observational astronomer of the day, unable to perceive stellar parallax? A new physics was needed, provided later by Galileo, as well as a new heuristic, furnished by Kepler. But it did not matter because there was plenty wrong with the Ptolemaic system as well. In fact, 70 percent of the conversion of laterborns to the Copernican system occurred before Galileo and his telescopic observations. The scientific evidence was strong enough to overturn Ptolemy for radical thinkers, but the ideological correspondences were too strong for the rest. Sulloway, in fact, estimates that at least 70 percent of the reason for the delayed acceptance of the Copernican system was ideological.[32] Copernicus created an ideological revolution, not just a scientific one, and the fact of its delayed impact is indicative of this because ideological revolutions are slower to be accepted than scientific revolutions. To understand why this is so we must return to the sixteenth century to glimpse a vision of the medieval worldview. It is difficult for us to understand the impact of the Copernican theory, or to realize the repercussions it must have had on the people of that age and culture, without knowing what it would have meant to them. Therefore let us think ourselves back into a twilight of the imagination to see the world as our ancestors did.

THE MEDIEVAL WORLDVIEW

The worldview of the sixteenth century individual was an Aristotelian/Ptolemaic earth-centered universe that was fixed, finite, small, and closed, making the entire cosmos knowable and the farthest distances of the universe within reach.

Aristotle laid the foundation of this cosmology, which was subsequently modified by Apollonius of Perga in the third century B.C.E., developed further by Hipparchus of Rhodes in the second century B.C.E., and completed by Ptolemy of Alexandria in the second century C.E.: "If one should next take up the question of the earth's position, the observed appearances with respect to it could only be understood if we put it in the middle of the heavens as the centre sphere," Ptolemy concluded in the *Almagest*. "All the observed order of the increases and decreases of day and night would be thrown into utter confusion if the earth were not in the middle." The Aristotelian/Ptolemaic system was passed down through the centuries, remaining relatively intact. The Venerable Bede's eighth-century *De Natura Rerum*, for example, gave a naturalistic account of earthquakes, thunder, and the like, while referring to the spherical form of the earth, the order of the seven planets in their orbs around the earth, and the spherical nature of heaven. In a twelfth-century work ascribed to William of Conches, the writer observed that "the earth is in the middle, as the yolk in the egg, and outside it is the water like the white round the yolk; around the water is the air like the skin round the white, and finally fire, corresponding to the eggshell." The twelfth-century bishop of Paris, Peter the Lombard, in his *Libri Sententiarum*, wrote of the spherical nature of heaven, the hierarchy of angels residing in the heavens and arranged in nine orders, and the descending order of spheres from the fixed stars to the centralized earth, the center of which is hell.[33]

The medieval illustration in FIGURE 20 depicts how this universe looked to the medieval mind. The earth was centralized and stationary. Everything revolved around it, including the nine solid but transparent crystal spheres in which the moon, sun, planets, and fixed stars were embedded. It was the spheres that rotated, not the heavenly bodies. The moon, sun, planets, and stars were carried along with their respective rotating crystal sphere. Since the earth was stationary in this system (no daily rotation), and yet the sun and stars reappeared each day, all spheres, including the sphere of the fixed stars (they too were fixed to a sphere), had to have a daily motion of their own. The ninth sphere, the *Primum Mobile*, controlled the cosmos and propelled the other spheres. God resided just outside the ninth sphere, where heaven and angels were also located. In this model each sphere was driven by God and the *Primum Mobile* through a *Resident Intelligence*, a type of angel that looks after each sphere. Actually, to account for "irregularities" in planetary motion, it was sometimes necessary to have a planet associated with four or five spheres. Aristotle, in fact, used fifty-four spheres altogether to account for the motions of the seven planets.

The space between the earth and the sphere of the moon was the critical dividing point between the sublunary/terrestrial region and the supralunary/celestial zone. Beneath the sphere of the moon everything consisted of a composition of the four "ordinary" elements of matter: earth, water, air, and fire. Above the sphere of the moon, the aggregate was made up of a more "perfect" representation of matter—the fifth element of ether. The inherent characteristic of the four ordinary elements was that they seek their "proper place" in the universe. For instance, earth, being the most "humble"of the elements (and heaviest), comes to rest at the center of the planet, while water, which is lighter than earth, rests atop earth. Air and fire, respectively still lighter, come to rest over the water.

If left alone, these four elements would sort themselves into spheres, similar to the crystal spheres of the celestial world. This does not happen, however, because the sphere of the moon rotates, stirring up the fire that in turn churns up the air, that disturbs the water, and so on. The atmosphere that fills the sublunary zone, as it approaches the sphere of the moon, can be burned (fire), and this burning produces comets and meteor showers. The interaction of the

Figure 20. A Medieval View of the Cosmos A medieval king representing Atlas holds aloft a three-dimensional representation of the geocentric cosmos. In the center are the four elements consisting of "yearth," "water," "aer," and "fier," surrounded by the planetary spheres, the "Cristalline Firmament," and the zodiac. The "Primum Mobile," or Primary Mover, sets the whole system in motion. (From William Cunningham, *The Cosmographical Glasse*, 1559.)

four elements produces transformations of all kinds, and this becomes the basis of change. So all terrestrial motions are ultimately controlled by those forces celestial which, in turn, are controlled by the *Primum Mobile*, itself directed by God. The hierarchy of the Roman Catholic Church was a perfect embodiment of this Aristotelian structure, ranking from the parish priest to the pope. In all, there were nine levels of the church, corresponding to nine levels of sacraments, all of which corresponded to the nine crystal spheres.[34]

Dante's *Divina Commedia*, published in 1307 (see FIGURE 21) is emblematic of this worldview, particularly since Dante was a learned man and had studied the Aristotelian/Ptolemaic system. In Dante's poem, the cosmos is a series of spheres beginning with hell in the center, and moving out to the furthest reaches of the heavens. Around the sloping sides of hell the places of punishment are ordered in circles of gradually decreasing diameter, depending on the severity of the sin. Lucifer, naturally, dwells at the very center of hell. Purgatory, similarly, is also spherical, and rises out of the ocean opposite Jerusalem, the navel of the dry land. Once the top rung of Purgatory is reached, one can pass upward and through the celestial spheres of the moon, Mercury, Venus, the sun, Mars, Jupiter, and Saturn. The eighth sphere is the fixed stars, the ninth the *Primum Mobile*, and the tenth, the *Empyrean*, or the dwelling of the Deity.

Since God's influence was a function of proximity, the closer the sphere the "purer" the body. The earth, being the farthest object from God, was the most corruptible. E. M. Tillyard, in his splendid little book *The Elizabethan World Picture*, describes this continuum from terrestrial to celestial, the higher the purer: "The farther the distance from the earth and the nearer to heaven, the purer and more brilliant was the atmosphere. Contrariwise the earth itself was gross and heavy and the more so towards its own centre. Far from being dignified, the earth in the Ptolemaic system was the cesspool of the universe, the repository of its grossest dregs."[35] Therefore the goal was to ascend out of this "cesspool repository" to heaven and God, as Milton expresses in *Paradise Lost*: "Well has thou taught the way that might direct Our knowledge, and the scale of Nature set from center to circumference, whereon In contemplation of created things By steps we may ascend to God."[36]

James Daly, in probably the best single paper on the subject, summarized this worldview: "From God down to the tiniest grain of sand or insect, each link in the chain had its particular virtue and claim to importance, its lesson for the different links in the social chain." Where was humanity in this picture? "His key position in the chain of being, his role as link between higher and lower, incorporeal and corporeal, angels and animals, meant that the whole universe

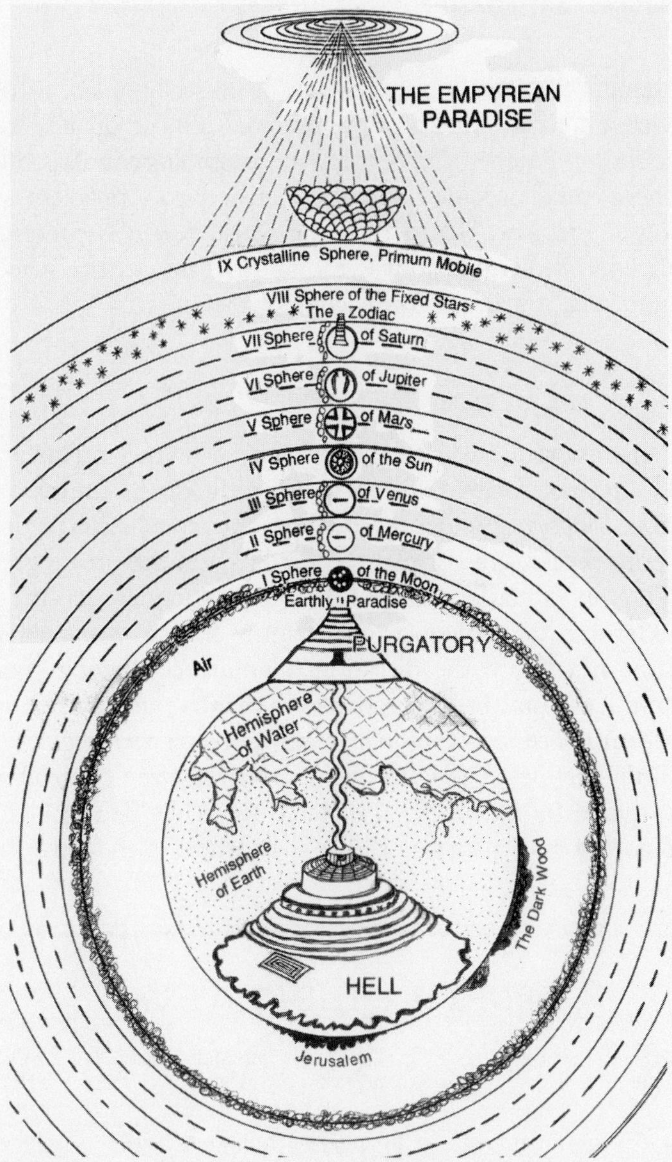
THE EMPYREAN
PARADISE
IX Crystalline Sphere, Primum Mobile
VIII Sphere of the Fixed Stars
The Zodiac
VII Sphere of Saturn
VI Sphere of Jupiter
V Sphere of Mars
IV Sphere of the Sun
III Sphere of Venus
II Sphere of Mercury
I Sphere of the Moon
Earthly Paradise
PURGATORY
Air
Hemisphere of Water
Hemisphere of Earth
The Dark Wood
HELL
Jerusalem

Figure 21. Dante's Fourteenth-Century Cosmos This worldview was developed in an attempt to synthesize ancient Greek science with twelfth-century church doctrine. In this Aristotelian/Ptolemaic cosmos, the earth is a stationary globe in the center of the universe, surrounded by nine translucent and revolving spheres. Between the earth and the moon is the *sublunary zone*. Beyond is the *supralunary zone*, consisting of eight spheres carrying the Moon, Mercury, Venus, the Sun, Mars, Jupiter, Saturn, and the fixed stars. The ninth sphere is the *Primium Mobile*, or *Prime Mover* of the entire mechanism. Outside this sphere is paradise where God and the angels reside. (From Dante Alighieri's *Divine Comedy*.)

revolved on him. . . ."[37] This worldview was historically linked to the ancients, whose knowledge and wisdom were considered almost divine. Religion was revealed in scripture, geometry in Euclid, and cosmology in Aristotle and Ptolemy. Since there could only be one correct answer to a problem, and the ancients had solved these problems, then the worldview was complete. If there were anomalies in the system, it was the fault of the scribes who must have made copying errors, not in the knowledge of the ancients.

MACROCOSM AND MICROCOSM

The stability of this system was further reinforced by the interactions between celestial and terrestrial forces. Within this tightly closed universe there were interconnections, or correspondences, between the crystal spheres, the earth, and man. Correspondences were correlations, or linkages that made the entire system interlocking and interdependent. These correspondences—the creation of Renaissance Hermetic and Neoplatonic magical traditions—were linked from one level to the next, with the direction of influence flowing generally from macrocosm to microcosm. For example, since the four elements of earth, water, air, and fire form concentric sublunary rings (albeit intermingled) around the centralized earth, these elements, in conjunction with the controlling forces of the stars (as studied by astrologers), dictated the daily life, health, and fate of our medieval ancestors. FIGURE 22 shows the "orders of the universe" and how

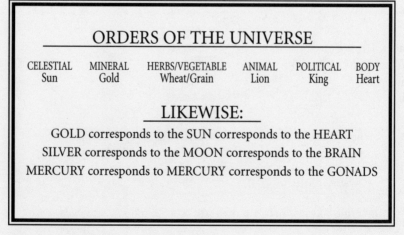

ORDERS OF THE UNIVERSE

CELESTIAL	MINERAL	HERBS/VEGETABLE	ANIMAL	POLITICAL	BODY
Sun	Gold	Wheat/Grain	Lion	King	Heart

LIKEWISE:

GOLD corresponds to the SUN corresponds to the HEART

SILVER corresponds to the MOON corresponds to the BRAIN

MERCURY corresponds to MERCURY corresponds to the GONADS

Figure 22. Orders of the Universe

important a geocentric structure was in this worldview. In this paradigm there were interactions between astronomical bodies, chemical elements, flora and fauna, and bodily parts.

In this system of orders every concentric ring, or crystal sphere of the cosmos, itself had a corresponding link to an element, a mineral, a vegetable, an animal, and a part of the physical body. The "yellowish" color of the sun corresponded to similar colors of gold, wheat, a lion's mane, and the king's crown. Kings, in fact, were said to have fiery eyes, for fire resembled the color of gold. Of King James I it was said that "His integrity resembled the whiteness of his robe, his purity the gold in his crown, his firmness and clearness the precious stones he wore, and his affection the redness of his heart."[38] An individual with a mental disorder was a "lunatic," corresponding to the influence of the moon, which was "silverish" in color. Medicine was also directed by this system. Mercury, for example, was used in the treatment of syphilis. Composites of gold and other elements were used in the treatment of heart disease.[39] The medical chemistry of Paracelsus was based on these connections between mineral agents and disease, with a goal of establishing specific causal agents of specified disorders.[40] Through this paradigm it can be seen how the controlling influence of the stars made astrology an important field of study. The chemical elements in this model were not perceived in the same manner as the modern mind interprets them. To the sixteenth-century iatrochemist (one who applies the knowledge of chemistry to medical practice), the elements were defined by their effect on the sublunary world (particularly the human body) in conjunction with the stars and God. Each of the four elements had some balance of heat, cold, moisture, and dryness:[41]

Earth: Cold and Dry
Water: Cold and Moist
Air: Hot and Moist
Fire: Hot and Dry

Physical health was a matter of striking a balance between the elements, while disease was an indication of an imperfect balance. The physician's job was to "tune the body" by balancing the elements. The four humours also corresponded to the four elements and their characteristics. The liver converts food into four liquid substances, called the humours, which are to the human body what the four elements are to the earth's body.

Thus, personality could be accounted for by the mixture of the elements and humours, as revealed in FIGURE 23. Differences in amounts of the four humours in the body resulted in personality differences. A person who had too much of the earth element, for example, was cold and dry, and therefore melancholic. Additional correspondences included a linkage between human physical anatomy and celestial bodies in such a way "as the noblest heavenly bodies are those highest in the sky, so man's noblest part, his head, is uppermost; and that as the sun is in the midst of the planets, giving them light and vigour, so is the heart in the midst of man's members."[42] Furthermore, the human body corresponds to the earth, as Sir Walter Raleigh described in his *History of the World* (1614):

> His blood, which disperseth itself by the branches of veins through all the body, may be resembled to those waters which are carried by brooks and rivers over all the earth, his breath to the air . . . the hairs of man's body to the grass which covereth the upper face and skin of the earth. The sun is the fundament of all heat and of all time, all in such wise as the heart of a man is the fundament by his valour that is in him of all natural heat.[43]

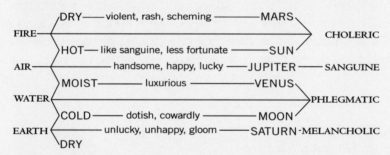

Figure 23. Medieval Personality Analysis. Personality traits correspondenced with the elements, planets, and humors.

Since correspondences are linked to one another, the medieval mind also ranked the living creations of God in a "chain of being" from inanimate objects to living creatures, and from living creatures to more complex life forms, illustrated by Robert Fludd in 1617 and Tobias Schutz in 1654 (see FIGURES 24 and 25). Richard Hooker, in his *Of The Laws of Ecclesiastical Polity* (1594), draws the links across a broad canvas, from stones to angels, and makes comparison judgements along the way.

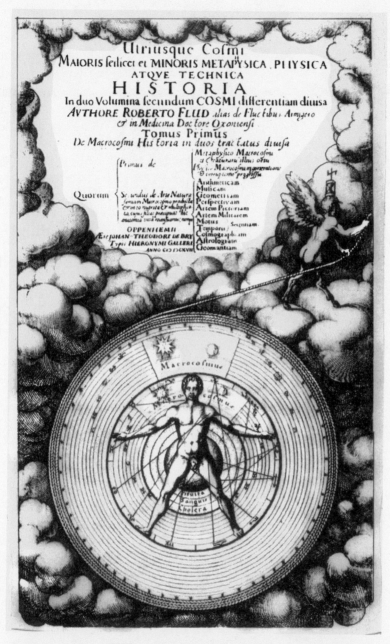

Figure 24. Correspondences Between Celestial and Terrestrial Objects Correspondences are found between the macrocosm (celestial/supralunary) and the microcosm (terrestrial/sublunary). Lines drawn from the signs of the zodiac show correspondences between organs and parts of the human body that are regulated by the spheres. From Robert Fludd's *Ultrisque Cosmic Maioris Scilicet Et Minoris Metaphysica* (1617).

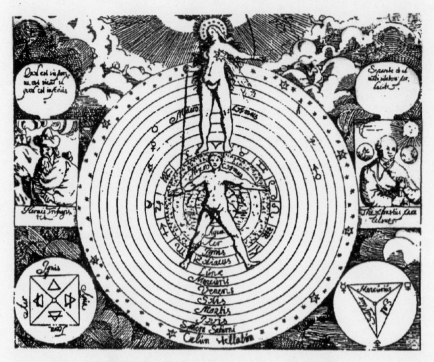

Figure 25. Microcosm and Macrocosm Top: Man as microcosm is joined to the spheres of the macrocosm, represented by a young woman, and the creator, all connected by the chains of nature. The portraits are of Hermes (left) and Paracelsus (right), two natural magicians. The four diagrams depict the three principles (right) and the four elements, (left). From *Harmonica Macrocosmi Cum Microcosmi* (1654), by Tobias Schutz. Below: a detail of a similar system of elaborate correspondences from Robert Fludd (1617).

Stones may be lowly but they exceed the class above them, plants, in strength and durability. Plants, though without sense, excel in the faculty of assimilating nourishment. The beasts are stronger than man in physical energy and desires. Man excels the angels in his power of learning, for his very imperfection calls forth that power, while the angels as perfect beings have already acquired all the knowledge they are capable of holding. The angels presumably excel God himself in the faculty of adoration, for perfection and infinitude have nothing to adore.

HEAVENLY BODIES, HUMAN BODIES, AND THE BODY POLITIC

It was the Greeks who first began construction of this worldview, and it was they who concluded that since the spheres moved, and movement created vibrations, the spheres' vibration caused music—a sort of cosmic music, audible to souls in tune with the cosmos. King James I, for example, was told that the earth and the spheres sang "God save the king." However, "Take but degree away, untune that string,/And hark, what discord follows."[44] Just as the physical body could be "out of tune," the political body could suffer similar maladies. Milton thought the British Commonwealth, for example, was "more divinely and harmoniously tuned" than either Rome or Sparta, because its government was more perfectly balanced.[45] Just as the physician tuned the physical body by balancing the elements, the sovereign tuned the political body. "The king could just as easily describe himself as husband to the whole island and head to that whole island's body," Daly explains. The king was, in this sense, the "Resident Intelligence" of the body politic. John Norden, in *A Christian Familiar Comfort*, compared the political state to the celestial spheres, and Queen Elizabeth to the *Primum Mobile*.[46] The political body is linked to the human body by Thomas Hobbes, in this classic passage from his great work *Leviathan*:

> By art is created that great leviathan called a commonwealth, or state which is but an artificial man, though of greater stature and strength than the natural for whose protection and defence it was intended and in which the sovereignty is an artificial soul as giving life and motion to the whole body. The wealth and riches of all the particular members are the strength. The people's safety its business.[47]

A moving earth and sun-centered cosmological structure challenged the political system of the day, where a monarchy believed it was dependent upon an earth-centered universe. The Aristotelian-Ptolemaic system was finite and closed, with humans near the center and God just outside, running the cosmos. Likewise, the body politic, or state, was compared to the prime mover running the country, with the king, God's chosen representative, at the center. A disruption of cosmic harmony could mean a disruption of political harmony.

Remove the earth and the king is uprooted. Set the earth in motion, and the entire political system, along with most other cultural elements, gets turned upside down.

A WORLD TURNED UPSIDE DOWN

In the medieval worldview all of human life, from the most sublime concepts of God and immortality to the most mundane aspects of daily living, was interpreted in terms of a geocentric universe. To alter this was to undermine the entire superstructure of man's life. If the earth moved, where was stability? If the earth moved, where was certainty? If the earth moved, could Aristotle, Ptolemy, and other ancient authorities be wrong about other matters? If the earth moved, what was the ultimate meaning of human existence?

No less a mind than Martin Luther immediately grasped the implications: "There is talk of a new astrologer who wants to prove that the earth moves and goes round instead of the sky, the sun and moon, just as if somebody moving in a carriage or ship might hold that he was sitting still and at rest while the earth and trees walked and moved. That fool wants to turn the whole art of astronomy upside-down. However, as Holy Scripture tells us, so did Joshua bid the sun to stand still and not the earth."[48] The *Homily on Obedience* below, dated 1547, would be what the typical Anglican church member might hear from the pulpit. The hierarchy of orderliness is pervasive throughout the sermon:

> Almighty God hath created and appointed all things in heaven and earth and waters in a most excellent and perfect order. In heaven, he hath appointed distinct and several orders and stages of Archangels and Angels. In earth he hath assigned and appointed Kings, princes, with other governors under them, in all good and necessary; order.

It was then made quite clear what the repercussions would be if this medieval order were to break down:

> For where there is no right order, there reigneth all abuse, carnall liberty, enormitie, sin, and Babylonical confusion. Take away kings, princes, rulers,

magistrates, judges, and such estates of God's order, no man shall ride or
go by the high way unrobbed, no man shall sleep in his own house or bed
unkilled, no man shall keep his wife, children, and possession in quietness,
all things shall be common, and there must needs follow all mischief, and
utter destruction.[49]

Richard Hooker lamented the consequences for man of an unstable cosmos that
would be the result of a moving earth:

Now if nature should intermit her course and leave altogether the observation of her own laws [and] lose the qualities which now they have . . . if
celestial spheres should forget their wonted motions . . . if the moon should
wander from her beaten way what would become of man himself, whom
these things now do all serve?[50]

Following his description of heavenly orderliness in *Troilus and Cressida*,
Shakespeare is clear on the consequences of a disruption of the cosmic harmony
(Act I, Scene 3):

> But when the planets,
> In evil mixture to disorder wander,
> What plagues and what portents, what mutiny
> What raging of the sea, shaking of the earth,
> Commotion in the winds, frights, changes, horrors,
> Divert and crack, rend and deracinate,
> The Unity and married calm of states
> Quite from their fixture! Oh, when degree is shaked,
> Which is the ladder to all high designs,
> The enterprize is sick!

In the Copernican system, God was no longer just outside the fixed set of
stars watching over a centralized earth. In time, with thinkers such as Giordano
Bruno, it became possible to conceive of an infinite universe, with the possibility
of infinite worlds, and infinite space.

THE HERETICAL IDEOLOGY OF COPERNICUS

It would appear from the historical evidence that Copernicus never intended his
research to result in either a scientific or ideological revolution. Rather, he saw

his work as an incremental improvement on the Ptolemaic system. He retained the use of epicycles, maintained that the heavenly bodies were embedded in crystal spheres as they moved about the sun, and made the structure of *De Revolutionibus* similar to that of Ptolemy's *Almagest*.[51] Proceeding cautiously at first, Copernicus considered the movement of the earth:

> It seemed an absurd notion. Yet I knew that my predecessor had been granted the liberty to imagine all sorts of fictive circles to save the celestial phenomena. I therefore thought that I would similarly be granted the right to experiment, to try out whether, by assigning a certain movement to the earth, I might be able to find more solid demonstrations of the revolutions of the celestial spheres than those left by my predecessors.[52]

In fact, Copernicus put the sun in the center with an appeal to the medieval correspondences of the sun with power and dignity (11): "In this most beautiful temple, who would care to put this lamp [the sun] in any other or better place than whence it can illuminate the whole at the same time? Thus some people not inaptly call [it] the lamp of the world, others [its] mind, others [its] governor. Indeed, residing on a royal throne, as it were, the sun rules the revolving family of the stars."[53] Further, Rosen[54] demonstrates that Copernicus's "orbs" were really "spheres" in which the planets, or "wandering stars," were still firmly embedded as they had been in the Ptolemaic system.[55]

In spite of this caution, and a claim that its purpose was only to discover more adequate "demonstrations" of celestial revolutions, Copernicus believed that the heliocentric system was real and his work, says historian of science Richard Olson, was not "devoid of philosophical implications." Nor were its readers "able to read it without seeing that if accepted, it would directly overturn virtually all traditional natural philosophy and cosmology and indirectly disrupt the rationales for such overarching schemes as those associated with macrocosm-microcosm correspondences."[56] As it turns out, Copernicus's methods of calculation did not yield results more in harmony with what was observed in the cosmos than Ptolemy's calculations. In fact, Copernicus had to add epicycles, and epicycles on epicycles, in order to reach an acceptable level of accuracy in his calculations. Indeed, Owen Gingerich, in comparing calculations between the Ptolemaic and Copernican systems, found that "the Copernican system is slightly more complicated than the original Ptolemaic system" it was to replace.[57]

For both scientific and ideological reasons the Copernican system was slow to be accepted. Edward Rosen demonstrates that when Rheticus attempted to persuade the University of Wittenberg to publish Copernicus's book in 1541, he was rejected "for ideological reasons, grounds already previously expressed by Luther and Melanchthon, who were number one and number two in the town. Therefore, Rheticus had to leave Wittenberg University to get another job elsewhere, and he also had to leave Wittenberg in order to find a printer for the Revolutions."[58] Robert Westman examined the diffusion of the Copernican theory and discovered that between 1543 and 1600, "I can find no more than ten thinkers who choose to adopt the main claims of the heliocentric theory. In view of the fact that Copernicus's book was widely known and discussed during this period, one can justly term this initial reaction conservative, but certainly not reactionary." Most did not completely reject Copernicus, rather they used, appropriately enough, "those parts of the theory which did not depend upon the claim that the earth moves. Interestingly, Westman argues that those who were most receptive to Copernicus were not scientists in the mainstream of traditional institutions playing the role of astronomer; rather Copernicus's strongest adherents were those "actively engaged in reformulating that role outside of those institutions."[59]

If Copernicus's work was not widely read, and it was not a superior method of astronomical calculation, then how did it eventually cause a revolution? Most people understood the implications of the challenge to the earth-centered system, despite Andreaus Osiander's disclaimer "To the Reader" of *De Revolutionibus*, which stated that the postulates were "not put forward to convince any one that they are true, but merely to provide a correct basis for calculation." Richard Olson notes that "most readers could not read Copernicus without seeing that his work denied . . . the fundamental presumption of a hierarchical spatial ordering of the cosmos from the highest Divine sphere to the lowest terrestrial sphere." Because of this, "the largest potential audience for Copernicus's work, the class of medical-astrologers, found Copernicanism incompatible with their most fundamental beliefs; and they naturally rejected it."[60]

Copernicus's initial attack upon the medieval worldview, followed by salvos from Bruno, Kepler, Galileo, and Newton, ultimately destroyed this grand picture, and along with it the interconnecting correspondences. Details of the cosmos and the earth were seen and studied for themselves, rather than as parts of a whole existing for man's use and God's purpose. There was a shift from the qualitative nature of the world to a quantitative one. The Copernican revolution was the true beginning of the type of systematic thinking and mechanistic philosophy that gave rise to the scientific method. In this new world-view, the old moral lessons were no longer to be learned. Just as Mars was no more noble than Venus, the lion was no more noble than the snake. Sublunary zones were no better than superlunary realms, therefore the heart was not superior to the spleen. As astronomy, physics, and chemistry became quantitative, so too did other fields of study. Qualitative relationships between spheres and elements gave way to quantitative relationships between mathematics and the empirical world.

Religion, politics, economics, literature, art, science, medicine, and all aspects of human life had been governed by and were contingent upon the stability of a geocentric system. Had Copernicus merely launched a scientific revolution, it might have gained more rapid acceptance once the new physics and mathematics were in place. The fact that *De Revolutionibus* was not put on the *Index of Prohibited Books* until 1616 is further indication that Copernicus's was an ideological revolution. And because of the immune resistances against such ideological challenges, it was several generations before the Copernican system was commonly accepted. The day the earth moved came to represent a new age and a new way of thinking—it changed the worldview from a provincial medieval egocentric model to the modern ever-expanding and ever-changing infinite universe, a universe very possibly without boundaries and without end.

7
HERETIC-PERSONALITY
Alfred Russel Wallace and the Nature of Borderlands Science

IN THE ANNALS OF SCIENCE one would be hard pressed to find a more affable individual (who is also controversial) than the nineteenth-century British naturalist and codiscoverer of natural selection, Alfred Russel Wallace. After getting to know the man through a thorough reading of his letters and correspondence, papers, manuscripts, and books (which I did as Wallace was the subject of my doctoral dissertation), one cannot help but like him. For example, recalling the generosity shown him as a young scientist by those already established (e.g., Lyell, Darwin, Hooker, Huxley), Wallace later in his life returned the favor to the next generation of budding naturalists. His friend and editor of his letters, James Marchant, recalled that "Wallace loved to give time and trouble in aiding young men to start in life, especially if they were endeavoring to become naturalists. He sent them letters of advice, helped them in the choice of the right country to visit, and gave them minute practical instructions how to live healthily and to maintain themselves. He put their needs before other more fortunate scientific workers and besought assistance for them."[1] Upon his death in 1913, Professor E. B. Poulton, writing in *Nature*, concluded: "The central secret of his personal magnetism lay in his wide and unselfish sympathy. It might be thought by those who did not know Wallace that the noble generosity which will always stand as an example before the world was something special— called forth by the illustrious man with whom he was brought in contact [Charles Darwin]. This would be a great mistake. Wallace's attitude was characteristic . . . to the end of his life."[2]

Wallace's humble origins and self-made career may help, in part, to explain this generosity and kindness. (See chapter 13 for a biographical sketch of Wallace's life.) He understood the difficulties of most people and could relate to their struggles. His background also shaped a separatist personality and created an independent thinker—good for creativity in breaking out of a paradigmatic mold (e.g., his discovery of natural selection), but making him more gullible to unusual claims (e.g., spiritualism), especially when his science failed to account for the anomalies of nature and humanity. When he could not explain the evolutionary significance of such human skills as mathematical reasoning, aesthetic appreciation, and spiritualism, for example, Wallace concluded that there must be a higher intelligence that guided and enhanced nature's selective hand (see chapter 8). This led Charles Darwin to call Wallace a "heretic."

As we shall see, however, Wallace's heretical view of the human mind was just one of many that he underwrote as part of the worldview of a heretic-scientist, bent on extreme independence of thought and maverick tendencies. He was willing (and quite scientifically able) to test any and all claims of the normal or paranormal. If he thought the evidence supported a claim, Wallace was willing to go to any length to lend his credibility and support. Where the evidence did not support a claim, Wallace was as vociferous in his rejection as any of his skeptical colleagues. This tendency was never more evident than when he accepted spiritualism (which he felt had substantial evidentiary backing) and defended it with all his literary and scientific power, while simultaneously taking on the Flat-Earthers in a battle that would span more than fifteen years of his life, causing him much anxiety and consternation. Taking Wallace as a case study in finding the essential tension between conservatism and openness in science, he usually erred on the side of the latter, preferring to risk being right rather than playing it safe and possibly missing out on an intellectual revolution. For Wallace, the decision was an intellectual one based on his perception of the boundaries of science and how its methodologies were defined. Of course, no one is purely objective in these matters, as personality can play a key role in shaping receptivity to ideas and preferences for where those boundary lines should be drawn. Thus, the life and personality of Wallace becomes a case study of the boundary problem in science.

In Wilma George's 1964 study of Wallace, she offers this disclaimer: "No attempt has been made to study Alfred Russel Wallace the man, nor to investigate the psychological reasons for his being both spiritualist and founder of zoogeography."[3] I have remedied that situation, and here I would like to examine Wallace's personality to see how his seemingly disparate intellectual interests and his scientific and spiritualistic beliefs relate to one another. This is not a "psychohistorical" analysis of the man in any Freudian sense, as I do not believe there is substantial evidence for the validity of most of the psychoana-lytically-based claims seen in much of the literature of psychobiography. Rather, I shall offer a general description and analysis of what Wallace actually did, without excessive speculation on internal cognitions and states of mind. My analysis constitutes more of a behavioral approach to personality, though in describing and analyzing his actions, I would offer one general personality de-scription of Wallace as a *heretic-personality*. After discussing this concept, I will review two bizarre incidents in Wallace's life, tangential to his science, and then examine two psychosocial explanations for these events and his many other heretical activities, making applications to other historical characters and events to gain further insight into the nature of heretical science.

THE HERETIC-PERSONALITY

A heretic is "One who maintains opinions upon any subject at variance with those generally received or considered authoritative," and personality is a "unique pattern of traits," where "a trait is any distinguishable, relatively en-during way in which one individual differs from others."[4] Personality, however, can be a fuzzy concept. Just what do we mean by personality, or a personality trait? The psychologist J. Guilford, for example, once noted: "One does not need to read very far in the voluminous literature on personality to be struck by the fact that there is a somewhat bewildering variation in treatments of the subject. One might even conclude that there is confusion bordering on chaos." Nevertheless, Guilford prefaces his definition of personality with "an axiom to which everyone seems agreed: each and every personality is unique." Uniqueness means differences from others (though "similar in some respects"), known as "individual differences," which allows Guilford to conclude: "An individual's personality is his unique pattern of traits." But what is a trait? "A trait is any distinguishable, relatively enduring way in which one individual differs from others."[5]

We may, based on this analysis, construct a composite (and modified) defi-nition of personality as: *The unique pattern of relatively permanent traits that makes an individual similar to but different from others*. Therefore a *heretic-scientist* or a *heretic-personality* is: *The unique pattern of relatively permanent traits that makes an individual maintain opinions upon any subject at variance with those considered au-thoritative*. In other words, a *heretic-scientist* or a *heretic-personality* is an individual who is different from others in his tendency to accept and support ideas con-sidered heretical, although similar to those who also maintain such antiauthoritarian, proradical tendencies. The assumption is that these traits, in being "relatively permanent," are not temporary "states," or conditions of the environment, the altering of which changes the personality. The heretic-personality, like any other type of personality, tends to be heretical in most environmental settings, throughout much of a lifetime. This definition fits Wallace, who routinely maintained opinions on a variety of subjects typically at odds with the received authorities.

Today's most popular trait theory is what is known as the Five Factor model, or just the "Big Five" — (1) *Extraversion* (gregariousness, assertiveness, excitement seeking), (2) *Agreeableness* (trust, altruism, modesty), (3) *Conscientiousness* (competence, order, dutifulness), (4) *Neuroticism* (anxiety, anger, depression), and (5) *Openness* (fantasy, feelings, values).[6] In evaluating Wallace's personality, social scientist Frank Sulloway and I used the 9-point scale devised and tested for the Five Factor model, on forty adjective pairs. For example:

> *I see Alfred Russel Wallace as someone who was:*
> Not dutiful 1 2 3 4 5 6 7 8 9 Dutiful
> Trusting 1 2 3 4 5 6 7 8 9 Suspicious
> Interested in the arts 1 2 3 4 5 6 7 8 9 Uninterested in the arts
> Aloof 1 2 3 4 5 6 7 8 9 Gregarious
> Self-conscious 1 2 3 4 5 6 7 8 9 Not self-conscious
> Prefers novelty/variety 1 2 3 4 5 6 7 8 9 Prefers routine
> Ambitious/hardworking 1 2 3 4 5 6 7 8 9 Lackadaisical
> Tough-minded 1 2 3 4 5 6 7 8 9 Tender-minded

Running correlations between the adjective pairs, and then comparing them to the general database of over 100,000 people Sulloway has in his lab at the University of California at Berkeley, Wallace's personality profile appears in Table 2 below.

WALLACE'S PERSONALITY
Ratings on the "Big 5" Personality Traits

Low 77th percentile • Conscientiousness	High
Low 99th percentile • Agreeableness	High
Low 99th percentile • Openness to Experience	High
Low 73rd percentile • Extroversion	High
Low 6th percentile • Neuroticism	High

The key traits are the first three: conscientiousness, agreeableness, and open-ness to experience. In chapter 10 we will see that Carl Sagan was high in con-scientiousness, very high in openness to experience, but low in agreeableness. Carl did not suffer fools gladly, as they say, but Wallace certainly did, and this is expressed in his expectionally high score on agreeableness—at the 99th per-centile, in fact. Good scientists must find that exquisite balance between being open-minded enough to accept radical new ideas, but not so open-minded that all manner of goofiness is embraced. The key ingredients of personality that help keep that balance is conscientiousness and agreeableness. Wallace was lower than Sagan on conscientiousness (but still relatively high compared to the general population), but he was agreeable to a fault. Wallace was simply far too con-ciliatory towards almost everyone whose ideas were on the fringe. He had a difficult time discriminating between fact and fiction, reality and fantasy, and he was far too eager to please, whereas his more tough-minded colleagues had no qualms about calling a foolish idea foolish.

Wallace became interested in heretical theories as a very young man, inves-tigating, for example, phrenology, and considered controversial biological prob-lems such as the mutability of species. This was not, however, a temporary flirtation with antiauthoritative ideas by a young, undisciplined mind. In mid-life, after codiscovering with Darwin their innovative (and at the time heretical) theory on the origin of species by means of natural selection, Wallace began experimenting with spiritualism and many other controversial beliefs. What es-tablishes Wallace as a genuine heretic-personality was that he demonstrated a unique pattern of relatively permanent traits that caused him to maintain opin-ions upon a variety of subjects throughout his life at variance with those con-sidered authoritative. The following two incidents, both of which occurred well into his later years, provide an exemplar of the heretic-personality in action.

CHALLENGING THE FLAT-EARTHERS

In January 1870, Alfred Wallace read the following advertisement in the journal *Scientific Opinion*:

> The undersigned is willing to deposit from £50. to £500., on reciprocal terms, and defies all the philosophers, divines, and scientific professors in the United Kingdom to prove the rotundity and revolution of the world from Scripture, from reason, or from fact. He will acknowledge that he has for-

feited his deposit, if his opponent can exhibit, to the satisfaction of any intelligent referee, a convex railway, river, canal, or lake.[7]

The undersigned was John Hampden, who had become convinced by Samuel Birley Rowbotham's book, *Earth Not a Globe,* that the earth is immovable and flat, with the North Pole at the center and the sun orbiting a toasty-warm 700 miles above the plane of the planet. All proofs of the earth's sphericity, such as the earth's rounded shadow on the moon in a lunar eclipse, were written off as scientific propaganda constructed by unskeptical Copernicans. (The rounded shadow, Flat-Earthers argue, is because the earth is both round and flat, like a saucer, but not spherical. Likewise, modern Flat-Earthers believe satellite images from space are merely photographs of the upper side of the flat, rounded plane.)

Hampden took up the Flat-Earth cause with proselytizing enthusiasm. He obtained permission from Rowbotham to arrange for the publication of a pamphlet extracted from his book, and printed by William Carpenter, who was soon to enter the story in another capacity. The advertisement soon followed the publications, and, unfortunately for Wallace, the challenge proved too tempting to resist. Soon after, Wallace became embroiled in an incident which, he later claimed, "cost me fifteen years of continued worry, litigation, and persecution, with the final loss of several hundred pounds." Wallace confessed that the blame was entirely his: "And it was all brought upon me by my ignorance and my own fault—ignorance of the fact so well shown by the late Professor de Morgan—that 'paradoxers', as he termed them, can never be convinced, and my fault in consenting to get money by any kind of wager." Wallace later admitted that this was "the most regrettable incident in my life."[8]

His sense of challenge piqued (and his pockets rather empty), Wallace wrote his friend, the renowned geologist Charles Lyell, "and asked him whether he thought I might accept it. He replied, 'Certainly. It may stop these foolish people to have it plainly shown them.' "[9] Hampden suggested using the Old Bedford Canal in Norfolk because it had a straight stretch of six miles between two bridges. Wallace agreed, suggesting that John Henry Walsh, editor of *Field* magazine, act as chief referee. Hampden agreed to Walsh as judge and witness, in addition to which each man would bring along a personal referee. Wallace was accompanied by one Dr. Coulcher, "a surgeon and amateur astronomer." Hampden brought with him none other than William Carpenter, the printer of Hampden's Flat-Earth literature. Walsh held the money to be given to the winner.

On the morning of March 5, 1870, Wallace set up three objects—a telescope,

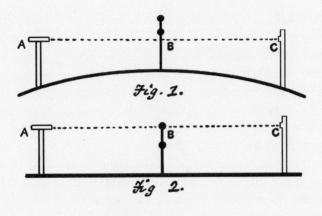

From Dr. Coulcher's report—"Signed by Mr. Carpenter."

The "Bedford Level Survey"—Sketches by the two referees.
Copied from the *Field* for March 26, 1870.

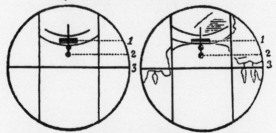

These two views, as seen by means of the *inverting* telescope, are exact
representations of the sketches taken by Mr. Hampden's Referee,
and attested by Dr. Coulcher as being correct in both cases:
first, from Welney Bridge; and secondly, from the Old Bedford Bridge.

Figure 26. Debunking the Flat-Earthers Wallace's test of the earth's sphericity in the Old Bedford Canal proved nothing to the Flat-Earthers.

a disc, and a black band—along the six-mile stretch of the Old Bedford Canal, eighty miles north of London, such that "if the surface of the water is a perfectly straight line for the six miles, then the three objects . . . being all exactly the same height above the water, the disc would be seen in the telescope projected upon the black band; whereas, if the six-mile surface of the water is convexly curved, then the top disc would appear to be decidedly higher than the black band, the amount due to the known size of the earth."

As the diagrams in FIGURE 26 show, Wallace's experiment clearly "proved that the curvature was very nearly of the amount calculated from the known dimensions of the earth." Not surprising, Hampden refused to even look through the telescope, trusting this to his personal referee, William Carpenter, who claimed that he saw "the three were in a straight line, and that the earth was flat, and he rejected the view in the large telescope as proving nothing." Walsh, on the other hand, as the official referee, declared Wallace the winner, and published the results in the March 26, 1870, issue of *Field*. Wallace's victory was well-deserved and he badly needed the money (throughout most of his life Wallace was short of funds and in search of work). But Hampden promptly wrote Walsh, "demanding his money back on the ground that the decision was unjust, and ought to have been given in his favour."[10] According to Wallace, the law in England at that time was that "all wagers are null and void" and "the loser can claim his money back from the stakeholder if the latter has not already paid it away to the winner. Hence, if a loser immediately claims his money from the stake-holder, the law will enforce the former's claim on the ground that it is his money."[11]

Wallace's loss of the newly won 500 pounds (a workingman's wages for one year) turned out to be the least of his problems. Hampden became a one-man nuisance in Wallace's life, initiating a series of abusive letters to the presidents and secretaries of the scientific societies of which Wallace was a member, such as the following to the president of the Royal Geographical Society on October 23 and 26, 1871:

> If you persist in retaining on your list of members a convicted thief and swindler, one A. R. Wallace, of Barking, I am obliged to infer that yr Society is chiefly made up of these unprincipled blackguards, who pay you a stipu-lated commission on their frauds, & secure the confidence of their dupes by their connexion with professedly respectable associations.
>
> In spite of the bluster of the whole English press, J. H. Walsh, of the Field and A. R. Wallace F.R.G.S. are still being posted as a couple of rogues

and swindlers, and will continue to be so if their insolent supporters were as thick as tiles on the houses. Pray inform them that no amt or kind of exposure that can possibly suggest itself, will hear cease till every Socty is ruined to which they respectively belong.[12]

Not content to libel Wallace in public, Hampden even wrote this remarkably caustic personal letter to his wife, Annie, which Wallace kept in order to bring suit against him:

> Madam—If your infernal thief of a husband is brought home some day on a hurdle, with every bone in his head smashed to pulp, you will know the reason. Do you tell him from me he is a lying infernal thief, and as sure as his name is Wallace he never dies in his bed. You must be a miserable wretch to be obliged to live with a convicted felon. Do not think or let him think I have done with him.

This was no bluff—nor was Hampden done. Wallace struck back with libel charges and lawsuits, for which Hampden was arrested and jailed on several occasions. But for the next fifteen years he tormented Wallace with letters, newspaper articles, leaflets, and the like. Wallace became concerned for more than just his reputation, as he indicated in a letter of May 17, 1871, to R. MacLachlau:

> I return Hampden's letters. I have actioned him for Libel, but he won't plead, and says he will make himself bankrupt & won't pay a penny. As the man is half mad I don't want to indict him criminally & infuriate him, & so I suppose he will continue to write endless torrents of abuse as long as he lives.

And so he did. On October 24, 1871, Hampden even got a petition read into the record at the Royal Geographical Society from his supporters, claiming, among other things, that "Mr. Hampden has been most urgent and importunate for the fullest and freest investigation of facts, the other side has been equally persistent in resisting and opposing all such practical tests as might in any way disturb the verbal decision of Mr. Walsh."

Wallace eventually recovered the 500 pounds, but "The two law suits, the four prosecutions for libel, the payments and costs of the settlement, amounted to considerably more than the 500 pounds . . . besides which I bore all the costs of the week's experiments, and between fifteen and twenty years of continued persecution—a tolerably severe punishment for what I did not at the time recognize as an ethical lapse."[13] The difference, of course, between Wallace and Hampden was the extent to which they would allow the evidence to answer a question of nature. Despite his scientism, however, Wallace's response to Hampden was one that would never have been made by his more conservative colleagues, Darwin and Lyell. Nevertheless, Wallace's personality dictated his need to take on such a radical claim. His were the actions of a heretic-personality. Traits usually trump states. Fascinated with all ideas on the radical edge, Wallace simply had to take up the cause regardless of the cost, which was substantial.

"LEONAINIE"—IN SEARCH OF THE LOST POEM OF POE

In digging through the stacks at the Honnold Library of the Claremont Colleges in search of any minutia on Wallace, I was surprised to find—under the section on Edgar Allan Poe—a 1966 publication entitled *Edgar Allan Poe: A Series of Seventeen Letters Concerning Poe's Scientific Erudition in Eureka and His Authorship of Leonainie*. The author of this tiny monograph (eighteen pages) was none other than Alfred Russel Wallace, who penned fifteen letters (and two extracts never mailed) to one Ernest Marriott, Esq., between October 29, 1903 and March 23, 1904. The incident in question—a rediscovered poem of Edgar Allan Poe supposedly written "at the Wayside Inn in lieu of cash for one night's board and lodging"—is emblematic of Wallace's vivid imagination and willingness to jump to conclusions on the scantiest of evidence.[14]

The story, as I have been able to reconstruct it, is as follows. Sometime around 1893, just seven years after his tour of America, Wallace received a letter from his brother living in California, which included a poem entitled "Leonainie," allegedly written by Poe. Wallace, however, was "occupied with other matters" and thus "made no enquiry how he got it, but took it for granted that he had copied it from some newspaper."[15] Ten years later, on November 3, 1903, Wallace wrote to Ernest Marriott (with no explanation offered of Marriott's role, other than he was an attorney) to inquire about confirmation of the claim: "I think you will agree with me that it is a gem with all the characteristics of Poe's genius." Wallace also made a bizarre reference in this letter about the last poems of Poe, "The Streets of Baltimore" and "Farewell to Earth," which Wallace believed were written *after* Poe's death "through another brain," and while they are "in my opinion fine and deeper & grander poems than any written by him in the earth-life . . . they are deficient in the exquisite music & rhythm of his best known work."[16]

With typical enthusiasm for all matters heretical, Wallace threw himself into an intense study of Poe's writings, obsessed with finding out if "Leonainie" was indeed his long lost, and perhaps last, poem (in *this* world anyway). One week later he told Marriott: "Since I wrote to you about 'Leonainie' I have read it many times & have it by heart, & on comparing it with the other poems by Poe which I have it seems to me to be in many respects the *most perfect* of all. The rhythm is most exquisite, and the form of verse different from any other I can call to mind in the double triplets of rhymes in each verse, carried on throughout by simple, natural and forcible expressions while the last verse seems to me the very finest in any of his poems." Wallace reprinted the poem for Marriott at the end of the November 2 letter:

Leonainie

Leonainie, angels named her, and they took the light
Of the laughing stars and framed her, in a smile of white,
And they made her hair of gloomy midnight, and her eyes of
    bloomy
Moonshine, and they brought her to me in a solemn night.
In a solemn night of summer, when my heart of gloom
Blossomed up to greet the comer, like a rose in bloom.
All foreboding that distressed me, I forgot as joy caressed me,
Lying joy that caught and pressed me, in the arms of doom.
Only spake the little lisper in the angel tongue,
Yet I, listening, heard the whisper; "songs are only sung
Here below that they may grieve you, tales are told you to
    deceive you,
So must Leonainie leave you, while her love is young."
Then God smiled, and it was morning, matchless and supreme,
Heaven's glory seemed adorning earth with its esteem,
Every heart but mine seemed gifted with a voice of prayer and
    lifted,
When my Leonainie drifted from me like a dream.

In response to Marriott's uncritical acceptance of Wallace's conviction that Poe had written from the "other side" (and thus warranting his praise), Wallace wrote: "Your letter about the 'Poems from the Inner Life' very much pleased as it shows you are open to conviction. I therefore send for your acceptance a copy of my little book—'Miracles & Modern Spiritualism'." Wallace's spiritualistic leanings are apparent in this and the following letter of December 19, 1903:

> Such teachings as these are in my opinion worth all the poems he wrote during life. The "Farewell to Earth" is such a favourite of mine that I know it by heart, & use it as an opiate if I lay awake. It really contains the essence of modern spiritualistic teaching, and such lines as—"Where the golden line of duty / Like a living pathway lies" strikes a higher note than anything in Poe's earthy poems.

With little evidence to go on, however, on the first day of 1904 Wallace noted that he still needed a scene and a motive for the poem: "I presume Poe was never in California, but I shall be glad to know if, at anytime, shortly before his death, he is known to have travelled anywhere in an almost penniless condition, where such an incident as his paying for a night's board & lodging with a poem *might* have occurred."[17] Undaunted by a lack of evidentiary support, however, and in his usual eagerness to get into print with an exciting new find, on January 6 Wallace told Marriott: "I think I can see when Leonainie was probably written & I shall now send it with a few preliminary remarks to the Editor of the *Fortnightly*, & its publication may possibly lead to its origin being traced in America."[18] Growing bolder by the day, on January 10 Wallace announced that the poem would be published and that "taking all the circumstances into consideration . . . I have come to the conclusion that this was the very last thing Poe wrote, & it was *probably* written only a few days before his death."[19]

Five days later Wallace was in print with the poem and, as usual, he found himself embroiled in controversy. Apparently someone identified the poem as a fake, written by one James Whitcomb Riley, but Wallace spin-doctored the accusation, setting a standard of proof he had not held for himself: "Till we have the alleged *proof* that Riley wrote 'Leonainie', it seems to me quite as probable that *he* found it, and on the suggestion of a friend made use of it to gain a reputation" (February 8). But then Wallace received a letter from a Mr. Law (reprinted in the February 8th letter to Marriott), implicating Riley as the perpetrator of the hoax, cajoled by friends who told him that if he could write like Poe he could achieve enough fame to establish himself as a poet of high caliber. Riley, speculates Wallace, then wrote "Leonainie," submitted it, and "after it had run the gauntlet of Poe critics and been pronounced genuine if not canonical Riley proved the authorship. This drew attention to his own works, and he has never since lacked for praise and pudding."[20]

On February 15, Wallace received more bad news, this from the "Librarian of the London Library," who "obtained a copy of Riley's 'Armazindy'—which contains 'Leonainie' & has sent it to me. The publishers say that this vol. 'contains some of Mr. Riley's latest and best work including 'Armazindy' & the famous Poe Poem.' " Despite the overwhelming evidence that "Leonainie" was a well-perpetrated hoax, Wallace was unable to recant. The remainder of the February 15th letter is a critique of Riley's other poetry, with Wallace's analysis that Riley did not have the skill to write "Leonainie," and his conclusion that the real hoax is that Riley *found* the Poe poem and pretended to have written it! On March 1, still obsessed with the problem, Wallace told Marriott: "I have now looked through 4 vols. of Riley & can find no sign of his being able to write *Leonainie* with all its defects." Wallace then conducted a careful line-by-line analysis and comparison of "Leonainie" with Riley's other poetry, and drew up a final summary of the whole affair: "The more I consider the matter the more I am convinced he did *not* compose the poem. It looks to me very much as if he really got hold of the poem in the form I have it or nearly,—that to cover himself from exact copying he made the alterations in words, which he might think would make it more like his own work, and the alteration in the arrangement of lines &c. so that it might be accepted as a bad imitation of Poe."[21]

The entire incident encapsulates Wallace's heretic-personality—his eagerness to investigate unusual claims, his thorough, almost obsessive analysis of a subject, his willingness to make a serious commitment to a position early in the absence of substantial evidence, and his resolution, regardless of contrary evidence, to maintain his original position (even using contradictory evidence in his favor). When he was right, as in his discovery of natural selection, these traits worked in his favor. But when he was wrong, as is most likely the case here as in his investigations of spiritualism, Wallace's heretic-personality brought down upon him the scorn and ridicule of scientists, skeptics, and more conservative personalities.

THE "ANTI-BODY"

From his letters one gets a sense of Wallace's feeling of being an outcast—a rebel of sorts—even amongst his closest colleagues. He turned down many honorary degrees and only reluctantly agreed to be admitted as a Fellow of the Royal Society after Sir W. T. Thiselton-Dyer invited him three times. After receiving the Order of Merit in 1908, the highest honor ever conferred upon him, Wallace wrote to his close friend, Mrs. Fisher:

> Is it not awful—two more now! I should think very few men have had three such honours within six months! I have never felt myself worthy of the Copley medal—and as to the Order of Merit—to be given to a red-hot Radical, Land nationaliser, Socialist, Anti-Militarist, etc., etc., etc., is quite astounding and unintelligible![22]

From the tone of such letters, it is obvious that Wallace was both aware of and proud of his autonomous stand on certain issues that his more conservative scientific colleagues would not vouchsafe to take. In a November 4, 1905, letter to Raphael Meldola, after the reviewers of his autobiographical *My Life* took note of his "faddish" interests in the fringes of science and spiritualism, he recalled: "Yesterday I got a notice in a paper called '*Reviews*' with a very fair bit . . . he says—'For on many subjects Mr. Wallace is an antibody. He is anti-vaccination, anti-state endowment of education, anti-land-laws, and so on. To compensate, he is pro-spiritualism, and pro-phrenology, so that he carries, as cargo, about as large a dead weight of fancies and fallacies as it is possible to float withal." A more conservative scientist might blanch at such a description, but not a heretic-personality. Three days later, to his friend Mrs. Fisher, Wallace actually boasted of his new title:

> The reviewers are generally very fair about the fads except a few. The Review invents a new word for me—I am an "anti-body"; but the Outlook is the richest: I am the one man who believes in spiritualism, phrenology, anti-vaccination, and the centrality of the earth in the universe, whose life is worth writing. Then it points out a few things I am capable of believing, but which everybody else knows to be fallacies, and compares me to Sir I. Newton writing on the prophets! Yet of course he praises my biology up to the skies—there I am wise—everywhere else I am a kind of weak, babyish idiot! It is really delightful![23]

Wallace's anti-body attitudes, which at times led him to greatness, occasionally took him down paths that led to dead ends. An unparalleled observer in some fields, he was almost blind in others.

BIRTH ORDER AND HERETICAL SCIENCE

Having examined Alfred Wallace as a heretic-personality, it might be constructive to consider the general origins and development of such a personality in all heretic-scientists throughout the history of science. As we saw in chapter 6, in a 1990 paper on "Orthodoxy and Innovation in Science," and in his 1996 book *Born to Rebel*, historian and social scientist Frank Sulloway tested and confirmed the importance of a series of factors in the development of a personality willing to explore and ultimately accept heretical ideas. But Sulloway tested far more than just the Copernican revolutionaries. He examined 28 different scientific controversies and revolutions over the past 400 years, and throughout he discovered that the likelihood of a laterborn accepting a revolutionary idea was 3.1 times higher than a firstborn, and for radical revolutions the likelihood was 4.7 times higher.

For example, Sulloway also assessed the attitudes of over 300 scientists toward the Darwinian revolution between 1859 and 1870. The criteria for acceptance or rejection of Darwinism were based on three premises: "(1) that evolution takes place, (2) that natural selection is an important (but not an exclusive) cause of evolution, and (3) that human beings are descended from lower animals without supernatural intervention" (3). Acceptance of all three makes one a Darwinian. The results for this particular controversy were consistent with those of the general model—83 percent of Darwinians were laterborns, and 55 percent of non-Darwinians were firstborns, statistically significantly different at p<.0001. Included in those who rejected Darwinian evolution by natural selection were Louis Agassiz, Charles Lyell, John Herschel, and William Whewell, all firstborns. Counted as full supporters were Joseph Hooker, Thomas Henry Huxley, Ernst Haeckel, and, of course, Charles Darwin and Alfred Russel Wallace, all laterborns.[24]

Plotting over 1,000,000 data points (that also confirm the effect in nonscientific revolutions such as the Protestant Reformation and the French and American Revolutions), as we also saw Sulloway found that the degree of radicalness of the new theory was also correlated with birth order. Laterborns prefer probabilistic views of the world, such as Darwin's and Wallace's theory of natural selection, to a more mechanical and predictable world view preferred by firstborns. Finally, Sulloway found that when firstborns did accept new theories, they were typically the most conservative of the bunch, "theories that typically reaffirm the social, religious, and political status quo and that also emphasize hierarchy, order, and the possibility of complete scientific certainty."[25] Louis

Agassiz, for example, opposed Darwin (predicted at a 98.5 percent probability by the model) but he supported glaciation theory, explained by the fact that relative to the ideological implications of Darwinism, glaciation was conservative, as well as being linked with the already accepted theories of catastrophism and creationism. The theory of evolution by means of natural selection, by contrast, did not reaffirm the status quo of any social institution, and was thus especially appealing to laterborn radicals like Wallace.

A deeper question to ask in this context of an analysis of heretical personality, is *why* firstborns are more conservative and influenced by authority, while laterborns are more liberal and receptive to ideological change? What is the causal connection between birth order and personality? One hypothesis is that firstborns, being first, receive substantially more attention from their parents than laterborns, who tend to enjoy greater freedom and less indoctrination into the ideologies of and obedience to authorities. Firstborns generally have greater responsibilities, including the care and liability of their younger siblings. Laterborns are frequently a step removed from the parental authority, and thus less inclined to obey and adopt the beliefs of the higher authority. Sulloway summarizes the birth order-personality connection:

> Sandwiched between parents and younger siblings, the firstborn child occupies a special place within the family constellation and, for this reason, generally receives special treatment from parents. Moreover, as the eldest, firstborns tend to identify more closely with parents, and through them, with other representatives of authority. This tendency is probably reinforced by the firstborn's frequent role as a surrogate parent to younger siblings. Consistent with these developmental circumstances, firstborns are found to be more respectful of parents and other authority figures, more conforming, and more conscientious, conventional, and religious. Laterborns, who tend to identify less closely with parents and authority, also tend to rebel against the authority of their elder siblings.[26]

There is independent corroboration of this hypothesis in the field of developmental psychology. J. S. Turner and D. B. Helms, for example, report that "firstborns become their parents' center of attention and monopolize their time. The parents of firstborns are usually not only young and eager to romp with their children but also spend considerable time talking to them and sharing their activities. This tends to strengthen bonds of attachment between the two."[27] Quite obviously this attention would include more rewards and punishment, further reinforcing obedience to authority and controlled acceptance of the "right way" to think. Adams and Phillips[28] and Kidwell[29] report that this excessive attention causes firstborns to strive harder for approval than laterborns. Markus has discovered that firstborns tend to be anxious, dependent, and more conforming than laterborns.[30] Hilton, in a mother-child interactive experimental setting with 20 firstborn, 20 laterborn, and 20 only children (four years of age), found that firstborns were significantly more dependent on, and asked for help or reassurance from their mothers, than the laterborn or only children.[31] In addition, mothers of firstborns were significantly more likely to interfere with their child's task (constructing a puzzle) than the mothers of laterborn or only children. Finally, it has been shown by Nisbet[32] that laterborns are far more likely to participate in relatively dangerous sports than firstborns, which is linked to risk taking, and thus "heretical" thinking.[33]

Sulloway's historical study—the most comprehensive ever conducted on birth order effects—will certainly provide psychologists with data confirming or rejecting any number of psychological theories about development and personality. Birth order alone, of course, does not determine the ideological receptivity to radically new ideas. Instead, Sulloway explains, birth order is a proxy for other influencing variables, such as age, sex, and socioeconomic class, which in turn influence openness to new ideas. For example, although "social class itself exerts absolutely no direct influence on the acceptance of new scientific ideas," Sulloway discovered "it is only through a triple-interaction effect with birth order and parental loss that social class plays a subtle but significant role in attitudes toward scientific innovation."[34]

This is not surprising, and it is the reason I have belabored this subject for so long. Alfred Wallace was laterborn (the eighth of nine children), was in the Middle/Lower Class (in Sulloway's classification system), and was separated from his parents at age fourteen—the triple-interaction effect that generates the greatest amount of support for radical scientific theories. According to Sulloway's multivariate model, which includes twelve predictors and their interaction effects, Wallace "possessed a 99.5 percent probability of championing that theory."[35]

Wallace's family background was working class, his formal education (and thus indoctrination into conserving traditional beliefs) was a minimal seven years, and his father went bankrupt when Wallace was only 13, at which time he went to live with his brother John, and then later with his brother William. Wallace rarely returned home, and by the time he completed his four-year trip to the Amazon, at age twenty-nine, his father had died. By Sulloway's analysis,

Wallace was almost destined to be a radical scientist because he had a heretic-personality. Other factors make this even more likely, as Sulloway explains:

> In my multivariate model, he is, of course, a laterborn, in the most liberal, political, and religious cohorts of the model, already somewhat acquainted with Darwin before 1859, and relatively young at that time (36). What differentiates him from Darwin is primarily his greater degree of political radicalness. Darwin falls in the third rather than the fourth category for political and religious beliefs, since he was a Whig and a deist (averaging about 3.6 on my 5-point scales), whereas Wallace was a clear radical and a deist (about a 4.4).[36]

On the Five Factor model of personality traits, *Openness* is most sensitive to birth order effects, and here Wallace and Darwin tie for third out of 2,458 scientists, scoring in the top 1/10th of one percent as defined by amount of travel, number of interests, and position on scientific controversies.

THE TEMPTATION OF HERETICAL SCIENCE

A final psychological component to consider in the heretic-personality of Wallace is what philosopher Paul Kurtz calls the *Transcendental Temptation*, discussed at length in his 1986 book of this title. In essence, this cognitive variable affects all who have thoughtfully considered the ultimate purpose of our existence. The temptation, says Kurtz, "lurks deep within the human breast. It is ever-present, tempting humans by the lure of transcendental realities, subverting the power of their critical intelligence, enabling them to accept unproven and unfounded myth systems."[37]

Specifically, Kurtz argues that myths, religions, and claims of the paranormal are lures tempting us beyond rational, critical, and scientific thinking, for the very reason that they touch something in us that is sacred and important—life and immortality: "This impulse is so strong that it has inspired the great religions and paranormal movements of the past and the present and goaded otherwise sensible men and women to swallow patently false myths and to repeat them constantly as articles of faith."[38] What drives this temptation? Kurtz believes it is creative imagination:

> There is a constant battle in the human heart between our fictionalized images and the actual truth. We fabricate ideal poetic, artistic, and religious

visions of what might have been in the past or could be in the future. There is a constant tension between the scientist and the poet, the philosopher and the artist, the practical man and the visionary. The scientist, philosopher, and practical man wish to interpret the universe and understand it for what it really is; the others are inspired by what it might become. Scientists wish to test their hypothetical constructs; dreamers live by them. All too often what people crave is faith and conviction, not tested knowledge. Belief far outstrips truth as it soars on the wings of imagination.[39]

If there is one lesson we can learn from experimental psychology it is that individual differences are the norm among humans. There is a range of variability in all behaviors and beliefs. Not everyone is equally tempted by transcendence, nor is everyone equal in the ability to look beyond the temptation. Heretic-personalities are significantly more tempted by the various transcendent claims, and, further, are less willing to analyze such beliefs with the same critical scrutiny as they might apply to other belief systems.

It is true that Wallace's belief in spiritualism had, in his mind, a certain amount of scientific evidentiary support behind it. But the sense one gets from reading the vast correspondence and literature Wallace produced on the subject is that the transcendental temptation was just too powerful for him to overcome. As in his involvement with the Flat-Earthers and Poe's last poem, his traits for radicalness overwhelmed his states of caution. Consider the following letter, written by Wallace in 1894 to a friend, upon the death of his beloved sister, Frances:

> Death makes us feel, in a way nothing else can do, the mystery of the universe. Last autumn I lost my sister, and she was the only relative I have been with at the last. For the moment it seemed unnatural and incredible that the living self, with its special idiosyncrasies you have known so long can have left the body, still more unnatural that it should (as so many now believe) have utterly ceased to exist and become nothingness. With all my belief in and knowledge of Spiritualism, I have, however, occasional qualms of doubt, the remnant of my original deeply ingrained scepticism; but my reason goes to support the psychical and spiritualistic phenomena in telling me that there must be a hereafter for us all.[40]

Such commentary in Wallace's writings is the norm, not the exception, and he received much support from the lay public and his fellow scientists, over many years and from around the world, in this temptation to transcendence.

A typical letter in a folio of Wallace's correspondence on matters spiritual is from Professor Theo D. A. Cockerell, in Las Cruces, New Mexico, who had corresponded with Wallace several times on biological matters. On September 24, 1893, Cockerell wrote Wallace regarding the death of his wife in childbirth:

> When a man speaks of his "better half,"—it is usually a form of speech, but I feel as if my better half was indeed taken away from me, and scarcely know how the other half can work to any purpose until they are reunited. The more one thinks, the more one sees every reason to hope and indeed believe that the present separation is only temporary. I am sure you will agree to this. The outlook in every way should make one cheerful, but it is impossible to be so philosophical as to ignore the present, which is hard enough. I find so few people who seem to have any clear notion about immortality. To me, it seems simply axiomatic just like the infinity of time and space—although in each case the conception eludes our mental capacity.[41]

Wallace responded with his usual sensitivity, as he did to all inquirers in such otherworldly matters, and by doing so he added the weight of scientific credibility to any who succumbed to this trancendental temptation. While there were many heretic-personalities in Wallace's time to whom the public might and did turn for comfort in such matters, Wallace was among the most important because of his stature as a scientist.

Those with heretic-personalities—scientists and nonscientists alike—must be more cautious than most, for while their boldness may lead them to extraordinary success in one field, it may occasionally turn to temerity and lead them down the road to deception and disaster in others. The rub in science is to find the right balance between being so open to heretical ideas that it becomes difficult to separate sense from nonsense; and so closed to heretical ideas that it becomes difficult to abandon the status quo. Heretic-personalities, so numerous among the various pseudosciences, need to temper their beliefs with a little caution. Skeptics, so numerous among the various sciences, need to moderate their skepticism with a little boldness. Where the heretic meets the skeptic a creative scientist will emerge.

8

A SCIENTIST AMONG THE SPIRITUALISTS
Crossing the Boundary from Science to Pseudoscience

HISTORIANS HAVE A MOST unusual task among seekers of truth. In order to think ourselves into the minds of our predecessors to understand how they thought, we must *forget* what we know because we would unfairly judge them by our standards—they did not know what we know. On the other hand, in order to glean lessons from the past to understand which ideas were dead ends and which led to the modern worldview, we must *remember* what we know and compare their ideas with ours in order to make history useful as well as entertaining. It is a tricky balance to maintain, especially when traveling along the borderlands of science where what we might today call pseudoscience, a different age would call science. A case study in exploring the boundary issues in the nature of science versus pseudoscience can be found in the investigations of spiritualism by the renowned nineteenth-century British naturalist Alfred Russel Wallace, best known for his codiscovery (with Charles Darwin) of natural selection.

Wallace merits our attention not only because he was honest and passionate (lots of people are, but that does not make them good investigators), but because he was considered one of the greatest scientists of his age. How does an eminent scientist, through a series of investigations (as opposed to compartmentalized religious or spiritual beliefs), come to accept suprascientific or supernatural ideas? The answer is not just a historical curiosity. There is a powerful social movement underfoot, driven largely by the Templeton Foundation, that distinctly and purposefully crosses the boundary between science and religion through attempts to prove previously faith-based beliefs such as God's existence, the efficacy of prayer on healing, or the relationship between guilt, forgiveness, and well-being. In like manner, in Frank Tipler's 1994 book, *The Physics of Immortality: Modern Cosmology, God and the Resurrection of the Dead*, the author claims "modern physics requires the God principle." By this Tipler means that the universe is structured in such a way that the laws of nature must give rise to intelligent life; and once formed, the resurrection of all intelligence — immortality — is inevitable. "Science now tells us," Tipler concludes, "how to go to heaven."[1] While Tipler's science is modern, his argument is not. It is Wallace's argument for the necessity of a higher intelligence clothed in modern physics. By examining the development of Wallace's ideas in the context of his culture we can, perhaps, gain an understanding of how scientists, then and now, flirt with and occasionally cross the boundary from science into pseudoscience.

WALLACE'S HERESY

The first public announcement of Wallace's scientific heresy on the evolution of man and mind, so influenced by his experiences with the spiritualism movement, can be dated to the April 1869 issue of *The Quarterly Review*, in a review of Charles Lyell's tenth edition of *Principles of Geology*. For Wallace, the problem of evolution was the failure of natural selection to explain the exceptionally large human brain:

> In the brain of the lowest savages and, as far as we know, of the prehistoric races, we have an organ . . . little inferior in size and complexity to that of the highest types. . . . But the mental requirements of the lowest savages, such as the Australians or the Andaman Islanders, are very little above those of many animals. How then was an organ developed far beyond the needs of its possessor? Natural Selection could only have endowed the savage with a brain a little superior to that of an ape, whereas he actually possesses one but very little inferior to that of the average members of our learned societies.[2]

Since natural selection was the only force Wallace knew of to explain evolution, and since he decided that it could not adequately account for the human brain, he concluded that "an Overruling Intelligence has watched over the action of those laws, so directing variations and so determining their accumulation, as finally to produce an organization sufficiently perfect to admit of, and even to aid in, the indefinite advancement of our mental and moral nature."[3]

Wallace's reasoning was sound and consistent. Natural selection does not select for organs that will be needed in the future. Natural selection operates on the here-and-now level of the organism. The usefulness or uselessness (or even harmfulness) of a given structure or function can only matter to the organism *now*. Nature did not know we would one day need a big brain in order to contemplate the heavens or compute complex mathematical problems; she merely selected amongst our ancestors those who were best able to survive in their environment. But since we *are* capable of such sublime and lofty mental functions, clearly natural selection could not have been the originator of a brain big enough to handle them. Only an "Overruling Intelligence" could have fashioned us in advance—a rational, if not natural, explanation for the phenomenon at hand.[4]

Lyell supported Wallace's new stance, telling Darwin in a letter of April 28, 1869, that "I rather hail Wallace's suggestion that there may be a Supreme Will and Power which may not abdicate its function of interference but may guide the forces and laws of Nature."[5] Lyell was an important ally for Wallace, bolstering his confidence in his decision to break from the Darwinian camp. Darwin, not surprisingly, was not so conciliatory. Anticipating his friend's reaction, Wallace wrote Darwin on March 24, 1869, to warn him that "in my forthcoming article in the 'Quarterly' I venture for the *first time* on some limitations to the power of natural selection." Knowing how this new scientific and ideological development would be received, Wallace continued: "I am afraid that Huxley and perhaps yourself will think them weak & unphilosophical. I merely wish you to know that they are in no way put in to please the *Quarterly* readers,—you will hardly suspect me of that,—but are the expression of a deep conviction founded on evidence which I have not alluded to in the article but which is to me absolutely unassailable."[6]

Darwin responded on March 27: "I shall be intensely curious to read the *Quarterly:* I hope you have not murdered too completely your own and my child."[7] After he read the article Darwin's response was predictably and understandably brusque. In the margin of Darwin's copy of the article, next to the passage on the inadequacy of natural selection to endow humans with a large brain (quoted above), he wrote a firmly pressed "NO," underlined three times with numerous added exclamation points. Darwin told Lyell that he was "dreadfully disappointed" in Wallace, and then he wrote him again, exclaiming: "If you had not told me, I should have thought that they had been added by someone else. As you expected, I differ grievously from you, and I am very sorry for it. I can see no necessity for calling in an additional and proximate cause in regard to Man."[8] Several months later, on January 26, 1870, with the not-so-subtle hint of a disappointed friend and mentor, Darwin lamented: "I groan over Man—you write like a metamorphosed (in retrograde direction) naturalist, and you the author of the best paper that ever appeared in the Anthropological Review! Eheu! Eheu! Eheu!—Your miserable friend, C. Darwin."[9]

Wallace's reaction to Darwin's disappointment was immediate and understanding. "I can quite comprehend your feelings with regard to my 'unscientific' opinions as to Man, because a few years back I should myself have looked at them as equally wild and uncalled for." In addition to his skepticism of the ability of natural selection to account for the human brain and other features, Wallace was now diverging from Darwin down a track about which he could only hint: "My opinions on the subject have been modified solely by the consideration of a series of remarkable phenomena, physical and mental, which I have now had every opportunity of fully testing, and which demonstrate the existence of forces and influences not yet recognised by science."[10]

Wallace, ever the heretic-scientist willing to explore any and all aspects of the mysterious world around him, had become enthralled with and caught up in the spiritualism renaissance that had become the rage of England and America over the previous two decades. Anticipating Darwin's less than enthusiastic response, Wallace marshaled his allies who had corroborated these findings, and then requested that Darwin delay an assessment of insanity: "This will, I know, seem to you like some mental hallucination, but as I can assure you from personal communication with them, that Robert Chambers, Dr. Norris of Birmingham, the well-known physiologist, and C. F. Varley, the well-known electrician, who have all investigated the subject for years, agree with me both as to the facts and as to the main inferences to be drawn from them, I am in hopes that you will suspend your judgment for a time till we exhibit some corroborative symptoms of insanity."[11]

HERETICAL EXPLANATIONS

Why did Alfred Wallace retreat from his own naturalistic interpretations in favor of supernatural intervention when it came to the origins and evolution of the mind? He was, after all, the self-styled defender of Darwinism who once confessed that "some of my critics declare that I am more Darwinian than Darwin himself and in this, I admit, they are not far wrong."[12] How can someone more Darwinian than Darwin break from this tradition?

Evolutionary biologist Stephen Jay Gould offers this explanation: "Wallace emerges from most historical accounts as a lesser man than Darwin for one (or more) of three reasons, all related to his position on the origins of human intellect: for simple cowardice; for inability to transcend the constraints of culture and traditional views of human uniqueness; and for inconsistency in advocating natural selection so strongly (in the debate on sexual selection), yet abandoning it at the most crucial moment of all." Though Gould's is a mono-causal explanation, he confesses "I cannot analyze Wallace's psyche, and will not comment on his deeper motives for holding fast to the unbridgeable gap between human intellect and the behavior of mere animals"; instead Gould seeks to assesse "the logic of [Wallace's] argument, and recognize that the traditional account of it is not only incorrect, but precisely backwards." Gould argues forcibly for *hyper-selectionism* as the sole agent of change: "Wallace did not abandon natural selection at the human threshold. Rather, it was his peculiarly rigid view of natural selection that led him, quite consistently, to reject it for the human mind. His position never varied—natural selection is the only cause of major evolutionary change."[13] Historian of science Malcolm Jay Kottler, by contrast, believes that "Wallace's spiritualist beliefs were the origin of his doubts about the ability of natural selection to account for all of man." Indeed, Kottler has good reason and historical evidence for such a conclusion, since Wallace states so himself in the letter to Darwin cited above, on the "series of remarkable phenomena" that "modified solely" his beliefs about the evolution of man and mind. Therefore, Kottler concludes, "Something happened between 1864 and 1869 to change his mind: the crucial event was Wallace's conversion to spiritualism."[14]

But a single quote does not an ideological shift make, even if it is from the ideologue himself. When we examine the whole man we see that this argument is too narrow in attributing Wallace's shift entirely to spiritualistic beliefs. It is obvious that these spiritual phenomena played a significant role in Wallace's thoughts, but so did many other factors. In the context of his entire corpus of writings, especially his correspondence, it is obvious that Wallace's hyper-selectionism was far more than a mere justification for a belief in spiritualism. Wallace's hyper-selectionism was potent, sustaining, and pervasive in his entire world-view, and led him into a number of scientific controversies. Historian Joel Schwartz, for example, disagrees with Kottler and argues that "After 1865 Wallace's religious views were responsible for widening the gulf between Darwin and himself."[15] Schwartz presents evidence that Wallace's interest in spiritualism antedated his papers on human evolution by several years. In an interview with

W. B. Northrop, published in the *Outlook* in 1913, for example, Wallace explained the origins of his interest in spiritualism:

> When I returned from abroad [the Malay Archipelago] I had read a good deal about Spiritualism, and, like most people, believed it to be a fraud and a delusion. This was in 1862. At that time I met a Mrs. Marshall, who was a celebrated medium in London, and after attending a number of her meetings, and examining the whole question with an open mind and with all the scientific application I could bring to bear upon it, I came to the conclusion that Spiritualism was genuine. However, I did not allow myself to be carried away, but I waited for three years and undertook a most rigorous examination of the whole subject, and was then convinced of the evidence and genuineness of Spiritualism.[16]

Curiously, Schwartz concludes from this passage that "Wallace was receptive to spiritualism because it filled a religious void in his life. He belonged to no organized church and, prior to his conversion in 1865, probably considered himself an agnostic. After 1865 his attitude changed: spiritualism was no longer a phenomenon that required investigation, it was his religion."[17] Unfortunately Schwartz never defines what he means by "religion," or what he thinks Wallace might have meant by it. Schwartz then ventures an explanation for why Wallace departed "from the Darwinian view of the origin of man," that being "his inability to bridge his scientific and moral beliefs." According to Schwartz, this "arose from his disenchantment with life in Victorian England and with the answers that the scientific community offered as an explanation of that world." Thus, in the end, Wallace's "split with Darwin also expressed his desire for a new and better world, which his evolutionary scheme could provide and the Darwinian mechanism could not."[18]

It is true that Wallace's evolutionary worldview was much broader in scope and more open to supernatural intervention than Darwin's, but the reason is far more complex than just a disillusionment with his culture or a personal religious need. Wallace's worldview was scientific and naturalistic to the core, with natural selection as the driving force behind all evolutionary change. If anything, Wallace was *too* committed to scientism and naturalism, leaving himself no room for the ambiguities of knowledge or the anomalies of nature. He was driven to find an explanation for *everything*, and herein lies the problem and the explanation for his bifurcation down the alternative path from Darwin.

THE NATURAL AND THE SUPERNATURAL

Wallace's scientific worldview is clearly and powerfully presented in his 1870 paper, "The Limits of Natural Selection as Applied to Man," in which he recognizes the heretical nature of his theory in proffering a force that is beyond those known to science: "I must confess that this theory has the disadvantage of requiring the intervention of some distinct individual intelligence. . . . It therefore implies that the great laws which govern the material universe were insufficient for this production. . . ."[19]

Wallace then sets out to argue the logical necessity for the existence of a higher intelligence. He cites Professor Tyndall's presidential address to the Physical Section of the British Association at Norwich, delivered in 1868, in which Tyndall poses the classic mind-brain problem: "How are these physical processes connected with the facts of consciousness? The chasm between the two classes of phenomena would still remain intellectually impassable."[20] Wallace begins with the materialist position that a collection of molecules, even if structured into levels of "greater and greater complexity, even if carried to an infinite extent, cannot, of itself, have the slightest tendency to originate consciousness." Consciousness, he argues, is a qualitative phenomenon, not quantitative. It cannot be spontaneously generated with just more molecules, as if there were some critical amount that when reached produces consciousness:

> If a material element, or a combination of a thousand material elements in a molecule, are alike unconscious, it is impossible for us to believe that the mere addition of one, two, or a thousand other material elements to form a more complex molecule, could in any way tend to produce a self-conscious existence. There is no escape from this dilemma — either all matter is conscious, or consciousness is, or pertains to, something distinct from matter, and in the latter case its presence in material forms is a proof of the existence of conscious beings, outside of, and independent of, what we term matter.[21]

The question Wallace is asking is a critical one for any attempt to explain the origins of consciousness: How do you go from zero to one — from no consciousness to even a little consciousness? "We cannot *conceive* a gradual transition from absolute unconsciousness to consciousness," Wallace argues, because "the mere rudiment of sensation or self-consciousness is infinitely removed from absolutely . . . unconscious matter." Furthermore, our own free will cannot be explained by any known natural force ("gravitation, cohesion, repulsion, heat, electricity, etc."); therefore, there must be another force that accounts for our free will.[22] Without this supernatural force, that is, if there were only the known natural forces, "a certain amount of freedom in willing is annihilated, and it is inconceivable how or why there should have arisen any consciousness or any apparent will, in such purely automatic organisms." Therefore, Wallace concludes, "it does not seem an improbable conclusion that all force may be will-force; and thus, that the whole universe is not merely dependent on, but actually *is*, the WILL of higher intelligences or of one Supreme Intelligence."[23]

Clearly Wallace is arguing from a naturalistic position to a supernaturalistic one, but he is doing so through what he considers consistent logic and scientistic rationality. Nowhere in this paper does he argue, or even discuss at all, spiritualism, religion, or God. Alfred Wallace perceived his methods to be purely scientific and his conclusions to be derived through unsullied rational arguments: "These speculations are usually held to be far beyond the bounds of science; but they appear to me to be more legitimate deductions from the facts of science than those which consist in reducing the whole universe . . . to matter conceived and defined so as to be philosophically inconceivable."[24]

For Wallace, at long last the ancient philosophical belief in the existence of a uniquely human spirituality has been proven through the enterprise of modern science: "Philosophy had long demonstrated our incapacity to prove the existence of matter, as usually conceived; while it admitted the demonstration to each of us of our own self-conscious, spiritual existence. Science has now worked its way up to the same result, and this agreement between them should give us some confidence in their combined teaching."[25]

Two decades later Wallace's marriage of the philosophical and the scientific had grown ever stronger. In his 1889 work, *Darwinism*, Wallace includes mathematical reasoning and artistic skills among those "outgrowths of the human intellect which have no immediate influence on the survival of individuals or of tribes, or on the success of nations in their struggles for supremacy or for existence."[26] As in the 1870 article, he never references spiritualism, phrenology, or any other paranormal phenomena. The reason is that these phenomena were only a *part* of a much grander scientific worldview that was derived through logical analysis. Wallace's various experiences in investigating spiritual events became evidentiary support for his larger scientific worldview. Natural selection and the Darwinian paradigm fit snugly into his scientific vision of man evolving into a higher state of physical, intellectual, and spiritual development:

The Darwinian theory . . . not only does not oppose, but lends a decided support to, a belief in the spiritual nature of man. It shows us how man's body may have been developed from that of a lower animal from under the law of natural selection; but it also teaches us that we possess intellectual and moral faculties which could not have been so developed, but must have had another origin; and for this origin we can only find an adequate cause in the unseen universe of Spirit.[27]

The remainder of Wallace's life was devoted to fleshing out the details of this global scientism that encompassed so many different issues and controversies. Despite his lifelong interest in spiritualism, Wallace called himself a "scientific skeptic," but clearly had a broader view of science than most of his contemporaries who held physics to be the queen of the sciences. He felt this was limiting, however, since "there are whole regions of science in which there is no such regular sequence of cause and effect and no power of prediction," as he wrote in an 1885 response to a criticism of his conclusions. "Even within the domain of physics we have the science of meteorology in which there is no precise sequence of effects; and when we come to the more complex phenomena of life we can rarely predict results and are continually face to face with insoluble problems; yet no one maintains that meteorology and biology are not sciences — still less that they are out of harmony with or opposed to science." If such accepted sciences as meteorology and biology lack "uniformity" and cannot predict "what will happen under all circumstances," then the study of spiritualism is not alone.[28]

For over half a century Wallace tried to reconcile his vision of science, his conviction to natural law, his own theory of evolution, and his belief in spiritualism. As we shall see, in the context of his culture and personality, this was not so ill conceived.

THE CULTURAL CONTEXT

The rebirth of interest in spiritualism and phrenology in the mid- to late nineteenth century, by both the general public and the scientific community, added to Wallace's polemics on the shortcomings of natural selection when applied to cognitive domains. Commingling his teleological thinking with the spiritual phenomena he was observing, Wallace understood the ultimate purpose of nature to be the development of the spirit — the final end of an immeasurably long evolutionary process.

The arrival of phrenology on the European continent preceded that of spiritualism by two decades. It was introduced in the 1790s by the Viennese physician Franz Joseph Gall, and picked up momentum in the 1820s. Phrenology is based on a few basic tenets: the mind is an aggregate of mental processes localized in specific brain areas (for Gall, it was a composite of 37 independent faculties, propensities, and sentiments, each with its own brain area); the larger the localized area the more powerful that specific mental process. Since the skull in infant development is plastic and malleable it ossifies (hardens) over the brain, forming external "bumps" or "valleys," indicating an individual's internal mental faculties. Gall's first protégé, Johann Gaspar Spurzheim, added the notion that certain personality characteristics and moral propensities, such as evil, were a result of an imbalance between the faculties. Where Gall sought to build a science of the mind through phrenology, Spurzheim hoped to expand the field's horizon beyond the individual and into the realm of the social and political. This approach attracted a Scottish lawyer George Combe to the phrenological movement, which would, in time, result in a seminal work read by Wallace.

Historian Roger Cooter has noted that before 1820 phrenology was widely criticized by both the general public and the intelligentsia. But it experienced a boom from approximately 1820 to 1840, supported at first by radicals in opposition to any form of established authority. In time it picked up advocates within the bourgeois class, as its proponents worked to emphasize the empirical and quantifiable quality of its claims (through a wide range of Rube Goldbergian devices placed over the head of the client, that gave an air of "hard science" at work). "Regarding phrenology," Cooter remarks, "it is indeed remarkable that men of such high intelligence should have given such nonsense any credence, but, for a time, it ensnared the nation, particularly its upper classes and is by no means the only instance of the gullibility of the public and the medical profession."[29]

From 1840 on, however, phrenology declined in credibility within the scientific community, though it remained popular in the working class, especially among the most radical, which was Wallace's social position. In 1844, in fact, the working-class naturalist first read about phrenology in a book entitled *Constitution of Man Considered in Relation to External Objects*, first published in 1839 and written by George Combe, the noted Scottish lawyer, phrenologist, founder of the Edinburgh Phrenological Society, and the ideological disciple of Spurzheim. Combe turned Spurzheim's phrenology into a natural philosophy of the mind, attempting to explain human emotion and suffering in the context of natural laws governing thought.

Wallace, always the social and political speculator in search of grand underlying causes, took immediately to Combe's philosophy. Linking phrenology to
mesmerism as well, Wallace began his lifelong quest for a scientific basis of such
phenomena in the mid 1840s with a number of "experiments." He recalled, for
example, "having my patient in the trance, and standing close to him, with the
bust [a phrenological skull] on my table behind him, I touched successively
several of the organs, the position of which it was easy to determine. After a
few seconds he would change his attitude and the expression of his face in
correspondence with the organ excited. In most cases the effect was unmistak-
able, and superior to that which the most finished actor could give to a character
exhibiting the same passion or emotion."[30] Wallace's belief in the basic premises
of phrenology never really attenuated throughout his life, and in old age he still
proudly exhibited a phrenological cranium reading done on himself by the same
individuals (E. T. Hicks and J. Q. Rumball) who measured Herbert Spencer's
head. (The social Darwinist Spencer—the man who coined the expression "survival of the fittest"—was one of Wallace's intellectual idols.)

A revivification of spiritualism began in mid-nineteenth-century America and
quickly spread across the Atlantic to England and the continent. Historian Henri
Ellenberger has written a history of this "dynamic psychology," in which he
identifies the beginnings of an interest in spiritualism around 1850, when the
words "telepathy" and "medium" were first used in print. Mediums holding
seances soon spread rapidly through centers of population, with divers claims
such as the ability to contact the dead, read the past and future, and produce
such psychic phenomena as rapping noises and appearances of ghosts.[31] A gullible general public was quickly swept up by the enthusiasm and excitement that
surrounded such mystical phenomena, especially when endorsements came from
respected members of the scientific community, some from the very highest
levels.

Although we dislike the notion that truths, especially scientific truths, might
be strongly influenced by who is doing the truth telling as much as by the
quality of the evidence, the fact is that who you are and who you know sometimes matters as much as the consistency of your arguments or the quality of
your evidence. Integrity, trust, reputation, fame, society memberships, and institution affiliations, all converge to construct the validity of a claimant, and
thus his or her claim. When Einstein spoke, people listened—no matter what
he said (e.g., in popular culture Einstein's views on war and peace, religion, and
other social issues are quoted far more than his scientific opinions). It was as
true in Wallace's time as it is today.[32]

As with phrenology, a considerable segment of the more staid and conservative scientific community also took an interest in spiritualism for a time. In 1882 the *Society for Psychical Research* was founded in London, with a membership centered at Cambridge University and a roster that included such renowned scientists as the physicists Sir William Crookes, Lord Rayleigh, and Sir Oliver Lodge, the noted eugenicist (and Darwin's cousin) Francis Galton, the mathematician Augustus De Morgan, the naturalist St. George Mivart, the physiologist Charles Richet, and the psychologists Frederic Myers and G. T. Fechner. Their goal, explains historian Ian Hacking in an interesting twist on what skeptics might expect from eminent scientists, "was not so much to challenge the validity of such claims, as it was to establish a scientific, naturalistic explanation for the phenomena which they assumed to be real."[33] Not surprisingly, they often found what they were looking for, and thus, claims became truth when sanctioned by such noted truth seekers.

As a member of the society, Wallace found himself at the epicenter of the spiritualism movement. Like the other members, once he was convinced of the validity of the claims he sought further verification and an explanation. More important, he pursued a deeper natural cause. In 1866 Wallace published a 57-page monograph (with the very nineteenth-century style title): *The Scientific Aspects of the Supernatural: Indicating the Desirableness of an Experimental Enquiry by Men of Science into the Alleged Powers of Clairvoyants and Mediums*. Confirming the social nature of science, Wallace explains:

> A little enquiry into the literature of the subject, which is already very extensive, reveals the startling fact, that this revival of so-called supernaturalism is not confined to the ignorant or superstitious, or to the lower classes of society. On the contrary, it is rather among the middle and upper classes that the larger proportion of its adherents are to be found; and among those who have declared themselves convinced of the reality of facts such as have been always classed as miracles, are numbers of literary, scientific, and professional men, who always have borne and still continue to bear high characters, are above the imputation either of falsehood or trickery, and have never manifested indications of insanity.[34]

GHOSTS, SPIRITS, AND MEDIUMS

In the 1860s, Wallace's interest in spiritualism that would reinforce his hyperselectionism was actually a revitalization of a curiosity that had begun nearly three decades previously. In July of 1865, Wallace's inquisitiveness was again piqued when he attended a seance at the home of a friend. The table moved and vibrated and rapping noises were heard. In November of 1866, he began to experiment at home with a well-known medium named Miss Nichol. Somewhat naively, Wallace claims that he entered the inquiry "utterly inbiased [sic] by hopes or fears, because I knew that my belief could not affect the reality."[35] But the levitation of the corpulent Miss Nichol, and the production of fresh flowers in the dead of winter, convinced Wallace that further investigation was necessary.

Unlike so many others driven by religious motivations to confirm the existence of a spiritual world, Wallace was in search of a *natural* explanation for the supernatural. *The Scientific Aspects of the Supernatural*, in fact, is one long attempt to prove that these phenomena are "not really miraculous in the sense of implying any alteration of the laws of nature. In that sense I would repudiate miracles as entirely as the most thorough sceptic."[36] Thorough scientist that he was, Wallace began his analysis of miracles with the classic skeptic David Hume, noting that "Hume was of opinion that no amount of human testimony could prove a miracle" because "a miracle is generally defined to be a violation or suspension of a law of nature" and "the laws of nature are the most complete expression of the accumulated experiences of the human race."[37] If these spiritual events are not miracles, then what are they? According to Wallace, "The apparent miracle may be due to some yet undiscovered law of nature."[38] Just because we cannot understand or explain these occurrences does not mean they lack causes, or that the causes are miraculous. It is just that we have yet to discover them: "A century ago, a telegram from 3000 miles' distance, or a photograph taken in five seconds, would not have been believed possible, and would not have been credited on testimony, except by the ignorant and superstitious who believed in miracles."[39] Thus, Wallace concludes, "it is possible that intelligent beings may exist, capable of acting on matter, though they themselves are uncognisable directly by our senses."[40]

These intelligent beings, however, are not in any way connected with divine providence or "acts of the Deity." In fact, Wallace argues in an interesting twist on the argument from miracles (for God's existence), "The nature of these acts is often such, that no cultivated mind can for a moment impute them to an infinite and supreme being. Few if any reputed miracles are at all worthy of a God."[41] Natural phenomena are to be explained by natural causes. Wallace's worldview was thoroughly scientistic. He was not the schizophrenic man of sense and nonsense, science and nonscience. If there were spiritualistic occurrences to be explained, the scientist could only do so through scientific means, "by direct observation and experiment," Wallace proclaimed in the empiricist mode of following the data wherever they may lead:

> It would appear then, if my argument has any weight, that there is nothing self-contradictory, nothing absolutely inconceivable, in the idea of intelligences uncognisable directly by our senses, and yet capable of acting more or less powerfully on matter. Let direct proof be forthcoming, and there seems no reason why the most sceptical philosopher should refuse to accept it. It would be simply a matter to be investigated and tested like any other question of science. The evidence would have to be collected and examined. The results of the enquiries of different observers would have to be compared.[42]

For the next forty years that is precisely what Wallace did—involving himself with the systematic examination of spiritualism, with such experiments as this one, described in a letter to a friend:

> Our seance came off last evening, and was a tolerable success. The medium is a very pretty little lively girl, the place where she sits a bare empty cupboard formed by a frame and doors to close up a recess by the side of a fireplace in a small basement breakfast-room. We examined it, and it is absolutely impossible to conceal a scrap of paper in it. Miss Cooke is locked in this cupboard, above the door of which is a square opening about 15 inches each way, the only thing she takes with her being a long piece of tape and a chair to sit on. After a few minutes Katie's whispering voice was heard, and a little while after we were asked to open the door and seal up the medium. We found her hands tied together with the tape passed three times around each wrist and tightly knotted, the hands tied close together, the tape then passing behind and well knotted to the chairback. We sealed all the knots with a private seal of my friend's, and again locked the door. A portable gas-lamp was on a table the whole evening, shaded by a screen so as to cast a shadow on the square opening above the door of the cupboard till permission was given to illuminate it. Every object and person in the room were always distinctly visible. A face then appeared at the opening, but dark and indistinct. After a time another face quite distinct with a white turban-like headdress—this was a handsome face with a considerable general likeness to that of the medium, but paler, larger, fuller, and older—decidedly a different face, although like. We were then ordered to release the medium. I opened the door, and found her bent forward with her head in her lap, and

apparently in a deep sleep or trance—from which a touch and a few words awoke her. We then examined the tape and knots—all was as we left it and every seal perfect.[43]

BELIEVING IS SEEING

Wallace's active involvement with the spiritualist movement postdated his theory of natural selection (1858) but predated his 1868 paper "Limitation of Natural Selection applied to Man." This sequence is important in understanding how a naturalist (in the methodological sense as well as the biological!) comes to believe in the supernatural. Wallace approached the study of spiritualism with his usual analytical enthusiasm. His first seance was 1865. By 1866, he had already published the pamphlet *The Scientific Aspects of the Supernatural*, and in 1875 he wrote an entire book on *Miracles and Modern Spiritualism*. Wallace sent the former to Darwin's tireless defender, Thomas Huxley, who confessed that he "could not get up any interest in the subject," later observing that spiritual manifestations might at least reduce suicides: "Better live a crossing-sweeper than die and be made to talk twaddle by a 'medium' hired at a guinea a seance."[44] Likewise Charles Darwin remained a skeptic. He attended one seance and wrote: "The Lord have mercy on us all, if we have to believe in such rubbish."[45]

But Robert Chambers, author of the *Vestiges of the Natural History of Creation*, received it with great "gratification" and wrote back to Wallace: "I have for many years known that these phenomena are real," and "My idea is that the term 'supernatural' is a gross mistake. We have only to enlarge our conceptions of the natural, and all will be right."[46] Inspired by Wallace, in fact, Chambers revised a later edition of the *Vestiges* to include spiritual phenomena.

Wallace was not alone in his fascination with the paranormal, and he accumulated what he believed to be much empirical evidence in support of these claims. One of the more bizarre confirmations of the spiritual world came from his own sister, Frances Sims (née Wallace), which I discovered in the archives at Oxford University when I came across a copy of *The Scientific Aspects of the Supernatural*. On the frontispiece, in the hand of his sister, is written the following (reprinted in FIGURE 27):

This book was written by my Brother *Alfred* and with 24 others was laying on my table. They had been there 4 days and I had not had time to give them away. One morning I had been sitting at my Table writing and left the room for a few minutes when I returned the paper parcel was opened

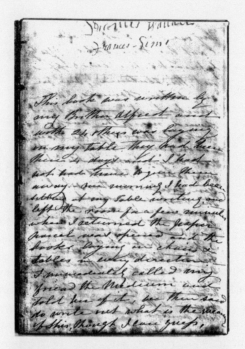

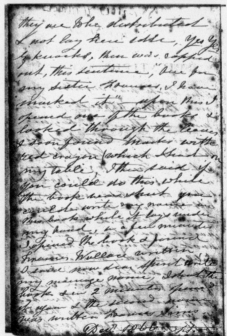

Figure 27. On the frontispiece of a copy of Alfred Wallace's *Scientific Aspects of the Supernatural*, owned by his sister Frances, appears this inscription in her hand describing a paranormal event that convinced her that her brother was right about the reality of spiritualism.

and the books laying on chairs & tables in every direction. I immediately called my friend the medium and told her of it, she then said to write out what is the meaning of this, though I can guess, they are to be distributed & not lay here idle. *Yes Yes* by knocks, then was rapped out, this sentence, "One for my Sister Frances. I have marked it"—upon this I opened one of the books & looked through the leaves & soon found marks with red crayon (which I had on my table). I then said if you could do this while the book was shut you could write my name in this book while it lays under my hand, in a few minutes I opened the book & found *Frances Wallace* written. I said now dear Spirit write my marriage name, I shut the book & in 2 minutes opened it again & the second name was written *Frances Sims*.

Dec. 1866. FS

One explanation for this curiosity is that Frances (or someone else) was attempting to perpetrate a hoax and wrote her name in the book, perhaps to enhance the publicity or add to the credibility of her brother's publication. There is a little similarity between the names at the top of the page and those in the text, though not enough to conclude this explanation is correct. It is possible her "medium" friend concocted a trick to reinforce the belief of Frances and/or Alfred, though the text of the passage is unclear where the medium was at the time. Perhaps the medium was in the other room and when Frances "left the room for a few minutes" the deed was done. How then could the medium have written the names while the book was under Frances' hand? This, of course, we will never know, but there are standard techniques used by magicians to do something very similar, so it seems reasonable to assume that Wallace and his sister were duped.

Wallace's American lecture tour in 1886 proved to be yet another testing ground for both the veracity and naturalism of spiritual phenomena. His journal from the trip is filled with entries that, in a most nonchalant way, mix botanical collecting, zoological exploring, public lecturing, and spiritual seances all in the same day. The entry for Saturday, December 18, 1886, for example, is reproduced in FIGURE 28. Wallace has drawn the room, indicating where he sat ("AW"), the cabinet where the medium was encased, and the sliding doors "privately marked with pencil & found untouched after." This is vintage Wallace—the man of science conducting a rigid experiment, complete with witnesses for corroboration and controls for fraud:

To library & Museum—called on Williams & McIntyre—Evening to Seance at Mrs. Ross. Remarkable Exd room carefully & rooms below . . .

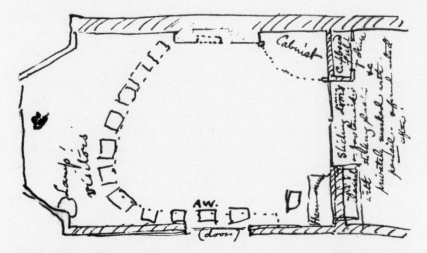

Figure 28. A drawing in Wallace's hand from his American journal, dated Saturday, December 18, 1886. The codiscoverer of natural selection attended a seance in which the medium was secured to prohibit sleight of hand (so Wallace thought). Wallace sat near the door ("AW") to further prevent trickery. Despite these precautions, female figures in white appeared, as well as a male figure recognized by a gentleman as his son, and a tall indian in white mocassins.

[diagram] Below the cabinet is the heating furnace & on the ceiling air pipes hot & cold clothed with cobwebs. Room carpeted up to walls, entire . . . walls solid. Cabinet a cloth curtain with cloth top 2 ft. below ceiling. Door to next room secured but gas-lights burning in it afforded perfect security. Ten visitors—Mr. & Mrs. Ross. Most striking phenomena.

1. A female figure in white came out with Mrs. Ross in black, and at the same time a male figure—to mid. of room.
2. Three female figures appeared together all in white of different heights—came 2 or 3 feet in front of cabinet.
3. A male figure came out recognised by a gentleman as his son.
4. A tall indian in white moccasins came out danced and spoke, shook hands with me & others—a large strong rough hand.
5. A female figure with a baby—to entrance of cabinet. Went up and felt baby's face, nose, and hair, & kissed it—a genuine soft skinned living baby as ever I felt. Other gentlemen & ladies agreed.

BEATEN BY THE FACTS

It was Wallace's certainty of the reality of such spiritual experiences that rein-forced his belief that natural selection would be inadequate as a causal agent to explain the origin of the human mind. In an 1874 *Fortnightly Review* article on "A Defence of Modern Spiritualism," Wallace argued that man is not just a physical being, but "a duality, consisting of an organized spiritual form. . . ."[47] In a letter to E. B. Poulton, February 22, 1889, he wrote (with a hint of doubt appropriate for a scientific attitude):

> I (think I) *know* that non-human intelligences exist—that there are *minds* disconnected from a physical brain,—that there *is*, therefore, a *spiritual world*. This is not, for me, a *belief* merely, but *knowledge* founded on the long-continued observation of facts,—& such *knowledge* must modify my views as to the origin & nature of human faculty.[48]

In the article on "Spiritualism" in *Chambers' Encyclopaedia*, Wallace defined spiritualism as "a science based solely on facts: it is neither speculative nor fan-ciful. On facts and facts alone, open to the whole world through an extensive and probably unlimited system of mediumship, it builds up a substantial psy-chology on the ground of strictest logical induction."[49] Wallace's brand of spir-itualism was strictly confined to his scientific worldview, as evidenced above and in this autobiographical passage from *On Miracles and Modern Spiritualism*:

> Up to the time when I first became acquainted with the facts of spiritualism, I was a confirmed philosophical sceptic, rejoicing in the works of Voltaire, Strauss, and Carl Vogt, and an ardent admirer (as I am still) of Herbert Spencer. I was so thorough and confirmed a materialist that I could not at that time find a place in my mind for the conception of spiritual existence, or for any other agencies in the universe than matter and force. Facts, how-ever, are stubborn things. My curiosity was at first excited by some slight but inexplicable phenomena occurring in a friend's family, and my desire to knowledge and love of truth forced me to continue the inquiry. The facts became more and more assured, more and more varied, more and more removed from anything that modern science taught, or modern philosophy speculated on. The facts beat me.[50]

Wallace's belief in the supernatural was caused by a number of important agents interacting over time from his youth to his final days. These include: a working-class background, self-education (and thus, lack of pressure to conform to the status quo), experiences in the Mechanic's Institutes (a type of night school for adults) with fringe and heretical ideas, youthful tinkering with mesmerism and phrenological readings, discovery of the then radical theory of natural selection and its subsequent payoff of scientific fame and confidence, taking his own theory to the extreme of hyper-selectionism that forced him to find a purpose for everything in nature, personal experiences with seances and spirit mediums that convinced him of the reality of a spiritual world, the need to incorporate these experiences into his scientific worldview, and the final leap from the natural to the supernatural when his science failed to explain by the laws of nature what he knew to be true. Subsequent observations, then, only served to confirm the validity of the knowledge claims, which led to more observations, and so on in an autocatalytic feedback loop: the variables influencing the development of ideas not only interact with each other with their potency changing over time, they become locked into an information system feedback loop with the thinker, where the ideas feed back into the culture and change the variables themselves, which in turn affect the ideas, that alter the variables, and so on. Throughout his long life and up to the very end, Wallace maintained this borderlands position between the natural and the supernatural, all the while believing he was doing good science in both worlds.

9

PEDESTALS AND STATUES
Freud, Darwin, and the Hero-Myth in Science

THE ESSENTIAL DICHOTOMY of history's heroes as "great men" or cultural products is succinctly summarized by Henry Wadsworth Longfellow's 1839 poem "A Psalm of Life:"

> In the world's broad field of battle,
> In the bivouac of Life,
> Be not like dumb, driven cattle!
> Be a hero in the strife!

Poets, being allowed to do more than merely reflect reality, can direct us to action, as Longfellow has done in his final couplet, which impels us to stand out as heroes amongst life's masses. Historians, however, do not share the poet's privilege. Their task is to understand the past "as it really happened," not as it should have been. And historians have gone to extremes in the matter of the role of the individual in history—from *hero as great man*, a seminal thinker, political leader, or military general, to *hero as myth*, an artifact of a *zeitgeist* that thrusts the ordinary person into fame that might just as well have gone to another.

The cardboard pennant of the first polemic is typically represented by the English philosopher Thomas Carlyle in his *Heroes, Hero-Worship, and the Heroic in History:* "Universal History, the history of what man has accomplished in this world, is at bottom the History of the Great men who have worked here."

And: "Worship of a hero is transcendent admiration of a great man."[1] Others concur with Carlyle. "It will never make any difference to a hero what the laws are," wrote Ralph Waldo Emerson. "His greatness will shine and accomplish itself unto the end, whether they second him or not."[2] And in the early nineteenth century Sydney Smith noted that "great men hallow a whole people, and lift up all who live in their time."[3]

By contrast, the cardboard pennant of the second polemic is traditionally carried by Marxists such as Joseph Stalin: "It is not heroes that make history, but history that makes heroes. It is not heroes who create a people, but the people who create heroes and move history forward."[4] And from Friedrich Engels: "That a certain particular man, and no other, emerges at a definite time in a given country is naturally a pure chance, but even if we eliminate him there is always a need for a substitute, and the substitute . . . is sure to be found."[5] Engels even went as high as a French emperor for an example. "That Napoleon—this particular Corsican—should have been the military dictator made necessary by the exhausting wars of the French Republic—that was a matter of chance. But in default of a Napoleon, another would have filled his place."[6]

Such extreme polarities as these, of course, are embraced by almost no historian today. These simplified caricatures are constructed by the opposition to magnify the rhetorical advantages of the dichotomy. Philosopher and historian Sidney Hook, in his classic study *The Hero in History*, offered this example: "Too many figures of history have been surrendered for exploitation to *belle-lettrists* and professional biographers who draw their subjects with one literal eye on earlier portraits and one imaginative eye on Hollywood."[7] Although such positions are not really held by anyone, the central claims of both sides are essentially correct. History's heroes are great individuals, but all individuals, ordinary or extraordinary, are and must be culturally bound—where else could they act out the drama of their times?

History may be modeled as a *contingency matrix*—a massively conditioned and crucially dependent multitude of linkages across space and time, where the hero is molded into, and helps to mold the shape of, those connections.[8] Traditional interpretations of history (e.g., the iconographies of ladders and cycles; the forces of culture and class) do not adequately reflect the rich and radiantly branching nature of contingent historical change and the role of the individual within that matrix. Hook is quick to emphasize that such historical forces as "Providence, Justice, Reason, Dialectic, and Zeitgeist" are no longer proffered as the sole triggers of historical change, where individuals are merely instruments of these forces. "Man is also an instrument who has something to say about what these purposes shall be. The Purpose he presumably serves is to be construed from the purposes he himself sets and realizes."[9] For Hook, a hero in history is defined as "the individual to whom we can justifiably attribute preponderant influence in determining an issue or event whose consequences would have been profoundly different if he had not acted as he did."[10] History is not merely contingent on the vagaries of the weather or geography, demographic trends or economic shifts, class struggle or scientific discoveries. There is no independent variable on which all or even most effects depend. A multiple analysis of variance, boundless in complexity, may more adequately reflect this matrix of change. If the present does not change in straight lines (note the complex multiple analyses of variance constructed by psychologists and sociologists to model even simple social phenomena), it would be naive to make such assumptions of the past. History, in this sense, is the present writ past. The contingent details and necessary forces will vary over time, but the interaction of these historical forces is fractal—the grain of complexity remains the same at any scale.

Where does the individual hero fit into this historical matrix? The hero is a great individual who altered history's contingencies unalterably. Hook, for example, makes a distinction between the *eventful* man, who was "at the right place at the right time," and the *event-making* man, who was an *eventful* man who helped to create the events themselves. "The event-making man finds a fork in the historical road, but he also helps . . . to create it. He increases the odds of success for the alternative he chooses by virtue of the extraordinary qualities he brings to bear to realize it."[11] The hero is a great person who was at the right place at the right time.

Even the two ends of what turns out to be a continuum, not a dichotomy, fail in their consistency to be extreme. Carlyle observed the impossibility of knowing everyone's biography thoroughly enough to glean a true understanding of the past: "Social life is the aggregate of all the individual men's Lives who constitute society; History is the essence of innumerable Biographies. But if one Biography, nay, our own biography, study and recapitulate it as we may, remains in so many points unintelligible to us; how much more must these million, the very facts of which . . . we cannot know!"[12] Even Marxists must step out of their paradigm of economic determinism when it comes to their own heroes—Marx and Lenin—whom they see as unique individuals who managed to step out of their cultural matrix to reshape their age. Hook notes: "Where [Marxism] paid adequate attention to the *work* of great historical figures—for instance, its own heroes, Marx and Lenin—its historical monism went by the board. Where it interpreted the historical activity of Alexander, Caesar, Cromwell, Peter the Great, Napoleon, as "expressions" of convergent social pressures or merely as instruments "of class interest, it often abandoned its scientific approach for the mystical *a priorism* which was part of its Hegelian heritage."[13]

The reason for these inconsistencies is that even the positivist historians of the nineteenth century could not ignore the influence of culture on the individual. Extremism equated with absolute certainty is a road filled with potholes. If this is so, then what is the origin of the *hero as great man* or *hero as myth*?

LARGE PEDESTALS FOR SMALL STATUES

In an essay entitled "From Hero to Celebrity," historian Daniel Boorstin observes that "since about 1900 we seem to have discovered the processes by which fame is manufactured."[14] Since the advent of what Boorstin calls the Graphic Revolution, "where we have developed the ability to make, preserve, transmit, and disseminate precise images," our thinking about what makes a hero has changed.[15] The Graphic Revolution has given us the "means of fabricating well-knownness," where a man can become a "great man" by virtue of his fame.[16] Heroes are now manufactured and packaged like so many mass-produced goods. It is the conspicuous consumption of artificially manufactured icons.

Only a celebrity can be created in this manner, not a hero. In Boorstin's delightful description, a celebrity is "a person who is known for his well-knownness. Their chief claim to fame is their fame itself. They are notorious for their notoriety."[17] Boorstin sardonically quips that "two centuries ago when a great man appeared, people looked for God's purpose in him; today we look for his press agent."[18] True heroes may attain celebrity status in today's world, but no celebrity can purchase greatness. Of celebrities, Boorstin notes that "despite the 'supercolossal' on the label, the contents are very ordinary."[19] He continues with this lucid distinction: "The hero was distinguished by his achievement, the celebrity by his image or trademark. The hero created himself, the celebrity is created by the media. The hero was a big man; the celebrity is a big name."[20] Recognition of greatness accrues only to true heroes who do not need to create an image for the public. Celebrities, in contrast, hire press secretaries and public relations consultants to construct, in the apt description by Edward Bernays, "large pedestals for small statues."[21]

Thus it is that occasionally the *hero as great man* in history is the creation of an image, both by the hero and by the hero's followers, and none fits this mold better than the founding father myths—Abraham as father of the Israelites, Romulus and Remus as parents of the Romans, Freud as the patriarch of psychoanalysis, Marx as the creator of communist philosophy, and Darwin as the progenitor of evolutionary theory—as if such creations had no incremental predecessors or sociocultural determinants. Yet the *hero as myth* distorts the legitimate role of founders no less than its counterpart. In fact, the *hero as great man* and *hero as myth* form the two ends of a false dichotomy, created, in part, by the heroes themselves. But such generalizations are hollow as abstract constructs. Sigmund Freud (1856–1939) and Charles Darwin (1809–1882) provide historical case studies to understand the termini of both ends of this historiographical dichotomy.

SIGMUND FREUD: THE CREATION OF A HERO MYTH

Frank Sulloway's 1979 biography, *Freud: Biologist of the Mind,* opens with the statement "Few individuals, if any, have exerted more influence upon the twentieth century than Sigmund Freud."[22] Despite this grand pronouncement, one of Sulloway's prime goals in his pathbreaking work is to destroy the image that Freud accomplished this impact almost entirely on his own—an image that he says Freud helped to create: "Historical 'decontextualization' is a prerequisite for good myths, which invariably seek to deny history. This denial process has followed two main tendencies in psychoanalytic history—namely, the extreme reluctance of Freud and his loyal followers to acknowledge the biological roots of psychoanalysis . . . and the creation and elaboration of the "myth of the hero" in the psychoanalytic movement.[23]

There is no question that Freud saw himself as a hero in history—a revolutionary scientist in the company of Copernicus and Darwin. According to Freud, humanity has suffered three great shocks to its narcissistic selfhood. The *cosmological* shock came when Copernicus placed us on "a tiny speck in a world-system of a magnitude hardly conceivable." The *biological* shock occurred when Darwin "robbed man of his peculiar privilege of having been specially created, and relegated him to a descent from the animal world."[24] The *psychological* shock, created by Freud's own psychoanalysis, was seen by the founder of the theory (in one of the least modest statements in the history of ideas) to be a scientific revolution "endeavoring to prove to the 'ego' of each one of us that he is not even master of his own house, but that he must remain content with the veriest scraps of information about what is going on unconsciously in his own mind."[25]

Freud's own blustering ego and visions of heroic greatness did not commence with the theory of psychoanalysis, but actually goes back to his childhood, where, according to Paul Roazen in his book *Freud and His Followers,* Freud had a "profound urge to become a mighty warrior." His own boyhood heroes included Hannibal, Cromwell, and Napoleon, whose lives of military intrigue and political machinations fired his imagination. A later figure of identification was the great explorer Columbus. In a letter to his fiancee at the age of 29, Freud wrote: "I have often felt as though I had inherited all the defiance and all the passions with which our ancestors defended their temple and could gladly sacrifice my life for one great moment in history." Later he would declare to his friend Wilhelm Fliess: "I am by temperament nothing but a *conquistador,* an adventurer . . . with all the inquisitiveness, daring, and tenacity characteristic of such a man." With burgeoning confidence that his successful career brought, Freud would go so far as to make this startling comparison in 1909 to Carl Jung: "If I am Moses then you are Joshua and will take possession of the promised land of psychiatry, which I shall only be able to glimpse from afar."[26]

Although Freud began his career in the medical and biological fields, his shift to the newly developing school of psychology would become his overwhelming "passion," his own choice of a word to describe his ambitious feelings for the new area of study: "A man like me cannot live without a hobby-horse, a consuming passion—in Schiller's words a tyrant. I have found my tyrant, and in his services I know no limits. My tyrant is psychology; it has always been my distant, beckoning goal and now, since I have hit on the neuroses, it has come so much the nearer."[27]

Freud saw himself as the founder and driving force behind this intellectual revolution. In *Sigmund Freud's Mission,* Erich Fromm writes: "Only if one understands this aspiration of Freud to bring a new message to mankind, not a happy but a realistic one, can one understand his creation: *the psychoanalytic movement.*"[28] Fromm makes this comparison of Freud with Marx and Darwin: "He was a great scientist indeed; but like Marx, who was a great sociologist and economist, Freud had still another aim, one that a man like Darwin did not have: he wanted to transform the world. Under the disguise of a therapist and a scientist he was one of the great world reformers of the beginning twentieth century."[29]

It was, in fact, the field of psychology toward which Freud gravitated, particularly with his early work on hysteria. His supposedly "novel" ideas on male hysteria were, according to Sulloway, "rudely rejected, thus marking the beginning of his 'isolation' from Viennese scientific life."[30] This had the function of legitimizing the myth of the hero as devising a new and innovative idea that is so revolutionary that the sterile and closed-minded scientists of German medicine were unable to realize its foresight.

According to Joseph Campbell, being "rejected" and "isolated" are key characteristics in the development of the myth of the hero. In *The Hero With a
Thousand Faces*, Campbell surveys hundreds of examples of hero myths in history, culling from them three main themes that Sulloway applies to the Freud
myth: *isolation, initiation,* and *return,* all of which, Sulloway notes, "appear
prominently in the Freud legend."[31] Through a variety of stimuli that trigger
isolation of the would-be hero, an initiation period ensues in which difficult
obstacles are presented to be overcome, so that the developing hero can reach
maturity, whereupon the return stage is reached and the hero is accepted back,
first in reluctance and finally in triumph. Sulloway aptly summarizes the process:
"Having undergone his superhuman ordeal, the archetypal hero now emerges
as a person transformed, possessing the power to bestow great benefits upon
his fellow man. Upon his return home, however, the hero usually finds himself
faced by nonunderstanding opposition to his new vision of the world. Finally,
after a long struggle, the hero's teachings are accepted, and he receives his due
reward and fame."[32]

As Sulloway shows in his "Catalogue of Major Freud Myths" with their appropriate function (as informed by Campbell's work), Freud and his followers
went to great lengths not only to see his career as fitting this heroic framework,
but to actually *mold* it to fit. This behavior is well suited to Hook's model of
hero worship, where, "for those who believe, the substance of things hoped for
becomes the evidence of things not seen."[33] Among the many hero-myths that
Freud projected in his career, two that he pursued relentlessly were *pristine empiricism* and *intellectual independence,* both for himself and for psychoanalysis.
Both myths are neatly expressed by Freud in his *Three Essays on the Theory of
Sexuality:* "I must emphasize that the present work is characterized not only by
being completely based upon psycho-analytic research, but also by being deliberately independent of the findings of biology. I have carefully avoided introducing any preconceptions, whether derived from general sexual biology or
from that of particular animal species, into this study."[34]

Because Freud saw himself as a great scientist in the heroic mode of
nineteenth-century science that emphasized pure empirical data collection that
was not to be cluttered with unnecessary hypothesizing, he portrayed himself
and his newly formed science as falling in line with this philosophy. (This pro-
clivity reached an extreme in the 1914 translation of Galileo's *Dialogues,* where
his translators portrayed Galileo as the "Father of Experimental Science" and
replaced in one passage the embarrassing words "without experiment" with the
archempirical words "by experiment," thus blotting out any sign that science

could have been conducted in any other fashion.) According to his biographer Ernest Jones, Freud was particularly sensitive to accusations that his theories were speculative and empirically unfounded. Freud complained in 1909: "If only one could get the better people to realize that all our theories are based on experience . . . and not just fabricated out of thin air or thought up over the writing desk. But the latter is what they all really assume."[35] In reality, speculative theorizing is precisely how Freud proceeded to develop many, if not most, of his ideas.

Freud fought passionately to keep neighboring fields of thought out of his own work—at least that was the image he wished to portray. "I try in general to keep psychology clear from everything that is different in nature from it, even biological lines of thought."[36] But it was biology, more than any other field, that influenced Freud from the beginning to the end of his career. From the earliest research on hysteria and childhood sexuality to the bizarrely metaphysical *Phylogenetic Fantasy,* Freud was so directed by biology and evolutionary theory that Jones called him the "Darwin of the mind," concluding that "if psychology is regarded as part of biology . . . then it is possible to maintain that Freud's work . . . signifies a contribution to biology comparable in importance only with that of Darwin's."[37] Sulloway, who calls Freud the "biologist of the mind," notes: "It is my contention that many, if not most, of Freud's fundamental conceptions were biological by *inspiration* as well as by implication. In my historical appraisal, Freud stands squarely within an intellectual lineage where he is, at once, a principal scientific heir of Charles Darwin and other evolutionary thinkers in the nineteenth century and a major forerunner of the ethologists and sociobiologists of the twentieth century."[38]

Just how far would Freud go to uphold the hero image of pristine empiricism and intellectual independence? Incredibly, in 1885 and 1907, Freud destroyed major portions of his personal papers, notes, letters, notebooks, private diaries, and manuscripts. Part of the cult of hero worship is to keep the hero shrouded in mystery. If you do not know the details of his past, it may appear that his thoughts and ideas were developed spontaneously out of the ether of genius. To his fiancee, Freud wrote this chilling account in 1885:

> I couldn't have matured or died without worrying about who would get hold of these old papers. . . . As for the biographers, let them worry, we have no desire to make it too easy for them. Each one of them will be right in his opinion of "The Development of the Hero," and I am already looking forward to seeing them go astray.[39]

The Freud hero-myth continues. A large collection of Freud's letters, including important notes and drafts to Fliess, was donated to the Library of Congress in 1970 by Freud's daughter Anna, with the restriction that they not be released to anyone until the year 2000. Sulloway's own attempts to gain access to the letters proved fruitless, and although he was told that the letters were being prepared for official publication, there was no sign of any progress in this project.[40] Sulloway believes that only through detailed analyses of such myths can we hope to clarify the truth behind the legend. However, he closes his biography with a touch of cynicism:

> Great "origins" myths are too much of a cultural art form, too good at conveying the "lesson of history," to die out simply because they are not true. Thus Ernst Kris was probably right when he proclaimed, "The story of Freud's self-analysis will, I believe, become part of that great store of accounts which in the history of science plays a peculiarly inspiring part— accounts which report how in sacrifice and loneliness the great scientist works."[41]

By contrast, if Charles Darwin shares any of the hero image that Freud portrays, it is because his ideas have stood the test (literally and figuratively) of time and experiment. Darwin was the consummate opposite to Freud in this respect and supplies a second case study in understanding the origin of the hero in science.

CHARLES DARWIN: CREDIT WHERE CREDIT IS DUE

Although they face each other from the opposite ends of the spectrum in regard to creating or denying the myth of the hero, Freud and Darwin did share a personal awareness of their importance in the history of their respective sciences. But where Freud saw himself as a "warrior" and "conquistador," and an enlightener of humankind, Darwin's hubris was cut with a mix of realism, modesty, and acknowledgment of credit where credit was due.

That Darwin knew of the great importance of his work and theory is not questioned. He was deeply entrenched in the academic culture of nineteenth-century England and communicated regularly with many of the great minds of his time. His personal letters and correspondence, currently being published out of the Darwin correspondence project at Cambridge University, reveal a rich network of contacts with his peers from around the world. But this self-awareness was tempered with a healthy dollop of modesty that reveals his antipathy to creating a hero image. In his *Autobiography*, Darwin writes: "My books have sold largely in England, have been translated into many languages, and passed through several editions in foreign countries. I have heard it said that the success of a work abroad is the best test of its enduring value. I doubt whether this is at all trustworthy; but judged by this standard my name ought to last for a few years."[42]

Some have used the *Autobiography* to bolster their claims that Darwin was not really a "great man" but happened along at the right place at the right time. For example, Jacques Barzun, in *Darwin, Marx, Wagner*, calls Darwin "a great assembler of facts and a poor joiner of ideas, a man who does not belong with the great thinkers."[43] Gertrude Himmelfarb, in *Darwin and the Darwinian Revolution*, writes that Darwin was "limited intellectually and insensitive culturally."[44] Based on a limited survey of the vast historical data on the man, these conclusions were generally based on a cursory reading of the *Autobiography*, which Darwin never intended for publication. Rather, Darwin "thought that the attempt would amuse me, and might possibly interest my children or their children."[45] A moral guide by the family patriarch, the project was laced with a strong dose of Victorian modesty: "I have no great quickness of apprehension or wit which is so remarkable in some clever men," reads a typical example. "I am therefore a poor critic: a paper or book, when first read, generally excites my admiration, and it is only after considerable reflection that I perceive the weak points."[46] But selective quoting can be misleading. Farther down the page Darwin defends such charges as he heard them in his time:

> Some of my critics have said, "Oh, he is a good observer, but has no power of reasoning." I do not think that this can be true, for the *Origin of Species* is one long argument from the beginning to the end, and it has convinced not a few able men. No one could have written it without having some power of reasoning.
>
> On the favourable side of the balance, I think that I am superior to the common run of men in noticing things which easily escape attention, and in observing them carefully. My industry has been nearly as great as it could have been in the observation and collection of facts.[47]

If Darwin was slow in "apprehension," he shows how patience is a virtue in avoiding regrettable mistakes made in haste: "I have as much difficulty as ever in expressing myself clearly and concisely; and this difficulty has caused a very great loss of time; but it has had the compensating advantage of forcing me to think long and intently about every sentence, and thus I have been often led to see errors in reasoning and in my own observations or those of others."[48]

Where Freud sought public acclaim and international fame through an orchestrated program of hero worship, Darwin avoided public notoriety and led an insulated life at Down House in the Kentish countryside. And where both Freud and Darwin coveted peer recognition (what scientist does not?), Darwin was far more concerned with the approbations of his immediate colleagues than with those of the public at large. "I did not care much about the general public," he penned in the *Autobiography*. "I do not mean to say that a favourable review or large sale of my books did not please me greatly; but the pleasure was a fleeting one, and I am sure that I have never turned one inch out of my course to gain fame."[49] To the botanist and his close friend Joseph Hooker, on hearing of the celebrated 1860 Huxley-Wilberforce debate (which Darwin missed while mourning the loss of a child), Darwin wrote: "Talk of fame, honor, pleasure, wealth, all are dirt compared with affection."[50]

Where Freud eradicated evidence of his intellectual heritage and ideological mentors, Darwin was generous in his acknowledgment of those whose ideas shaped his thinking. As a young man before entering Cambridge, he befriended Professor John Henslow, whom "I had heard of . . . from my brother as a man who knew every branch of science, and I was accordingly prepared to reverence him." Henslow held a weekly open house for students to gather and discuss the scientific issues of the day, by invitation only. Darwin eagerly sought and obtained a solicitation, and he attended regularly. "Before long I became well acquainted with Henslow, and during the latter half of my time at Cambridge took long walks with him on most days; so that I was called by some of the dons 'the man who walks with Henslow'."[51]

While at Cambridge, Henslow persuaded Darwin to study geology with the highly regarded geologist Adam Sedgwick, "and Henslow asked him to allow me to accompany him." An incident with Sedgwick convinced the young Darwin of the importance of the empirical methods of science. In a gravel pit near Shrewsbury, Darwin discovered a tropical shell and "was then utterly astonished at Sedgwick not being delighted at so wonderful a fact as a tropical shell being found near the surface in the middle of England. Nothing before had ever made me thoroughly realise, though I had read various scientific books, that science consists in grouping facts so that general laws or conclusions may be drawn from them."[52]

On the *Beagle*'s five-year circumnavigation of the globe, Darwin read Charles

Lyell's *Principles of Geology*, where Lyell expounded Hutton's principle of "uniformitarianism," which states that geologically uniform processes of today can be inferred to have always been operating that way in the past. Darwin explains the importance of this idea to his own thinking: "I had brought with me the first volume of Lyell's *Priniciples of Geology* which I studied attentively; and this book was of the highest service to me in many ways. The very first place which I examined, namely St. Jago in the Cape Verde islands, showed me clearly the wonderful superiority of Lyell's manner of treating geology, compared with that of any other author, whose works I had with me or ever afterwards read."[53]

Darwin's 1842 publication of *The Structure and Distribution of Coral Reefs* was, in fact, a clever and original application of Lyellian uniformitarianism, which he also applied in his own studies of organic nature, giving Lyell his due credit: "It appeared to me that by following the example of Lyell in Geology, and by collecting all facts which bore in any way on the variation of animals and plants under domestication and nature, some light might perhaps be thrown on the whole subject." Darwin continues, acknowledging Bacon's development of inductive reasoning: "My first note-book was opened in July 1837. I worked on true Baconian principles, and without any theory collected facts on a whole-sale scale."[54] The following year Darwin read the English economist Thomas Malthus's *Essay on Population,* crediting him for contributing a significant step in the development of the theory of natural selection: "In October 1838 . . . I happened to read for amusement 'Malthus on Population,' and being well prepared to appreciate the struggle for existence which everywhere goes on from long-continued observation of the habits of animals and plants, it at once struck me that under these circumstances favourable variations would tend to be preserved and unfavourable ones to be destroyed. The result of this would be the formation of a new species."[55]

In the *Origin of Species,* in reference to populations struggling for limited food supplies, Darwin notes that "it is the doctrine of Malthus applied with manifold force to the whole animal and vegetable Kingdoms."[56] After the publication of the *Origin* in 1859, Darwin credited Spencer's coinage of the now overused (and improperly applied) description, "survival of the fittest," as superior to his own phraseology: "I have called this principle, by which each slight variation, if useful, is preserved, by the term Natural Selection, in order to make its relation to man's power of selection. But the expression often used by Mr. Herbert Spencer of the Survival of the Fittest is more accurate and sometimes equally convenient."[57]

Even after the overwhelming success of the *Origin* and the praise heaped upon himself and his work, Darwin modestly lamented to Hooker: "If I lived twenty more years and was able to work, how I should have to modify the *Origin*, and how much the views on all points will have to be modified! Well it is a beginning, and that is something."[58]

Where Freud saw his theory as the third great blow to humanity's egocentrism, Darwin barely mentions in passing, on the final page of the *Origin*, that "light will be thrown on the origin of man and his history." In later editions he added the modifier "much" to the beginning of the sentence. It was not until 1871, after several more editions of the *Origin* and the publication of two more specific works supporting his theory (*On the Various Contrivances by Which British and Foreign Orchids Are Fertilized by Insects* in 1862, and *The Variation of Animals and Plants Under Domestication* in 1868), that Darwin ventured forth with his applications of the theory to humanity in *The Descent of Man*.

THE HERO OF THE MYTH

Some writers have suggested that Darwin's contribution has been overrated, that he was merely reiterating what others had already brought to the forefront. An example can be seen in Francis Hitching's *Neck of the Giraffe*, where the author states: "It is one of the less pleasing sides of Darwin's otherwise affable and scholarly nature that he could never bring himself to acknowledge a debt to the many predecessors in his field who were puzzling about the origin of species."[59] This statement follows a lengthy discussion by Hitching of Lyell's influence on Darwin and of Darwin's *willingness* to credit him, with such acknowledgments as "I feel as if my books came half out of Sir Charles Lyell's brain" and "I saw more of Lyell than of any other man, both before and after my marriage."[60]

Following this odd inconsistency in Darwin's allegedly niggardly generosity, Hitching cites an example in Loren Eiseley's book *Darwin and the Mysterious Mr. X* to establish Darwin's unwillingness to give appropriate credit. The mysterious "Mr. X" is the London chemist Edward Blyth, who in 1835 and 1837 published his theories about competition among species in the *British Magazine of Natural History*. Eiseley chronicles passages "that are almost word-for-word identical between Darwin and Blyth," with nothing more than "a cryptic reference in a letter" to the fact that Darwin had read these articles.[61] After castigating Darwin for this violation, Hitching now "forgives" him because "it was not the Victorian fashion to acknowledge predecessors and give references in the way it is mostly done in science today."[62] This is true, in part, but there is a deeper misunderstanding of Darwin that a cursory review of the literature can produce. Hitching, for example, quotes Blyth:

> Among animals which procure their food by means of their agility, strength, or delicacy of sense, the one best organized must always obtain the greatest quantity; and must, therefore, become physically the strongest and be thus enabled, by routing its opponents, to transmit its superior qualities to a greater number of offspring. The same law therefore, which was intended by providence to keep up the typical qualities of a species can be easily converted by man, into a means of raising different varieties.[63]

Several years ago, in preparing lectures for a course in evolution, I succumbed to a similar misunderstanding when I came across this passage in Charles Lyell's *Principles of Geology,* quoted in the text I was using for the course and presented, incorrectly as it turns out, as a precursor to Darwin: "Nature is constantly at war with herself and thus there will always be individuals who perish by disease or by the actions of predation. In a variable species it would not be the typical individual who succumbed to death, but the deviant ones—too great, too small, too thin legged, too thick legged—that would most often be the victims of predation or accident."[64]

On first blush this passage sounds like an anticipation of natural selection and the process of eliminating the unfit, and that is, in fact, how I read it. Indeed, it does describe an eliminating process, as does Blyth's "law," but this is not for the creation of new species. Lyell and Blyth (and others) were promulgating the accepted dogma of the day that the forces of nature were acting to preserve created kinds, not change them into new and different kinds. This *essentialistic* belief in the fixity of species was a vital part of the understanding of nature since the time of Aristotle. As Stephen Jay Gould explained: "Your quotation from Lyell is not an anticipation of Darwinian selection, but an expression of the very standard pre-Darwinian view that natural selection is a negative force acting to preserve the created form or essence of the species by eliminating those individuals that fall far from the mean. It was Darwin's particular genius that he converted a well-known mechanism thought to act for the preservation of stability into a theory of organic change."[65] In fact, reading Lyell deeper one comes across another passage that explains just that: "varieties have strict limits, and can never vary more than a small amount away from the original type."[66]

It seems highly probable that to the extent Darwin did not reference certain writers, it was *not* because they had anticipated him or that he plagiarized their ideas (or that referencing was not done in Victorian England, therefore we can forgive him), but because these ideas were in the air and well-known to most naturalists and, in fact, were orthogonal to what Darwin was proposing. Ernst Mayr stresses just how radical was Darwin's mechanism of change in the context of this creationist paradigm: "The fixed, essentialistic species was the fortress to be stormed and destroyed; once this had been accomplished, evolutionary thinking rushed through the breach like a flood through a break in a dike."[67]

A thorough review of the literature, rather than selected quoting, reveals a picture of Darwin that finds him neither demigod nor dotard. Rather, to the extent that Charles Darwin is a hero, that status arises from the impact of his ideas rather than an image created by himself or his followers. Darwin did virtually nothing to promote any sort of hero image, and at times even discouraged it. And if his followers were to be examined in depth for such contriving one would find many inconsistencies. For example, although Thomas Huxley ("Darwin's bulldog") proclaimed that the *Origin* was "the most potent instrument for the extension of the realm of knowledge which has come into man's hands since Newton's Principia," Darwin's close friend Lyell held back his support for Darwin's theory a full nine years, and even then hinted at a providential design behind the whole scheme.[68] These are hardly the actions of a coconspirator. Where Ernst Haeckel promoted evolution in Germany, and Asa Gray ("the American Huxley") supported Darwin's theory in the United States, the astronomer John Herschel called natural selection the "law of higgledy-piggledy." Adam Sedgwick (who was an Anglican cleric as well as a geologist) proclaimed that natural selection was a moral outrage, and penned this ripping harangue:

> There is a moral or metaphysical part of nature as well as a physical. A man who denies this is deep in the mire of folly. You have ignored this link; and, if I do not mistake your meaning, you have done your best in one or two cases to break it. Were it possible (which thank God it is not) to break it, humanity, in my mind, would suffer a damage that might brutalize it, and sink the human race into a lower grade of degradation than any into which it has fallen since its written records tell us of its history.[69]

Henry Fawcett in *Macmillan's Magazine* wrote of the great divide surrounding Darwin's book: "No scientific work that has been published within this century has excited so much general curiosity as the treatise of Mr. Darwin. It has for a time divided the scientific world with two great contending sections. A Darwinite and an anti-Darwinite are now the badges of opposed scientific parties."[70]

In his time much criticism accompanied the support Darwin received. Darwin is a hero because his ideas were so powerful that they survived his contemporaries and are still a vital force in twenty-first-century biology. Where Freud is pondered as a historical curiosity and as a launching platform for other, more sophisticated (and testable) theories of human behavior, Darwin is studied as a contemporary thinker nearly a century and a half after the publication of the *Origin*. Where only a small handful of psychoanalysts call themselves Freudian, only a small handful of biologists do *not* call themselves Darwinian (in some modified form, at least). Where Freud created a *hero as myth* image that has faded, Darwin refused to create a hero myth image, and he has come down to our time as a *hero as great man*.

10

THE EXQUISITE BALANCE

Carl Sagan and the Difference Between Orthodoxy and Heresy in Science

IN HIS CLASSIC 1959 WORK *The Logic of Scientific Discovery*, the philosopher of science Karl Popper identified what he called "the problem of demarcation," that is "the problem of finding a criterion which would enable us to distinguish between the empirical sciences on the one hand, and mathematics and logic as well as 'metaphysical' systems on the other." Most scientists and philosophers use induction as the criterion of demarcation—if one reasons from particular observations or singular statements to universal theories or general conclusions, then one is doing empirical science. Popper's thesis, presumably not derived through induction, is that induction does not actually provide empirical proof ("no matter how many instances of white swans we may have observed, this does not justify the conclusion that *all* swans are white") and that, de facto, scientists actually reason deductively—from the universal and general to the singular and particular.[1] But in rejecting induction as the preferred (by others) criterion of demarcation between science and nonscience, Popper was concerned that his emphasis on deduction would lead to an inevitable fuzziness of the boundary line. If a scientific theory can never actually be proven, then is science no different from other disciplines of knowledge?

Popper's ultimate solution to the problem of demarcation was the criterion of falsifiability. Theories are "never empirically verifiable," but if they are falsifiable then they belong in the domain of empirical science. "In other words: I shall not require of a scientific system that it shall be capable of being singled out, once and for all, in a positive sense; but I shall require that its logical form shall be such that it can be singled out, by means of empirical tests, in a negative sense: it must be possible for an empirical scientific system to be refuted by experience."[2]

The theory of evolution, for example, has been accused by creationists as being nonscientific because there is no way it can be ultimately proven to be true. No one was there to see it happen and biologists cannot observe it in the laboratory because it takes too long. But, in fact, by Popper's criterion of falsifiability the theory of evolution would be doomed to the trash heap of bad science if, say, human fossil remains suddenly turned up in the same geological bedding planes as 505-million-year-old Burgess Shale fossils. No such falsification of evolution has ever been found, and although by Popper's criterion this does not mean that the theory has been proven absolutely, it does mean that it has yet to be falsified, thus placing it firmly in the camp of solid empirical science, creationists notwithstanding. (Ironically, it was Karl Popper who mistakenly identified natural selection as a tautology—those who survive are most fit, those who are most fit will survive—a belief he later retracted. I own a copy of a 1971 anti-Darwinian book by an attorney named Norman Macbeth, *Darwin Retried*, endorsed on the back jacket by Popper, who called it "a really important contribution to the debate . . . a truly valuable book." Such misunderstandings will continue to be propagated as long as philosophers and scientists fail to recognize that the historical sciences are no less rigorous than the experimental sciences, even though their methodologies differ. See my book *Denying History* for a defense of this position.)

But Popper's attempt to solve the problem of demarcation—similar to what I have been calling the boundary problem—between science and nonscience begins to break down in the borderlands of knowledge. Consider the theory that extraterrestrial intelligent life exists somewhere in the cosmos. If we find one by making radio contact through the SETI program then the theory will have been proven absolutely (assuming the signal was verified to be truly of an extraterrestrial source, the signal was repeated, etc.). Physical contact would be even better, although this is far more unlikely, UFOlogists and alien abductees notwithstanding. But how could this theory ever be falsified? One might argue that it is falsified every day that SETI fails to find a signal, and critics of the SETI program make just that point. But the SETI folks remain undaunted for a very simple fact of nature: the cosmos is a very, very big place and we have only just begun our search. Our galaxy alone contains roughly 400 billion stars, and there may be as many as 400 billion galaxies in our universe, itself perhaps one of a near infinite number of bubble universes. How can the search through a few thousand stars be a criterion of falsifiability for this theory? Like Popper's white swans analogy, at what point could one ever say that the ET theory has been falsified, short of searching every star in every galaxy in the entire cosmos?

Clearly there are statements and theories that are not quite science, not quite nonscience, and these I have called borderlands science. SETI is one such example that I wish to explore further in this chapter, through the eyes of the man who nearly single-handedly put it on the map as a legitimate field of study for scientists—Carl Sagan. When Sagan began his studies in astronomy, the idea of searching for extraterrestrial life of any kind, intelligent or otherwise, was considered to be the province of fringe cranks. By the time of his death in 1996, however, Sagan's heresy had become orthodoxy. In early 2000, for example, a book was published entitled *Rare Earth*. The authors, Peter Ward and Donald Brownlee, were featured in the *New York Times* as radicals for daring to challenge orthodox science that the cosmos is probably teaming with complex life: "Now, two prominent scientists say the conventional wisdom is wrong."[3] In three decades Sagan changed the theory from heresy to orthodoxy, even though there still exists not one iota of concrete evidence of any life, simple or complex, intelligent or not, anywhere beyond Earth.

How did this belief shift happen? More importantly, how is it that so much can be said and written about an as-yet-to-be-detected phenomenon? The short answer is that in the borderlands between science and nonscience the mind fills in the gaps left by nature. The long answer is in understanding how scientists—or any of us for that matter—find that exquisite balance between heresy and orthodoxy, particularly when the evidence cannot lead us to a definitive conclusion. Why are some people more open to heretical ideas and others more inclined toward defending orthodoxy? The question is most relevant (and interesting) in the borderlands where definitive conclusions are not forthcoming. To find an answer we must turn to the social sciences, for it is here where we must take measure not of the phenomenon under question, but of the life doing the measuring.

THE MEASURE OF A LIFE

What is the measure of a life when it is gone? A newspaper obituary? A magazine story? A potted television biography? How shall we capture the essence of that life? A list of accomplishments? Highlights and lowlights? Interviews with family, friends, colleagues, and critics? A womb-to-tomb narrative? And if that life was an epoch-shaping life, how is a contemporary biographer to put that life in perspective before the epoch is over?

What tools should we use? Oral history interviews? Demographic and statistical data? Document analysis? What fields should we consult? Psychology? Sociology? Cultural history? Does the measure of a life depend as much on who is doing the measuring as it does on the measured life itself? Can we even get to the true core of a person? Can there be a science of biography?

Humans are storytelling animals. Our greatest stories are about ourselves, our lives, and how they play out within the larger context of culture and history. From Moses to Michener, narrative has been the vehicle of biography for three millennia. Thus, my aim is not to tear down the citadel but to build upon it. In considering how narrative biographies are constructed I thought of Barbara Tuchman—one of our era's most eloquent storytellers—and the frustration she experienced in facing the vast panorama of human variability and apparent contradictions in her attempt to generalize about the Middle Ages:

> Contradictions, however, are part of life, not merely a matter of conflicting evidence. I would ask the reader to expect contradictions, not uniformity. No aspect of society, no habit, custom, movement, development, is without cross-currents. Starving peasants in hovels live alongside prosperous peasants in featherbeds. Children are neglected and children are loved. Knights talk of honor and turn brigand. Amid depopulation and disaster, extravagance and splendor were never more extreme. No age is tidy or made of whole cloth, and none is a more checkered fabric than the Middle Ages.[4]

No life is tidy or made of whole cloth, and few form a more checkered fabric than Carl Sagan's. Nonetheless, this has not, nor will it, stop biographers from trying to capture that life in a comprehensive narrative biography. Less than three years after Sagan's death on December 20, 1996, two biographies hit the market—Keay Davidson's *Carl Sagan: A Life*, and William Poundstone's *Carl Sagan: A Life in the Cosmos*—with more on the way. Both Davidson and Poundstone did a remarkable job of getting their minds around such a larger-than-life figure, especially given the pressures they were under from their publishers to be first into print. But the problem of all narrative biography (and here I do not fault biographers for not writing biographies I would write) is in determining whether a particular action, a quote from a speech, an excerpt from a book, or a description of one's subject by a colleague or friend represents a passing fancy or a deep interest, a whim or a passion, a long-term personality trait or a short-term temporal state (trait v. state theory in personality psychology).

Was Sagan a tender-minded liberal or a tough-minded careerist? Was he a feminist or a misogynist? Was he really obsessed with the possibility of extra-terrestrial intelligence, or was this just a flighty avocation that happened to generate a lot of media attention? Was he a scientist of the first rank or merely a media-savvy popularizer? How can we tell? It is easy to start off with a hunch and then comb through books, papers, notebooks, diaries, interview notes, and the like, pick out the quotes that best support the hypothesis, and draw the anticipated conclusion. In statistics this is called "mining the data." In cognitive psychology it's termed "confirmation bias," a powerful explanatory concept that accounts for many thinking foibles where we tend to focus on information that confirms what we already believe and ignore disconfirming evidence.[5] Or as I like to say about psychic readings, we remember the hits and forget the misses.

TOWARD A SCIENTIFIC BIOGRAPHY

How can we avoid the confirmation bias in writing biography? One way is to apply the tools of the social sciences. Fortunately for any would-be scientific biographer, Sagan's curriculum vitae (c.v.) is, to say the least, comprehensive. Weighing in at 4.5 pounds, it totals 265 single-spaced typed pages. An analysis of it allows us to answer certain questions and to test specific hypotheses.

For example, how productive a career did Sagan have? FIGURE 29 presents his 293 advisory groups, professorship and lectureships, and professional societies by type. FIGURE 30 displays Sagan's 89 fellowships, awards, and prizes by type, offering insight into what he was most recognized for by society and his professional colleagues: first and foremost as a humanitarian and science popularizer, next for his scientific research, and last for his scientific writing (but one was the Pulitzer).

Such data alone, however, tells us little without a context. Was Sagan a world-class scientist or a mediocre scientist and a world-class popularizer? Since he was rejected by the National Academy of Science, the most prestigious scientific organization in America, I thought it would be instructive to compare Sagan's statistics to those of the average NAS member. Unfortunately such comparative data are not available. But even by NAS standards Sagan was no ordinary scientist, so I decided to compare him to a handful of scientists who represent the créme de la créme: Jared Diamond, Stephen Jay Gould, Ernst Mayr, and E. O. Wilson.

FIGURE 31 shows Carl's honorary degrees in comparison: Gould's 41 towers above the rest, but Sagan's 23 is nestled firmly between Wilson and Mayr

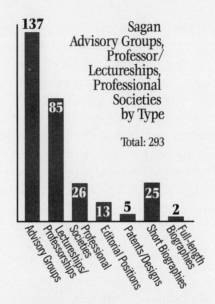

Figure 29. Sagan's Advisory Groups by Type.

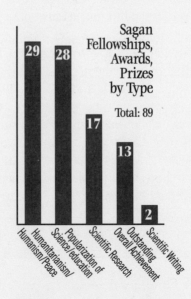

Figure 30. Sagan's Fellowships by Type.

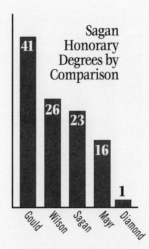

**Figure 31. Sagan's
Honorary Degrees
in Comparison.**

(although Ernst was quick to point out that his "are from the very best universities—the cream of the crop—such as Oxford, Cambridge, Harvard, Yale, and Bologna, the world's oldest university"). Diamond's single honorary doctorate actually helps us understand the meaning of the others. Honorary doctorates are one of several means of keeping score for driven careerists. Of these five, in my opinion Diamond is the most modest and unassuming. As he told me: "I only have one because they don't mean that much to me and they take time away from my family."

Sagan's book production is also telling—in totality, in content, and in comparison. FIGURE 32 shows Sagan's 31 books by content, indicating his primary professional interest in planetary science (from his first book in 1961, *The Atmospheres of Mars and Venus* with W. W. Kellogg, to *Pale Blue Dot* in 1994), as well as his pioneering efforts in the exotic science of exobiology and (at the time) the mildly radical SETI (from the classic *Intelligent Life in the Universe* with I. S. Shklovskii to his novel *Contact*). Under general science I included such books as *Cosmos* and *The Demon-Haunted World*, and under the category of evolution fall *The Dragons of Eden* and *Shadows of Forgotten Ancestors*, coauthored with Ann Druyan and considered by Keay Davidson to be Sagan's greatest work (because of her influence). But Sagan's most controversial books dealt with the topics of nuclear winter, disarmament, and the environment, especially *A Path Where No Man Thought* with Richard Turco.

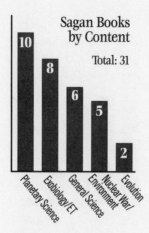

Figure 32. Sagan's Books by Content.

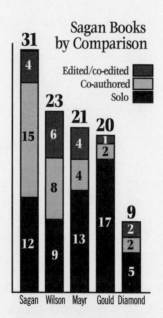

Figure 33. Sagan's Books in Comparison.

FIGURE 33 shows that Sagan's book productivity was the highest in my comparison group, out-generating Mayr by 10 in 35 fewer years, Wilson by eight in 10 fewer years, Gould by 11 in only five more years, and Diamond by 22 in the same time frame. Interestingly, for all his alleged arrogance and self-centeredness, Sagan has the highest ratio of coauthorships and coeditorships of this elite group (eight of his 15 coauthored books had four or more authors or were large group collaborations, artificially inflating his book total but demonstrating his ability and willingness to work with others).

In FIGURES 34, 35, and 36 we get to the meat of Sagan's c.v.—scientific output by content and in comparison. FIGURE 37 presents a content analysis I conducted on Sagan's 500 published scientific papers, revealing that planetary science was by far and away his greatest professional interest, with two-thirds more than all other papers combined. Nevertheless, nearly a third (31.6 percent) of the total were in the (then) controversial field of exobiology, and another 9

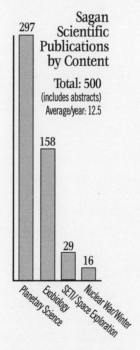

Figure 34. Sagan's Scientific Articles by Content.

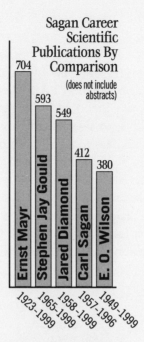

Figure 35. Sagan's Overall Scientific Productivity in Comparison.

percent in such career-hampering fields as SETI and nuclear winter. To many scientists, these washed out Sagan's remarkable 67 (13.4 percent) papers that appeared in the prestigious journals *Science* (37) and *Nature* (30). By comparison, Diamond had 13 in *Science*, 128 in *Nature* (with 120 of them as his regular "News and Views" column), and through 1996 Gould had 45 articles total published in *Science* and *Nature*.

Edward Teller was the most publicly vitriolic critic of Sagan, sputtering to Davidson, "Who was Carl Sagan? He was a nobody! He never did *anything* worthwhile. I shouldn't talk with you. You waste your time writing a book about a nobody."[6] Even though Teller had his own agenda, he was not alone in such criticisms. Many have claimed that Sagan was little more than a popularizer, forcing us to accept a crude binary taxonomy where one is *either* a scientist or a popularizer, but never both. These critics strengthen their case by citing Sagan's inability to get tenure at Harvard, or the National Academy of

Science's rejection of his bid for membership. When asked by David Swift to characterize Sagan's role as a SETI pioneer, Melvin Calvin put it bluntly: "He's a publicist."[7] Philip Morrison said of Sagan "I don't think he's actually done very much directly bearing on the technical problems," but that "There's no doubt that he's had an impressive impact on the public about the whole question."[8] When I asked a number of his colleagues to describe Sagan's "strengths and weaknesses as a scientist," one astronomer (who wished to remain anonymous) wrote: "I don't think about Carl Sagan's strengths and weaknesses as a scientist. I think of him as the most successful mass media promoter of science yet. Some of his own research, public pronouncements, and priorities were compromised by his personal vision and style, but that is how it is in science."[9]

Can we scientifically assess the relative value of Sagan's scientific contribution versus his popularization? We can. And we can make quantitative comparisons to other world-class scientists. In FIGURES 35, 36, and 37 we see Sagan's overall scientific production and annual rate of publication comparable to my select eminent group (in FIGURES 35 and 36 Sagan's total and average do not include

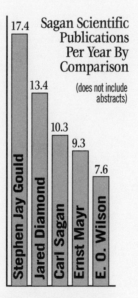

Figure 36. Sagan's Average Scientific Productivity in Comparison.

abstracts, which the others did not include in their c.v.'s). The data speak for themselves: by quality and quantity Sagan stands toe to toe with these giants of science. (Note: had Sagan lived to Mayr's age of 95 and continued publishing papers at his average pace, his total would have been 751 articles. If Diamond continues at his present rate to 95, his lifetime total will top out at 1,004, bettered only by Gould who, if he makes it to 95, will peak at 1,219. Gould's figures include his 288 essay columns in *Natural History*.)

FIGURE 37 shows that with the exception of a dip during the years *Cosmos* was under production in the late 1970s (and a subsequent messy divorce), Sagan's scientific productivity never wavered. In fact, he turned out more than a scientific paper per month from 1983 until his death in 1996, during the off-the-chart years of media exposure and popular writing. (Note: included in these 1,380, so labeled in his c.v. as "General Works, Interviews, Speeches, Policy Analyzes, Book Reviews, Television Writing, etc.," are articles written not by Sagan or even about Sagan, but by journalists who interviewed Sagan, along with others for an article on a subject on which Sagan was an

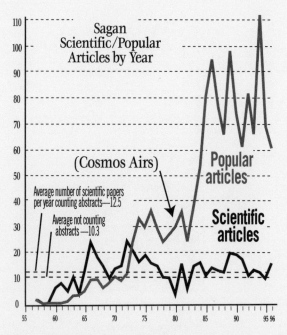

Figure 37. Sagan's Scientific and Popular Productivity Total and Average over Time.

expert. No one in my comparison group has anything comparable in their c.v.)

What Sagan was most famous for, and what got him in the biggest trouble with the academic establishment, was his Brobdingnagian outpouring of popular articles and interviews. For FIGURE 38 I conducted an eye-blurring content analysis of all 1,380 items, revealing that public and professional perceptions of Sagan as the ET go-to guy were not misdirected, with SETI and space exploration topping the list. The number two most popular subject, interestingly and tellingly for this analysis, was Sagan himself, with no less than 263 interviews and profiles of the man (and many with his wife and professional collaborator Ann Druyan).

Was Sagan politically and socially liberal? The data give us an unequivocal answer: one-third of everything he wrote or said was on nuclear war, nuclear winter, environmental destruction, women's rights, reproductive rights, social freedoms, free speech, and the like (and this figure does not include Op Ed pieces, which were not given titles in the c.v.).

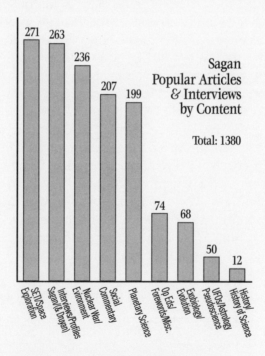

Figure 38. A Content Analysis of Sagan's Popular Writings.

But in looking beyond the raw data one finds a definite tension in Sagan between his liberal/feminist ideals and his career ambitions. Although he was already a social activist in his early 20s, according to his first two wives Sagan was no liberal or feminist in the home. As Davidson described it: "Sagan's liberalism, while sincere, had an abstract aspect; it was the clever, witty, after-dinner-speaker liberalism of Adlai Stevenson, not the passionate, heart-wrenching, take-to-the-streets liberalism of Martin Luther King Jr. Like so many aloof intellectuals, the young Sagan seemed to think in terms of People rather than people, of Humanity rather than humans."[10] His first wife, Lynn Margulis (herself a world-class scientist and a member of the National Academy of Science), recalled that Sagan "never changed a diaper in his life, he never cleared the table of his dishes, he never washed the dishes. . . . He needed ten thousand people to be raving about him all the time. I was just one young woman, trying to go to school and take care of kids and run a household. Every distraction he considered personal."[11]

As a consequence, his first marriage disintegrated. In a response emblematic of a man so committed to science and rationality as to not see its boundaries, Sagan tried to persuade Margulis to come back through what Davidson called "a very Saganish sales pitch—big on career, tongue-tied on love." Margulis recounted the breakup: "We were walking on the street and he told me how I was crazy because he was such an important person, and he was going to be much *more* important, and that I was really married to a fantastic guy and I was crazy to even think about leaving."[12] More than anyone else Ann Druyan showed Sagan that you have to *live* the principles, not just talk about them, even in the home. Even here, however, my wife Kim pointed out that it is much easier for one to be a liberal and a feminist later in life, when one is established and well off, where day-to-day chores can be hired out, when careers are flourishing and the children are grown. Sagan's first son with Margulis, Dorion, wrote Sagan a contentious letter to point out what he perceived to be hypocrisy (summarized by Davidson, Dorion is reflecting on the pain of being largely abandoned by his then excessively careerist father):

"His understanding of markets, which I had been studying, was simplistic. I remember being up at the Ritz Carlton . . . with his friends and his new wife [Annie]. Top floor of the Ritz Carlton, getting all kinds of perks—and they were going on about the virtues of communism. And that's classic champagne socialism, you know?" Dorion wrote his dad a letter implying that his left-leaning economic views were hypocritical—a letter that was,

Dorion admits, a pretext for his own inner hurts. "In the letter I said stuff like, 'You say that we should have an equal allotment of wealth. . . . Okay, why don't we cap [the maximum allowable wealth] at your earnings last year and we call the unit 'one sagan,' and nobody can make more than one sagan. While we're doing it, let's cap the number of books that anybody can write."[13]

Along similar lines, Poundstone properly nuances Sagan's conflicting feelings about, and attitudes toward, homosexuality. When Dorion was in high school he befriended a gay classmate, triggering Sagan to sit him down for a lecture explaining that homosexuality was not how a species can propagate itself. Nevertheless, Poundstone gainsays Dorion's stories (and we would do well to remember that, however understandable, Dorion still harbors a fair amount of ill will toward his father that may cloud his judgment) with examples of how Carl's closest scientific collaborator, Jim Pollack, was openly gay; how Carl came to the defense of Pollack's lover in a problem with obtaining treatment at the university health service emergency room, and that "in no visible way did Pollack's homosexuality impede Sagan's long and productive collaboration with him."[14]

SETI AS SCIENCE AND RELIGION

What made Sagan a pioneer in the search for extraterrestrial intelligence? To attempt an answer to this question, FIGURE 39 presents data I culled from David Swift's 1990 book *SETI Pioneers*. Not surprising, none believed that UFO sightings represent actual visitation by extraterrestrials. Equally unsurprising was their universal agreement that extraterrestrial intelligences (ETIs) probably exist somewhere in the cosmos (why else would they be involved in SETI?) I included these columns because, although interest in and the study of exobiology and the possibility of ETIs is certainly not mainstream science, then or now, it is nowhere near as fringy as belief in UFOs. In a way, SETI is elitist, UFOs populist; SETI is highbrow, UFOs are lowbrow; SETI is dominated by Ph.D. astronomers, physicists, and mathematicians, UFOs are predominantly the domain of noncredentialed amateurs. As revolutions go, SETI is on the conservative side. This observation will become important when we turn to the role of personality in science.

Swift asked each of the SETI pioneers about their parents' religiosity, but oddly did not ask about their own beliefs. Nevertheless, I was able to glean

CARL IN CONTEXT: THE SETI PIONEERS

NAME	BIRTH ORDER/ SIBSHIP SIZE	RELIGIOUS UPBRINGING	RELIGIOUS ATTITUDES	BELIEF IN ETI	BELIEF IN UFOS
Billingham	1/2	Church of England	Nonbeliever	Yes	No
Bracewell	1/2	Anglican Orthodox	Nonbeliever	Yes	No
Calvin	1/2	Jewish	Nonbeliever	Yes	No
Cocconi	2/2 (older sister)	?	?	Yes	No
Drake	1/3	Baptist	Nonbeliever	Yes	No
Dyson	2/2 (older sister)	Church of England	Nonbeliever	Yes	No
Kardashev	1/2	None (parents died young)	?	Yes	No
Kraus	2/2 (older sister)	Strong Methodist	Believer?	Yes	No
Morrison	1/2	Jewish	Nonbeliever	Yes	No
Oliver	Only	Nondenominational	Atheist	Yes	No
SAGAN	**1/2**	**JEWISH**	**NONBELIEVER**	**YES**	**NO**
Sakurai	2/7	?	?	Yes	Unlikely
Seeger	1/7	?	?	Yes	No
Shklovskii	1/2	Jewish	?	Yes/No	No
Tarter	Only	Protestant/Presbyterian	Nonbeliever	Yes	*
Troitskii	1/3	?	?	Yes	**

* Unlikely/Unexplained
** Unexplained/Possible

Figure 39. A Data Analysis of the SETI Pioneers.

most of that information from the interviews—enough to make the generalization that most were raised in a religious household but that not one believes in anything like the traditional Judeo-Christian God (although I am missing some data). What is the significance of this observation? Astronomer Frank Drake, ostensively *the* SETI pioneer if there was one (and creator of the infamous "Drake equation" for computing the probability of ETIs), who was raised "Very strong Baptist. Sunday school every Sunday," made this observation: "A strong influence on me, and I think on a lot of SETI people, was the extensive exposure to fundamentalist religion. You find when you talk to people who have been active in SETI that there seems to be that thread. They were either exposed or bombarded with fundamentalist religion. So to some extent it is a reaction to firm religious upbringing."[15] Similarly, John Kraus recalled: "We were very strong churchgoers, members of the Methodist church. I was brought up in a very religious atmosphere . . . there was never any thought of conflict between science and religion in my thinking or in my upbringing. Science and religion were simply both seeking ultimate truth but using different ways of going at it."[16]

But there were exceptions. Melvin Calvin's parents were Russian Jews who "didn't keep any religious practices. When I grew up I was without religion;

a-religious, not anti-religious."[17] And Bernard Oliver's parents "belonged to no orthodox church of any sort. I think my father had been christened a Congregationalist by his mother when he was very little, but he never went to church; it didn't interest him. My mother, however, had this strong interest—a philosophical interest, let's say, in life: what was life? And she believed that there was a soul. And the reason was that material things were far too gross to in any way hold this marvelous quality called life."[18]

Does religion play a role in attitudes toward ETIs? Philip Morrison gave his considered opinion: "Well, it might, but I think that it's just one of the permissive routes; it isn't an essential factor. My parents were Jewish. Their beliefs were conventional but not very deep. They belonged to the Jewish community; they went to services infrequently, on special occasions—funerals and high holidays."[19]

One might speculate that SETI, as a highbrow, elitist revolution, contains within it quasi-religious and spiritual overtones, in the sense that these scientists, while not believing in God, do believe in ETIs, uniformly portrayed as higher intelligences who, having survived what might be a tendency in species toward self-destruction once advanced technologies are created, and who are virtually certain to be significantly more advanced in every way than us, must also be morally superior. To the extent that religion involves belief in and hope for transcendence or transcendent beings, SETI is a high-cultural form of religion, and UFOs a low-cultural form of religion.

Melvin Calvin said as much about the impact of first contact: "It would have a marked effect. It's such a broad, major subject of concern to everyone, no matter where they are, that I think people would listen. It's like introducing a new religion, I suppose, and having it picked up by a lot of people."[20] Philip Morrison compared it to the Copernican revolution: "Up till now a great many people have the happy view that we are unique, the green footstool of creation, and that there is nothing else like us." Discovering ETIs "will have an impact over the long run comparable to the notion that the Earth is not the center of the Solar System."[21] And Bernard Oliver returned to the problem his mother posed: "My mother was involved in quasi-religious or metaphysical things. The question is really, 'Is life a negligible and extremely rare phenomenon in the universe—intelligent life, that is—or is it so prevalent that the universe can be considered to be somewhat efficient in producing it?' . . . life could, in the course of time, become an important force in the late evolution of the universe. I can imagine, though I can't tell you how, that this life, in a network of communication, could form a sort of superconsciousness throughout the galaxy that, in ways we can't foresee now, might modify the history of it."[22]

Although Sagan did not believe in God, he nevertheless said this about SETI's importance: "It touches deeply into myth, folklore, religion, mythology; and every human culture in some way or another has wondered about that type of question. It's one of the most basic questions there is."[23] In fact, in Sagan's novel/film *Contact,* described by Keay Davidson as "one of the most religious science-fiction tales ever written,"[24] Ellie discovers that pi—the ratio of the circumference of a circle to its diameter—is numerically encoded in the cosmos and this is proof that a super-intelligence designed the universe:

> The universe was made on purpose, the circle said. In whatever galaxy you happen to find yourself, you take the circumference of a circle, divide it by its diameter, measure closely enough, and uncover a miracle—another circle, drawn kilometers downstream of the decimal point. In the fabric of space and in the nature of matter, as in a great work of art, there is, written small, the artist's signature. Standing over humans, gods, and demons, subsuming Caretakers and Tunnel builders, there is an intelligence that antedates the universe.[25]

SAGAN'S EXQUISITE BALANCE

The left column in FIGURE 39 presents the birth orders and sibship size of each of the SETI pioneers. David Swift identified an apparent overabundance of firstborns in his population, including Sagan. But is it a statistically significant overabundance? Swift did not test for this, but U.C. Berkeley social scientist Frank Sulloway and I did, applying what is known as the Greenwood-Yule rule for expected number of firstborns. For the SETI pioneers, eight is the expected number of firstborns based on the number of siblings they had, but 12 is the observed number. This difference (four) is statistically significant at the .05 level of confidence.

What does this significant number of firstborns mean? In Sulloway's book *Born to Rebel* he presents a summary of 196 controlled birth-order findings classified according to the Five Factor model of personality, or the "Big Five" personality dimensions (discussed in chapter 7):

—*Conscientiousness*: Firstborns are more responsible, achievement oriented, organized, and planful.

—*Agreeableness:* Laterborns are more easygoing, cooperative, and popular.
—*Openness to Experience*: Firstborns are more conforming, traditional, and closely identified with parents.
—*Extraversion:* Firstborns are more extroverted, assertive, and likely to exhibit leadership.
—*Neuroticism/Emotional instability:* Firstborns are more jealous, anxious, neurotic, fearful, and likely to affiliate under stress.[26]

To measure Sagan's personality Sulloway and I requested a number of his family members, friends, and colleagues to rate him on a standardized Big Five personality inventory of 40 descriptive adjectives using a 9-step scale. For example: *I See Carl Sagan as Someone Who Was . . .*

—Ambitious/hardworking 1 2 3 4 5 6 7 8 9 Lackadaisical
—Tough-minded 1 2 3 4 5 6 7 8 9 Tender-minded
—Assertive/dominant 1 2 3 4 5 6 7 8 9 Unassertive/submissive
—Organized 1 2 3 4 5 6 7 8 9 Disorganized
—Rebellious 1 2 3 4 5 6 7 8 9 Conforming

FIGURE 40 presents the results for Sagan in percentile rankings relative to Sulloway's database of 7,276 subjects. (To measure the consistency of ratings between raters, Sulloway computed an interrater reliability score for the six raters who participated of .73, more than acceptable by social science standards.)

Most consistent with his firstborn status, Sagan was exceptionally high ranking on *conscientiousness* (ambitiousness, orderliness, dutifulness) and is strikingly

SAGAN'S PERSONALITY
Ratings on the "Big 5" Personality Traits

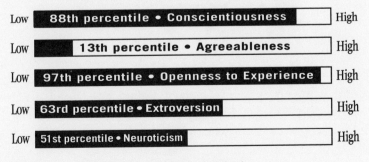

Low	88th percentile • Conscientiousness	High
Low	13th percentile • Agreeableness	High
Low	97th percentile • Openness to Experience	High
Low	63rd percentile • Extroversion	High
Low	51st percentile • Neuroticism	High

Figure 40. Sagan's Personality Analysis.

low ranking on *agreeableness* (tender-mindedness, easy-goingness, modesty). *Extroversion* and *Neuroticism* were non-descript, but Sagan's *openness to experience* (preference for novelty, variety, adventurousness) was nearly off the scale, significantly higher than what one would expect from a firstborn.[27] How can we reconcile this disparity? In Sulloway's family dynamics model he reveals a number of variables that shape personality:

Social Attitudes: "People who are socially liberal are more open to radical change." As we saw in FIGURE 9 for his publications and interviews about social issues, Sagan was extremely liberal.

Parental Social Attitudes: "People having liberal parents tend to be liberal and hence to support radical change." According to his biographers, Sagan's parents were both socially liberal. Plus, members of many minority groups (Jewish in Sagan's case) tend to support liberal causes and are more open to experience.

Personal Influences: "Mentoring and friendship influence the adoption of radical ideas."[28]

On the latter influence, as a graduate student at the University of Chicago Sagan befriended Joshua Lederberg, whom Keay Davidson calls "the godfather of exobiology," the meeting of which launched "the most high-profile dynamic duo of the early days of exobiology, the science of extraterrestrial life."[29] Sagan also worked closely with Nobel laureate H. J. Muller, and in 1976 wrote Lederberg: "if not for the encouragement by H. J. Muller and yourself, I might not have had the courage to seriously pursue what later has come to be called exobiology." As Davidson described it: "The older scientist did more than talk; he escorted Sagan into the corridors of power." Lederberg characterized the relationship as such: "I was often his protector and defender from folks who thought he was wild. He had a lot of offbeat ideas. They were always at some level not illogical, and some of them could prove to be right; and I would point out [to others] the value of listening closely to someone who has that degree of rigor and imagination at the same time."[30]

Sagan's high conscientiousness occasionally clashed with his high openness. Lederberg recalls: "He didn't stick to things very long. I think part of his reputation for not being 'solid' has less to do with lack of rigor on any one item than that he didn't build a body of work on one particular topic. His interests were so catholic."[31] Actually this is what Sulloway's family dynamics model predicts. Sagan's openness to experience led him to gamble on a number of revolutionary ideas, but his conscientiousness prevented him from taking these ideas too far into crankdom.

If we view SETI as high culture and UFOs as low culture, then we should not be surprised to see a personality like Sagan's support the former and reject the latter. This is the "essential tension" described by Thomas Kuhn in his apt distinction between normal science and revolutionary science, between tradition and change.[32] Science is normally conservative, yet to progress it must occasionally relinquish ground to revolutionaries who have built enough of a foundation to grab a foothold. Sagan was masterful at finding that "exquisite balance," as he noted (in a quote that serves as the epigram for my book *Why People Believe Weird Things,* from a 1987 lecture he gave in Pasadena on "The Burden of Skepticism"):

> It seems to me what is called for is an exquisite balance between two conflicting needs: the most skeptical scrutiny of all hypotheses that are served up to us and at the same time a great openness to new ideas. If you are only skeptical, then no new ideas make it through to you. On the other hand, if you are open to the point of gullibility and have not an ounce of skeptical sense in you, then you cannot distinguish useful ideas from the worthless ones. If all ideas have equal validity then you are lost, because then, it seems to me, no ideas have any validity at all.

In this context, in a 1998 study Sulloway and I found that *openness* was significantly correlated with lower levels of religiosity ($r = -.14$, $p < .0001$) and higher levels of religious doubt ($r = .18$, $p < .0001$). Moreover, *openness* was significantly correlated with change in religiosity, with higher *openness* scores being associated with lowered piety with increasing age ($r = -.09$, $p < .01$), as well as with lower rates of church attendance ($r = -.11$, $p < .01$). Not surprisingly, we also found a strong correlation between *openness* and political liberalism ($r = .28$, $p < .0001$).[33] These findings gel with Sagan's personality and attitudes toward SETI and religion, where he was a passionate believer in the former and a skeptic of the latter.

This hypothesis is supported by answers offered by our raters when asked to describe Sagan's unique thinking style. Astronomer David Morrison wrote: "Analytical, big-picture, great with students, excellent team member, probing and curious, ready to pursue unconventional paths, highly original thinker." Sagan's brother-in-law, Les Druyan, recalled: "In discussions Carl was scrupulously honest, always willing to admit to any flaws in his own argument; he had an uncanny memory for facts; his logic was wonderfully simple and clearly explained; he had the ability to discuss things on the same level as the person(s) he was with, from seven-year-olds to rocket scientists."[34] Poundstone well captured the essential tension:

Strengths: He had an almost phenomenal ability to look at a raw data set and see connections that no one else did. He was thereby able to pose the interesting "big" questions that are the starting point of most important research. *Weaknesses*: Lack of follow-through. He had no patience for hands-on experimental work or the high-powered math that is the basis of most theoretical work. Having posed an interesting question, his impulse was to move on, to pose other interesting questions.[35]

Thomas McDonough, who was a graduate student of Sagan's at Cornell and later his colleague through the Planetary Society, noted the strain between Sagan's rebelliousness and his ambitions:

> He was not afraid of controversy, indeed, he seemed to thrive on it. Thus he advocated SETI when it was highly disapproved of by the scientific establishment. It threatened to abort his career because it was so disreputable. [Yet he] sometimes he went out of his way to insult his enemies even when it was irrelevant. For example, once at a public lecture on space exploration for The Planetary Society, he attacked Republicans even though it had nothing to do with the talk and undoubtedly alienated a good part of the audience.
>
> He treated science virtually as his religion. He had an almost messianic passion for science. I have often suspected that science held the same emotional position in his mind as religion has for others.
>
> He used his verbal skills to help him accomplish much more than most people could have, by dictating whenever possible. He kept two secretaries working full-time just transcribing his tapes. . . . [36]

Some of the most interesting observations of Carl's personality I received came from e-mail conversations with Ann Druyan. Although he had no belief in God whatsoever, and considered most of the tenets of religion a tissue of illusions, Druyan set the record straight on how important Sagan's Jewish heritage was to him (in part, countering the suggestion by his biographers that Sagan hid his Jewishness in the interest of career ambitions):

> Carl was always completely out front about being Jewish. (And believed that his face was a gloriously unsubtle declaration of his origins.) It was his primary cultural identity. All three of his wives were Jewish, each wedding presided over by a rabbi. Our homes, replete with menorahs, yearly seders, etc. identifiably so. One of Carl's few unrealized lifelong goals was the writ-

ing of a new Haggadah. His conversation was dotted with Yiddish words and phrases. A check of his remarkable vitae will reveal that he was repeatedly honored by Jewish organizations, and went to considerable effort when he was gravely ill to be included in a *Life* magazine book and feature on American Jews of distinction.[37]

Sulloway has shown that firstborns are more parent-identified, and Sagan's biographers go on at length about his adoring and dominating mother. Druyan shows that the relationship with his father was no less special:

> To live with Carl and his father, Sam, was to witness the most tender and unambivalent father/son relationship I have ever known. I never once saw Carl be disrespectful or even slightly testy with his garment-cutter father, or with mine. He adored Sam and tried his best to be as much like him as he could.

As a tough-minded firstborn, Sagan preferred right over nice, putting conscientiousness above agreeableness, as Druyan recalled:

> Carl never participated in anything so shabby or short-sighted as the desire to pass for something other than he was. See his refusal to absolve Werner von Braun, one who presumably a Carl neutered of his Jewishness would have otherwise lionized. I believe Carl was the only major figure in that community to take him on in print and even more courageously, in Huntsville. And Carl hired Frank Kameny, a man who couldn't get a job because he was the first declared homosexual to sue the Federal government over his dismissal stemming from his sexual preference.
>
> As for Carl "the driven careerist," if it was careerism that motivated him, surely he wouldn't have turned down three dinner invitations to the Reagan White House. That's like landing on Boardwalk for careerists. No, this was the man who routinely turned down invitations to dine with the rich and powerful and curry their favor. Instead, he's the guy who schleps to the inner city kindergartens and citizenship inductions and jury duty. No careerist would have resigned from the Air Force Science Advisory Board and surrendered his top security clearance in protest over the Vietnam War. He would have played the game at Harvard, and believe me, if that's what he wanted to do, he would have done it brilliantly. I never knew him once to keep his mouth shut about a matter of principle when it was in his self-interest. How do we square Carl as "the driven careerist" with his consistent

lifelong pattern of choosing a course that would be problematical for his career?[38]

We square it by recognizing and analyzing Sagan's exquisite balance between orthodoxy and heresy dictated, in part, by the essential balance between his high conscientiousness and high openness. No one is all of one personality trait all of the time. These traits are tendencies, and when they are in conflict we see such seemingly paradoxical behavior. But the paradox is resolved when put into the context of this personality dynamics model.

A WONDERFUL LIFE

Scientific biography can only take us so far. Humans are storytelling animals, and narrative biography is storytelling in the personal mode. Not only was Carl Sagan the preeminent scientific storyteller of our time, his life right up to the end was heroic in the best Homeric mode, as Druyan expressed it to me so poetically: "Even facing death and excruciating physical torture, Carl remained heroically rational. His samurai-like conduct, his grace throughout his harrowing two year illness and three bone marrow transplants—two years on the rack—is a demonstration of the authenticity of his perspective and character."[39]

How fleeting is our tenure on Earth, Carl might have said. We must make the most of it. Sagan certainly did. To quote George Bailey's guardian angel, Clarence Oddbody, from It's a Wonderful Life: "Strange, isn't it? Each man's life touches so many other lives, and when he isn't around he leaves an awful hole, doesn't he?"

Whether Carl Sagan's life is measured qualitatively (through narrative biography) or quantitatively (through scientific biography), he really had a wonderful life.

PART III
BORDERLANDS
HISTORY

I acknowledge that it is much more difficult to understand human history than to understand problems in fields of science where history is unimportant and where fewer individual variables operate. Nevertheless, successful methodologies for analyzing historical problems have been worked out in several fields. As a result, the histories of dinosaurs, nebulas, and glaciers are generally acknowledged to belong to fields of science rather than to the humanities. I am thus optimistic that historical studies of human societies can be pursued as scientifically as studies of dinosaurs—and with profit to our own society today, by teaching us what shaped the modern world, and what might shape our future.

—Jared Diamond, final paragraph,
Guns, Germs, and Steel: The Fates of Human Societies,
1997

II

THE BEAUTIFUL PEOPLE MYTH

Why the Grass is Always Greener in the Other Century

LONG, LONG AGO, in a century far, far away, there lived beautiful people coexisting with nature in balanced ecoharmony, taking only what they needed, and giving back to Mother Earth what was left. Women and men lived in egalitarian accord and there were no wars and few conflicts. The people were happy, living long and prosperous lives. The men were handsome and muscular, well-coordinated in their hunting expeditions as they successfully brought home the main meals for the family. The tanned bare-breasted women carried a child in one arm and picked nuts and berries to supplement the hunt. Children frolicked in the nearby stream, dreaming of the day when they too would grow up to fulfill their destiny as beautiful people.

But then came the evil empire—European White Males carrying the disease of imperialism, industrialism, capitalism, scientism, and the other "isms" brought about by human greed, carelessness, and short-term thinking. The environment became exploited, the rivers soiled, the air polluted, and the beautiful people were driven from their land, forced to become slaves, or simply killed.

This tragedy, however, can be reversed if we just go back to living off the land where everyone would grow just enough food for themselves and use only enough to survive. We would then all love one another, as well as our caretaker Mother Earth, just as they did long, long ago, in a century far, far away.

ENVIRONMENTAL MYTH-MAKING

There are actually several myths packed into this fairy tale, proffered by no one in particular but compiled from many sources as mythmaking (in the literary sense) for our time. This genre of mythmaking, in fact, tucks nicely into the larger framework of golden age sagas and has a long and honorable history. The Greeks believed they lived in the Age of Iron, but before them there was the Age of Gold. Jews and Christians, of course, both believe in the golden age before the fall in the Garden. Medieval scholars looked back longingly to the biblical days of Moses and the prophets, while Renaissance scholars pursued a rebirth of classical learning, coming around full circle to the Greeks. Even Newt Gingrich has his own version of the myth when he told the *Boston Globe* on May 20, 1995, that there were "long periods of American history where people didn't get raped, people didn't get murdered, people weren't mugged routinely."

I first encountered this—what I call the Beautiful People Myth (BPM)—in a graduate seminar co-taught by an anthropologist and a historian in the late 1980s, when both fields were being "deconstructed" by so-called postmodern literary critics and social theorists. Anticipating the kind of anthropology done in the 1970s when I last studied the science—the customs, rituals, and beliefs of indigenous pre-industrial peoples around the world—I was shocked, and soon dismayed, to find myself bogged down in such books as Michael Taussig's *The Devil and Commodity Fetishism in South America,* with such chapters as "Fetishism and Dialectical Deconstruction" and "The Devil and the Cosmogenesis of Capitalism."[1] I couldn't figure out what was going on until the anthropologist announced his was a Marxist interpretation of history that sees the past in terms of class conflict and economic exploitation (the Beautiful People lived before capitalism). Taussig's anthropology of indigenous peoples of South America proclaims:

> Marx's work strategically opposes the objectivist categories and culturally naive self-acceptance of the reified world that capitalism creates, a world in which economic goods known as commodities and, indeed, objects themselves appear not merely as things in themselves but as determinants of the reciprocating human relations that form them. Read this way, the commodity labor-time and value itself become not merely historically relative categories but social constructions (and deceptions) of reality. The critique of political economy demands the deconstruction of that reality and the critique of that deception.[2]

This was as clear as the waters of the Rio Negro to me. I gleaned from my professors' commentary on the book (for I simply could not get through it) that indigenous peoples lived in relative harmony with their environment until you-know-who came along. Fortunately our seminar was provided with some readable material that brought balance to the discussion, such as William Cronon's *Changes in the Land: Indians, Colonists, and the Ecology of New England*. Cronon restates the Beautiful People Myth and why we must resist its temptation:

> It is tempting to believe that when the Europeans arrived in the New World they confronted Virgin Land, the Forest Primeval, a wilderness which had existed for eons uninfluenced by human hands. Nothing could be further from the truth. . . . Indians had lived on the continent for thousands of years, and had to a significant extent modified its environment to their purposes. . . . The choice is not between two landscapes, one with and one without a human influence; it is between two human ways of living, two ways of belonging to an ecosystem. . . . All human groups consciously change their environments to some extent—one might even argue that this, in combination with language, is the crucial trait distinguishing people from other animals—and the best measure of a culture's ecological stability may well be how successfully its environmental changes maintain its ability to reproduce itself.[3]

In the early 1990s when I was team-teaching courses in cultural studies at Occidental College, I encountered two more versions of the BPM, both of which push the blame further back in time and place it in the hands of other entities. In Carolyn Merchant's *The Death of Nature: Women, Ecology and the Scientific Revolution,* the author points her finger at science when, "between the sixteenth and seventeenth centuries the image of an organic cosmos with a living female earth at its center gave way to a mechanistic world view in which nature was reconstructed as dead and passive, to be dominated and controlled by humans."[4] The prescientific organic model of nature, says Merchant, was as a nurturing mother, "a kindly beneficent female who provided for the needs of mankind in an ordered, planned universe."[5] Then the Dead White European Males (DWEMs) destroyed this organicism, and with it egalitarianism. Hierarchy, patriarchy, commercialism, imperialism, exploitation, and environmental degradation soon followed. To prevent doom, Merchant concludes, we will have to adopt a new lifestyle: "Decentralization, nonhierarchical forms of organization, recycling of wastes, simpler living styles involving less polluting 'soft' technologies, and labor-intensive rather than capital-intensive economic methods, are possibilities only beginning to be explored."[6]

In Riane Eisler's *The Chalice and the Blade*, the author goes back 13,000 years to find another bogeyman. Instead of DWEMs perhaps they should be called DMOACs: Dead Males of All Colors. Before the DMOACs there was "a long period of peace and prosperity when our social, technological, and cultural evolution moved upward: many thousands of years when all the basic technologies on which civilization is built were developed in societies that were not male dominant, violent, and hierarchic."[7] As paleolithic hunting, gathering, and fishing gave way to neolithic farming, this "partnership model" of equality between the sexes gave way to the "dominator model," and with it came wars, exploitation, slavery, and the like. The solution, says Eisler, is to return to the *equalitarian* partnership model where "not only will material wealth be shared more equitably, but this will also be an economic order in which amassing more and more property as a means of protecting oneself from, as well as controlling others will be seen for what it is: a form of sickness or aberration."[8] The Beautiful People Myth lives. Why?

The grass seems greener in the other century for the same reason it does on the other side: the very human tendency to want what we don't have, reinforced by a distorted and unfair comparison of the realities (warts and all) of what we do have. The BPM is just one manifestation of the greener-grass psychology, but it has special appeal to us because of a conjunction of two historical circumstances: (1) we know more about our past than anyone in history and, with the aid of mass communication and visual technologies, can envision that past—or at least a fantasy of it—as never before; (2) this normal historical fantasizing is exaggerated by the realities of modern overpopulation and environmental pollution. In other words, pressures on the environment are higher now than they were in the past, but that past was no idyllic Eden.

THE ENVIRONMENTAL BOUNDARY PROBLEM AND THE BEAUTIFUL PEOPLE MYTH

The problem with examining environmental claims of diverse sorts is that they range, in my classification system, from solid science to mushy pseudoscience, with plenty falling into the borderlands in between. The reason is that while there may be plenty of quality data available for any given problem, the interpretations and conclusions from those data typically carry a lot of political and ideological baggage, especially when policy decisions may be based on the find-ings and billion-dollar industries potentially effected. This boundary problem is especially vital to solve, which is why finding balance between experts is so vital. Left-wing scientists argue that the data indicates human-caused global warming and recommend harsh restrictions on industry. Right-wing scientists with their own data (and often with the same data as the left-wing scientists), draw the opposite conclusion with corresponding policy recommendations. What are nonexperts to think? Short of finding scientists with no wings, we need open and public debate so that we may draw the most reasonable conclusion or push both ends of the debate to the middle.

While political demogogues and apologists for certain business interests may want to debunk the BPM as a way of cutting off discussion of the environmental pressures of modern industrial society and its impact on indigenous cultures and peoples, that is definitely *not* my purpose. Rather, I want to examine the an-thropological and historical evidence that disproves the BPM and then show how holding to the myth in the face of contrary evidence actually stands in the way of our effectively solving environmental and societal problems.

In a fascinating 1996 study, University of Michigan ecologist Bobbi Low used the data from the Standard Cross-Cultural Sample to test empirically the prop-osition that we can solve our ecological problems by returning to the BPM attitudes of reverence for (rather than exploitation of) the natural world, and opting for long-term group-oriented values (rather than short-term individual values).[9] Her analysis of 186 Hunting-Fishing-Gathering (HFG) societies around the world showed that their use of the environment is driven by eco-logical constraints and not by attitudes (such as sacred prohibitions), and that their relatively low environmental impact is the result of low population density, inefficient technology, and the lack of profitable markets, not from conscious efforts at conservation. She also showed that in 32 percent of HFG societies not only were they not practicing conservation, environmental degradation was severe.

In his 1996 book, *War Before Civilization: The Myth of the Peaceful Savage*, University of Illinois anthropologist Lawrence Keeley examines one element of the BPM—that prehistoric warfare was rare, harmless, and little more than rit-ualized sport.[10] Surveying primitive and civilized societies, he demonstrates that prehistoric war was, relative to population densities and fighting technologies, at least as frequent (as measured in years at war versus peace), as deadly (as measured by percentage of conflict deaths), and as ruthless (as measured by killing and maiming of noncombatant women and children), as modern war.

One prehistoric mass grave in South Dakota, for example, yielded the remains of 500 scalped and mutilated men, women, and children—a full fifty years before Columbus ever left port.

UCLA anthropologist Robert Edgerton, in his 1992 book *Sick Societies: Challenging the Myth of Primitive Harmony*, surveys the anthropological record and finds clear evidence of drug addiction, abuse of women and children, bodily mutilation, economic exploitation of the group by political leaders, suicide, and mental illness in indigenous preindustrial peoples.[11]

Anthropologist Richard Wrangham, in his 1996 book *Demonic Males: Apes and the Origins of Human Violence* (coauthored by Dale Peterson), traces the origins of patriarchy and violence across cultures and through history all the way back to our hominid origins millions of years before the Neolithic Revolution.[12]

Finally, evolutionary biologist Jared Diamond, in his Pulitzer Prize-winning 1997 book *Guns, Germs, and Steel*, surveys history's broadest patterns over the past 13,000 years, showing that the relative differences in the development of environmentally exploitative technologies and attitudes were a function of bio-geographical differences between environments.[13] Those peoples who lived in environments with domesticable plants and animals were able to develop food-producing technologies when their population numbers exceeded the carrying capacity of their environment and their traditional hunting and gathering methods. Hunting, fishing, and gathering alters the environment, but not as much as herding, ranching, and farming; and small hunter-gatherer Pleistocene groups exploited their environment, but not as much as large agricultural neolithic groups. In terms of environmental exploitation and destruction, the only difference between us and our ancestors is the size of our numbers and the effectiveness of our technologies. There is nothing noble about Rousseau's "noble savages." Give them the plants, animals, and technologies (and the need through population pressures) to exploit their environment, and they would; indeed, those that had that particular conjuncture of elements, did just that.

In other words, centuries before and continents away from modern economies and technologies, and long before DWEMs, humans dramatically altered their environments. As we shall see below, the Beautiful People turned rich ecosystems into deserts (Southwestern America), caused the extinction of dozens of major species (North America, New Zealand), and even committed mass eco-suicide (Easter Island and possibly Machu Picchu).

The Beautiful People have never existed except in myth. Humans are neither Beautiful People nor Ugly People. *Humans are only doing what any species does to survive; but we do it with a twist—instead of our environment shaping us through natural selection, we are shaping our environment through human selection.* Since we have been doing this for several million years the solution is not to do less selecting, but higher-quality selecting based on the best science and technology available. Demythologizing the BPM is one place to start.

THE ECO-SURVIVAL PROBLEM

The story begins two to three million years ago when ancient hominids in Olduvai Gorge in eastern Africa began chipping stones into tools. Archaeological evidence reveals an environmental mess of bones of large mammals scattered amongst hundreds of stone tools probably abandoned after use—in other words, our hominid ancestors littered the place. It is not for nothing that the Leakeys called this hominid *Homo habilis*—the handy man.[14]

Around one million years ago *Homo erectus* added controlled fire to our technologies, and between half a million and 100,000 years ago *Homo neanderthalensis* and *Homo heidelbergensis* developed throwing spears with finely crafted spear points, lived in caves, and had elaborate tool kits. It appears that many hominid species lived simultaneously and at this point we can only guess what speciation pressures these changing technologies put on natural selection. Perhaps human selection was already at work on itself in a type of autocatalytic feedback loop.

Sometime around 30,000 to 35,000 years before present (BP), Neanderthals went extinct (for reasons hotly debated amongst paleoanthropologists) and Cro Magnons flourished. By now the tool kits were complex and varied, clothing covered bodies, art adorned caves, bones and wood formed the structure of living abodes, language produced sophisticated communication, and anatomically modern humans began to wrap themselves in a blanket of crude but effective technology.[15] The pace of technological change, along with the human selection and alteration of the environment, took another quantum leap. The leap, in fact, may well account for the demise of the Neanderthals and the triumph of Cro-Magnons.

From 35,000 to 13,000 BP humans had spread to nearly every region of the globe and all lived in a condition of hunting, fishing, and gathering. Some were nomadic, while others stayed in one place. Small bands grew into larger tribes, and with this shift possessions became valuable, rules of conduct grew more complex, and population numbers creeped steadily upward. Then, at the end of the last Ice Age roughly 13,000 years ago, population pressures in numerous places around the globe grew too intense for the HFG lifestyle to support. The result was the Neolithic, or Agricultural Revolution. The simultaneous shift to farming was no accident; nor does it appear to be the invention of a single people from whence it diffused to others. Farming replaced HFG in too many places too far apart for diffusion to account for the change. The domestication of grains and large mammals produced the necessary calories to support the larger populations. In other words, overpopulation triggered another quantum leap in human selection and alteration of the environment.[16]

Now an autocatalytic (self-generating) feedback loop was established—a 13,000-year-long complex interaction between humans and the environment, with the rate of change accelerating dramatically beyond what stone tools, fire, and HFG could ever do. Around the globe peoples of all colors, races, and cultures altered their environments to meet their needs. And these altered environments, in turn, changed how humans survived: some continued destroying their ecosystems, some moved, some went extinct.[17] By the first year of the Common Era (2,000 years ago) the globe was filled with humans living in one of five conditions depicted in FIGURE 41: (1) complex stratified agriculturalism, (2) simple peasant agriculturalism, (3) nomadic pastoralism, (4) general HFG, and (5) specialized HFG.

From the earliest civilizations at Babylon, Ur, Mesopotamia, and the Indus Valley, to Egypt, Greece, and Rome, all the way to the Early Modern Period, most people lived similar lifestyles: over 90 percent were farmers. They used the barter system or had crude forms of money. Small numbers of the elite had access to goods and services, but the majority did not.[18]

The problem these neolithic farmers faced was, at the most basic level, the same one their paleolithic HFG ancestors faced and, for that matter, the same one we face. I call this the Eco-Survival Problem: *since humans need environmental products to survive, how can we meet our needs without destroying our environment and causing our own extinction?* In other words, how can human selection continue without selecting ourselves into oblivion?

THE TRADE-OFF

One of the problems in shattering the myth of the Beautiful People is that the alternative would seem to imply that civilization was a universally progressive step in cultural evolution. If *they* weren't the Beautiful People, then *we* must be. Not necessarily. One of the mysteries for archaeologists and environmental historians to solve is why our ancestors made the shift from HFG to farming. Scientists in the 1960s, like Jacob Bronowski, see this step as the first great

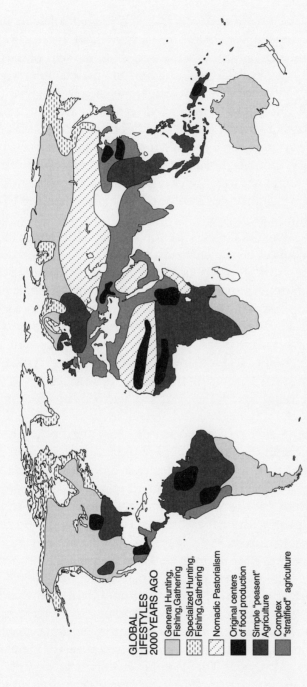

GLOBAL
LIFESTYLES
2000 YEARS AGO

General Hunting,
Fishing, Gathering

Specialized Hunting,
Fishing, Gathering

Nomadic Pastorialism

Original centers
of food production

Simple "peasant"
Agriculture

Complex
"stratified" agriculture

Figure 41. Global Lifestyles 2,000 years ago as agriculture gradually displaced hunting, fishing, and gathering and the rate of environmental change accelerated. Also noted are the original centers of food production. (Adapted from Roberts, 1989, 121.)

achievement in the "ascent of man."[19] In reality, if judged solely by health and longevity, paleolithic people were taller, bigger boned, ate better, lived longer, and had more free time than anyone from 13,000 BP to the start of the twentieth century. The average height of HFGers at 13,000 BP, for example, was 5'10" for men, 5'6" for women. By 6,000 BP the average dramatically dropped to 5'3" for men, 5'1" for women. Not until the twentieth century have heights again approached these marks, but still fall short.[20] Studies from modern HFG societies also show that they have more free time than neolithic farmers (or any farmers up to the Industrial Revolution). Kalahari bushmen, for example, invest between twelve and nineteen hours per week in food gathering and production, with an average of 2,140 calories and 93 grams of protein, higher than the FDA's Recommended Daily Allowance.[21]

If HFG is so great, why did humans take up farming? It requires more hours per day of work, it produces dependency on a narrower base of a less dependable food supply, and generates greater populations through which diseases spread more rapidly.[22] For starters, in many parts of the world the Neolithic revolution was really an *evolution*. According to Esther Boserup, "It apparently took ancient Mesopotamia over four thousand years to pass from the beginning of food production to intensive, irrigated agriculture, and it took Europe still longer to pass from the introduction of forest fallow to the beginning of annual cropping a few hundred years ago."[23] Still, in the grand historical sweep of the last 100,000 years, something big happened at the start of the neolithic that demands an explanation.

Archeologist Kent Flannery concludes from his digs in 10,000-year-old Mesopotamian village sites that humans turned to farming not to improve their diets or the stability of their food supply (since it did neither), but to increase the carrying capacity of the environment in response to larger populations. Small, local populations had grown large enough that they had exceeded the carrying capacity of their ecosystem and so turned to farming in order to produce enough calories to survive.[24] In his book *The Food Crisis in Prehistory: Overpopulation and the Origins of Agriculture*, Mark Cohen argues that the planet held as many people as could be supported by paleolithic technology.[25] Another way to say this is that the Beautiful People overpopulated and exploited their environment and so were forced to turn to technology to save themselves. Or as Alfred Crosby put it so well: "*Homo sapiens* needed, not for the only time in the history of the species, to become either celibate or clever. Predictably, the species chose the latter course."[26] The neolithic evolution was simply a human selection response to an Eco-Survival Problem. Whether the trade-off was worth it or not is irrelevant. We had to take that road to survive. We face a similar problem today.

ECOCIDE

Environmental history is the study of the effects of large-scale natural forces and contingent ecological events on human history, how human actions have altered the environment, and how these two forces interact.[27] This is not drum-and-trumpet history—wars and politics, generals and kings—the *proximate causes* of history. This is the study of the currents and eddies upon which we ride like flotsam and jetsam on the historical sea of change—the *ultimate causes* of history (see FIGURE 44).

Reconstructing environmental history reveals that the great rise and fall of civilizations, previously credited to "great men" or "class conflicts," was more often the result of human environmental exploitation or destruction. In each of four geographical locales we find a form of ecological suicide—*ecocide*—where the inhabitants failed to solve the Eco-Survival Problem; they were unable to meet the needs of their population without destroying their environment and thereby caused their own extinction. These four examples not only demonstrate that the BPM is wrong, they show what could be in store for us if the global rate of population growth is not checked, and if solutions to environmental problems caused by human selection are not found.

1. *New Zealand*. If anyone fits the image of a Beautiful People it is Polynesians, as portrayed in films living in an Eden-like condition of endless summers and timeless love. Environmental historians, however, paint a different portrait. When Europeans arrived in New Zealand in the 1800s, the only native mammal was the bat. But they found bones and eggshells of large extinct moa birds. From feathers and skeletal remains, we know moas were an ostrichlike bird of a dozen different species, ranging in size from three feet tall and forty pounds, to ten feet tall and 500 pounds. Preserved moa gizzards containing pollen and leaves of dozens of plant species give us a clue to the environment of New Zealand, and archeological digs of Polynesian trash heaps reveal that ecocide was well under way before the DWEMs arrived.[28]

Moas are believed to have evolved their flightless condition in New Zealand over millions of years of a predatorless environment. Their sudden extinction around the time of the arrival of the first Polynesians—Maoris—offers a causal clue. Although many biologists have suggested a change in climate as the cause, or Maori hunting as the last straw to an already drastically changing environment, Jared Diamond makes the case that when the extinction occurred New Zealand was enjoying the best climate it had in a long time.[29] If anything, the preceding Ice Age would be a more logical choice for an extinction trigger. Also, C^{14}-dated bird bones from Maori archaeological sites prove that all known moa species were still present in abundance when the Maoris arrived around 1000 C.E. By 1200 C.E. — six centuries before the arrival of Europeans — they were all gone. What happened?

Archaeologists have uncovered Maori sites containing between 100,000 and 500,000 moa skeletons, ten times the number living at any one time. In other words, they had been slaughtering moas for many generations, until they were all gone.[30] How did they do it so easily? As Darwin and hungry sailors discovered on the Galapagos Islands, animals that evolved in an environment with no major predators often have no fear of newly introduced predators, including humans. It would appear that moas were to the Maori what buffalos were to armed American hunters: sitting ducks. The result was that the Beautiful Maori People exterminated one of their major resources.

2. *Native America*. When anatomically modern humans crossed the Bering Strait from Asia into the Americas some 20,000 years ago (estimates vary considerably), they found a land teeming with big mammals: elephantlike mammoths and mastodons, ground sloths weighing up to three tons, one-ton armadillolike glyptodonts, bear-sized beavers, and beefy sabertooth cats, not to mention Native American lions, cheetahs, camels, horses, and many other large mammals. They are now all extinct. Why?

C. A. Reed has suggested that these species were unable to adapt during the period of rapid climactic change at the end of the last ice age.[31] But the weather was getting warmer, not colder, meaning that as the glaciers receded there were more niches to fill, not less; plus, comparable extinctions at the termini of previous ice ages resulted in no comparable megafaunal extinctions. Paul Martin and Richard Klein point to massive archeological "kill" sites where huge numbers of animal bones are found, accompanied by spear points buried in the rib cages of such animals as mammoths, bison, mastodons, tapir, camels, horses, bears, and others — the obvious remains of multiple species hunted into oblivion.[32] Since mammals adapted to both cold and warm weather went extinct, climate is an unlikely cause. G. S. Krantz, in balance, argues for a combination of climate and hunting as the trigger of the megafaunal extinctions, showing how human hunters also took over the niche of the carnivores they hunted, and in the process threatened the niche of such herbivores as the now extinct American Shasta ground sloth.[33] Either way — hunting alone, or hunting plus environmental change—the Beautiful Native American People were the ultimate cause since without the intervention of these sapient hunters such mass extinctions almost surely never would have occurred.

Archaeologists are also discovering that these indigenous Americans were no less destructive of their botanical resources. When the DWEMs first arrived in the American Southwest they found gigantic multistory dwellings (pueblos) standing uninhabited in the middle of treeless desert. Like many travelers, when I first visited these sites in Arizona, Colorado, and New Mexico I could not help but wonder how the Anasazi (Navajo for "Old Ones") could have survived in this desolate landscape. Pueblo Bonito in Chaco Canyon, New Mexico (FIGURE 42), is one of the most striking examples. Here you find a "D" shaped structure that was originally five stories high, 670 feet long, 315 feet wide, containing no less than 650 rooms and supporting thousands of people, all nestled in a dry, arid, treeless desert. How did these people make a living here?

Construction at Pueblo Bonito began around 900 C.E., but occupation terminated a scant two centuries later. Why? Well-meaning tour guides there will tell you that a drought drove the Anasazi out. David Muench dramatically closes his volume by concluding that the Anasazi "were a people escaping—from precisely what we cannot be sure—and they fanned out across the land like so many gypsies, carrying a few possessions on their backs and the cultural heritage of a thousand years in their heads."[34] We now have a fairly good idea of *precisely* what they were escaping from: self-generated ecocide. Archaeologists have calculated, for example, that the Anasazi would have needed well over 200,000 16-foot wooden beams to support the roofs of the multistoried Pueblo Bonito. Paleobotanists Julio Betancourt and Thomas Van Devender used packrat "middens" in Chaco Canyon to identify the fauna of the region before, during, and after the Anasazi occupation.[35] C[14] dating of the pollen and remains of plants in the middens reveals that when the Anasazi arrived in Chaco Canyon there was a dense pinyon-juniper woodland, with ponderosa pine forest nearby. This explains where the wood came from for building. As the population grew, the forest was denuded and the environment destroyed, leaving the desert we see today. As they destroyed their environment the Anasazi built an extensive road system to reach further for trees—upwards of fifty miles—until there were no more trees to cut. In addition, they built elaborate irrigation systems to channel water into the valley bottoms, but the erosion following the deforestation gouged out the land until the water table was below the level of the Anasazi fields, making irrigation impossible. Then, when the drought did hit, the

Figure 42. Pueblo Bonito in Chaco Canyon, New Mexico, represents the magnificent monumental architecture of the Anasazi, or "Old Ones," of North America. Today this structure lies in a dry, arid, treeless desert. What happened to the Old Ones? Unable to solve the Eco-Survival Problem, it appears they committed ecocide.

Anasazi were unable to respond and their civilization collapsed. Their ecocide was a direct result of their failure to solve their Eco-Survival Problem.

3. *Machu Picchu*. The closest I have ever come to having a mystical experience was a 1986 trip to Machu Picchu, the so-called "Lost City of the Incas" in the Andes Mountains in south central Peru. Nestled in a narrow saddle between two peaks at 9,000 feet altitude and fifty miles northwest of the 12,000-foot world's highest city, Cuzco, Machu Picchu is four and a half hours by train (or several days on foot), where you then take a circuitously steep dirt road to a tiny plateau hanging on the edge of a cliff. Clouds swirl around the adjacent peaks, and when dusk descends upon the stark ruins and the fog rolls in, you can almost sense the presence of the people who once carved out a living in this magnificent but harsh environment.

(The experience was enhanced by the fact that a terrorist organization, the Shining Path, had started a major prison riot to free their comrades. Many were killed on both sides. The head of the Peruvian military convinced the terrorists to surrender, whereupon he had over fifty of them murdered. In response, the Shining Path blew up the train to Machu Picchu the day after I was on it. Military troops roamed Cuzco and surrounded the airport. In order to leave I had to bribe an official to let me on the plane for which I was scheduled. Adding to the experience, New Agers believed that there was a planetary "harmonic convergence" that summer, so they converged on Machu Picchu, forming circles and chanting New Age mantras. Finally, to my shock I discovered too late that I was not booked at the pleasant Machu Picchu Hotel at the top of the mountain, but rather I was at the corrugated tin-roofed Hotel Machu Picchu down in the Urubamba River Valley next to the train station. My wife and I were rescued by a schoolteacher with retinitis pigmentosa who offered her spare bed to us if I would assist her to the top of the treacherously steep Huayna Picchu peak adjacent to Machu Picchu, from where the photographs in FIGURE 43 were taken.)

No wonder it took the Yale archeologist Hiram Bingham so long to find the place in 1911. He spent decades attempting to determine if this was the famed sixteenth-century last citadel of the escaping Inca leaders—Vilcambamba—concluding, over the opposition of most archaeologists, that it was. It appears that it was not. So what was Machu Picchu and what happened to the people who lived there?

What Hiram Bingham discovered was a five-square-mile city with a temple, citadel, about 100 houses, and agricultural terraces linked by more than 3,000 steps and an elaborate irrigation system carved into granite, for what appears

Figure 43. Machu Picchu, the "Lost City" of the Incas. What happened to the 1,000 + inhabitants of this remote city in the Andes of Peru? No one knows. It remains one of the great historical mysteries of our time. One possibility is that the population exceeded the carrying capacity of this very limited ecosystem and the Incas, unable to solve their Eco-Survival Problem, were forced to leave. Note the rather limited terraced farming that rises up the steep cliffs surrounding Machu Picchu. (Note also the encircled New Agers dressed in white chanting mantras in celebration of the "harmonic convergence" of the Earth's energies that day.)

to be an extremely limited system of farming.[36] The Incas had no cattle, horses, domestic pigs, poultry, or sheep, and most of their meat came from small animals such as guinea pigs, rabbits, and pigeons. Llamas were mostly beasts of burden and sources of wool rather than protein. The Incas were, therefore, highly dependent on agriculture. Yet here they had no native wheat or other cereals, no olives, rice, or grapes, and few green vegetables. Maize and potatoes were the primary source of calories so we can assume that this is what the Incas grew on these agricultural terraces surrounding Machu Picchu.[37] Strangely, of the 173 skeletons found there, about 150 of them were women (although others say the configuration is 135 skeletons, 102 females).[38] Whatever the exact numbers, it is unlikely that the people of Machu Picchu were the victims of war because of the natural defenses of the geography. The Spanish never knew about Machu Picchu, and archeologist Paul Fejos believes no defense was necessary as this was most likely a sacred religious city, not a military outpost.[39]

According to J. Hemming, Machu Picchu was not a last refuge, but an older city that flourished at the height of the Inca empire.[40] If so, and this point remains highly debated, what happened to the people of Machu Picchu? In light of what we have seen happen around the globe, particularly in places of limited agricultural and animal resources, it seems reasonable to consider the possibility that the extremely limited carrying capacity of Machu Picchu was exceeded by population and environmental pressures, and the people there were forced to abandon this resplendent outpost.

4. *Easter Island.* In 1722, the Dutch explorer Jakob Roggeveen came upon the most isolated hunk of real estate on the planet—an island 2,323 miles west of Chile, 4,000 miles east of New Zealand, and so remote that the nearest island is 1,400 miles away (Pitcairn, the desolate rock where the Bounty mutineers took refuge). What Roggeveen found when he arrived on Easter Sunday (hence the name), were hundreds of statues weighing up to 85 tons and standing as tall as 37 feet. It appeared that these statues were carved from volcanic quarries, transported several miles, raised to an upright position without metal, wheels, or even an animal power source. Oddly, many of the statues were still in the quarries, unfinished. The whole scene looks as if the carvers quit in the middle of the job.

How and why did these Polynesian peoples carve and raise these statues and, more importantly, what happened to them? Thor Heyerdahl was shown by modern islanders that their ancestors used logs as rollers to transport the statues, and then as levers to erect them.[41] Piecing together the history of the Easter Islanders from archeological and botanical remains, it appears that around the time of the fall of Rome—400 C.E.—eastward migrating Polynesians discovered an island covered by a dense palm forest that they gradually but systematically proceeded to clear in order to create usable land for farming, use logs for boats, and for transporting statues from the quarry to their final destination.[42] Between 1100 and 1650 C.E. the population had reached 7,000, all living in a fairly dense 103 square miles. The Easter Islanders had carved upwards of 1,000 statues, 324 of which had been transported and erected. By the time Roggeveen arrived the forests had been destroyed and not a single tree stood. What happened in between?

As archeologist Paul Bahn and ecologist John Flenley conclude in their provocatively-titled book *Easter Island, Earth Island*, the Easter Islanders committed ecocide. "We consider that Easter Island was a microcosm which provides a model for the whole planet."[43] Initial deforestation led to greater population, but this triggered massive soil erosion, resulting in lower crop yields. Palm fruits would have been eaten by both humans and rats (initially introduced for food), attenuating the regeneration of felled palms. Without palms and palm fruits, the rats would have raided seabird nests, while humans would have eaten both birds and eggs. No logs for boats meant less fishing and this, coupled to limited land, led to starvation, internecine warfare, and cannibalism. At that point a warrior class took over, battle spear points were manufactured in huge quantities and littered the landscape. The defeated peoples were enslaved, killed, and some were even eaten. With no logs or ropes there would have been no point in carving additional statues, or finishing the ones already started. The statue cult lost its appeal, rival clans pulled down each other's statues, and the population crashed, leaving only a handful by 1722.[44]

The lesson is clear but especially disturbing when you consider that on an island 10 by 11 by 13 miles it would have been impossible to miss the obvious fact that the last palm trees were about to disappear forever. They had to know it was the end of their most important resource, yet they did nothing to stop it. The Easter Islanders were not the Beautiful People, but neither were they any worse than the DWEMs. It would appear this is a very human problem. Is Easter Island a microcosm for Earth Island?

WHAT WILL WE DO?

Change in physical, biological, and human systems is inevitable, and the science of history records this change. Humans have been altering their environment for millions of years. As soon as a stone tool was chipped or a wooden spear carved, the step toward ecological change by human selection began. Living in hunting, fishing, and gathering societies, as populations grew the pressures and changes on the environment increased. It is not that there was just more change, but that the *rate* of change accelerated. Humans caused the extinction of large numbers of species for tens of thousands of years. Civilization has accelerated the *rate* of change even more, and for the last 10,000 years peoples of all races and geographic locales have significantly altered their environments.

Humans have been successful in changing the environment for productive uses that have led to a higher standard of living and a richer, more diverse lifestyle. We have also changed it for destructive uses leading to the extinction not only of whole species, but of whole peoples. Change cannot be stopped without stopping history, because *change is history*. And as chaos and complexity theory have shown, small changes early in a historical sequence can trigger enormous changes later. So many quirky contingencies construct determining necessities, making it virtually impossible to reverse the change once it is under way.[45] Once the channel of change is dug, it is almost impossible to jump the berm to another channel. The question is: what type of change will be triggered by human actions, and in what direction will it go? (See FIGURE 44.)

As for our future, it is very difficult to legislate historical change because of the impossibility of determining the consequences of legislative actions. Which change do we prohibit and which do we allow? Since all human actions cause change in the environment, once we start down the road of prohibiting actions that cause change, where do we stop? Obviously the majority of us have no desire to return humanity to a hunting, fishing, and gathering state. Nor could the environment support our population under such a condition. We are animals, it's true, but we are *thinking* animals. All technologies alter the environment—from stone tools to nuclear power plants. Once we start down the road of technologically-triggered change, there is no turning back. But we can go forward in a new direction.

One solution to environmental problems is more and better science and technology, and the application of them to solve problems older science and technologies caused. My libertarian inclinations make me resistant to supporting government intervention to solve the problem. Free-market solutions are already available to many of these free-market-caused problems, if only the market could be allowed to act in a truly free manner. Yet my historical and scientific training leads me to fear that unrestricted free markets could result in a planetary ecocide—Easter Island writ large. Perhaps the risks run too high for us to place our trust in the free market.

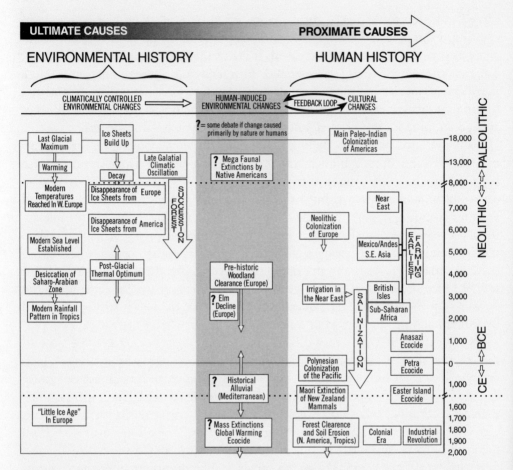

Figure 44. Twenty-two thousand years of environmental and human changes. Time flows from top to bottom, noting the two major breaks at the Neolithic and the Common Era. Climatically-controlled environmental changes are a part of environmental history and are the deeper, ultimate causes of change. Human-induced environmental changes are a part of both environmental history and human history, and this section, in a feedback loop with cultural changes, denotes the shift to proximate causes of change. (Adapted from Roberts, 1989, 183.)

Are we on the verge of imitating our ancestors who were unable to solve their Eco-Survival Problems? There is compelling evidence that overpopulation, pollution, global warming, the ozone hole, chemical poisoning, and many more could threaten our very existence. But there is equally good evidence that we can adapt and solve problems. There is no need to cry doom yet. But we should be alert. If we are going to legislate change, it should be based on the best scientific knowledge available. The politics of this most political of sciences has muddied the issues. We need more data. It would seem that a truly "wise use" of government funding would be for more and better environmental science in order to determine with high confidence what needs to be legislated, where, and when. Will we be like the Easter Islanders standing there staring at the last palm tree, and say "screw the future, let's cut the damn thing down"? Or will we heed the lessons of history and find a solution to our own Eco-Survival Problem? There is a difference between us and all those who failed to find this solution. We are the first to realize the consequences of our actions in time to do something about them. The question is, what will we do?

12

THE AMADEUS MYTH
Mozart and the Myth of the
Miracle of Genius

BEFORE READING ANY FURTHER try to answer these questions: what five-letter word do all college graduates spell wrong? How is it possible that our basketball team won a game last week by the score of 73–49, and yet not one man on the team scored as much as a single point? How can it be that a man in a certain town in the United States married 20 women from the town, yet he broke no law, he is not a Mormon, and they are all still alive?

If you answered W-R-O-N-G for the first question, that the basketball team was an all-woman's team for the second question, or that the man is a minister for the third question, you likely experienced what is called the "aha" reaction of insight in problem solving. This is one theory of genius—the answer inexplicably pops into the ingenious mind from on high, a mental miracle from the muses. Einstein stumbles onto relativity theory while dreaming of riding on a beam of light. Kekule discovers the structure of the benzene ring by dreaming of a snake biting its tail. Darwin suddenly becomes an evolutionist while visiting the Galapagos Islands. Wallace discovers natural selection while in a feverish fit of malaria in the Malay Archipelago. Evariste Galois, out of fear of a foreshortened life, pens the entirety of his mathematical group theory the night before he was killed in a duel over a woman. Newton flashes onto universal gravitation when beaned by an apple. Coleridge creates his brilliant poem *Kubla Khan* one afternoon during an opium-induced altered state of consciousness. And, perhaps best known of all, Mozart composes perfect symphonies on first draft—no corrections, additions, or deletions needed—a miraculous masterpiece.

The only problem with this scenario of the development of genius is that none of these stories are true. Creative genius simply does not happen this way. This is what I call the *Amadeus Myth*, patterned on the portrayal of Mozart as a mysterious musical miracle in Peter Shaffer's and Milos Forman's film *Amadeus*. The Amadeus Myth is *the belief that genius and original creations are produced by mysterious mental miracles limited to a special few.* Its source is ancient, dating back to the ancient Greeks who believed that the Muses, or gods, *breathed* creativity into certain individuals, *inspiring* them (Latin for *breathe in*) to genius. One of my graduate school professors—Harry Liebersohn—told us first-year history students that when we write most of us will be Beethovens, rewriting draft after draft until we get it right; few of us could ever hope to be Mozarts, with a perfectly crafted paper the first go around. The Amadeus Myth persists because it provides a parsimonious explanation and makes for a more compelling story than a more complex and nuanced model allows. And, anecdotely, it is the experience of most of us that when we read about, or even personally witness a genius in action, we rarely if ever see the haulting steps from neophyte amature to polished professional.

There is, in fact, neither scientific nor historical evidence to support the Amadeus Myth. Studies by cognitive psychologists do not corroborate the claim that somehow geniuses are qualitatively different from the rest of us. The above anecdotes are cast in doubt once historical details fill in the gaps of faulty memory and selective presentation. These gaps in our scientific and historical knowledge are filled with myths. The "aha" solution is the exception, not the norm. Or at the very least, the "aha" comes at the end of a very long train of thought—the darkened room slowly illuminated by the gathering dawn of knowledge, not instantly lit by the flip of an ingenious switch.

Consider the following brainteaser: *When the music stopped the girl died. . . .* Asking only questions that can be answered yes or no, one is to explain what the sentence means. This problem takes most people 30 to 45 minutes to solve. The solution is that the girl is a blind tightrope walker performing in a circus. Her cue that she is at the end of her (literal) rope is the termination of the music, at which time she steps off onto the platform. Her murderer stopped the music early and she stepped off to her demise. This puzzle circulated through the dorms at Pepperdine University, and as a student of psychology there I kept informal tabs on times and mental strategies used to solve the problem. Most people begin with questions that attempt final explanations. When these fail, they begin to organize their questions into categories of career, location, number of people involved, the link between the music and what the girl was doing, the nature of her health, and so forth. Once the general categorical answers are established the questions become more specific until the solution begins to unfold—health problems? She was blind. Career? She works in a circus. Specific action? She was walking a tightrope. The final "aha" comes only after all the pieces are in place, and the final connection is made between the girl, the music, and her death.

This *continuous*, step-by-step process of problem solving (as opposed to a *discontinuous*, leap-by-leap process called for by the Amadeus Myth) appears to be the way all of us operate, dullard and genius alike. The difference is to be found more in the *quantity* of mental power and less on the *quality*. That is, a genius is not so much *qualitatively* different, with some brain module or cognitive process absent in the rest of us; rather, the genius is *quantitatively* different—faster, more efficient, with a refined and systematic memory, focused attention, and especially years of honed practice invisible to all but a few of the genius's closest circle. This is not to deny the existence of something that might reasonably be called genius. Like great art, it may be difficult to define but we know it when we see it. Einstein, Newton, Darwin, Mozart—these were geniuses. But did they have something the rest of us do not, or just a lot more of what we all have? Evidence from both history and the cognitive sciences lead us to conclude that it is more of the latter than the former. Another way to look at it is that geniuses have so much more of what others have, and they used it so much more effectively, and do so in a cultural setting that would allow its expression and appreciation. Thus, one way to define a genius is: *an individual with a quantitative difference in ability so great that he or she appears to be qualitatively different.* That is, a genius is an individual of such high ability that he or she seems wholly different from and better than the rest of us. The *mystery* of how a genius operates is what creates the Amadeus Myth, and the deconstruction of the Amadeus Myth helps us to better understand how, in particular, science and scientists work, and, in general, how thinking and thinkers operate. This insight into the scientific and thinking process aids us further into our explorations of the difference between science, borderlands science, and nonscience, because making the transition from one level to another often requires a spark of creative genius, a new way of thinking about a problem.

THE SCIENCE OF GENIUS

A good place to start on the scientific analysis of genius is Robert Weisberg's 1986 *Creativity: Genius and Other Myths*.[1] Weisberg presents scientific evidence from the most recent studies by cognitive psychologists on the nature of creativity and genius as a quantitative difference. Weisberg cites studies on chess masters in which their cognitive styles and thought patterns were discovered not to differ in kind from average players. Intuitively we assume that chess masters consider more moves than average players. Psychologist Adriaan DeGroot, however, discovered just the opposite: chess masters actually consider significantly fewer moves, but the ones they examine are the most important moves.[2] And how do they select the key moves? After a pivotal move in a championship match that confounded the experts into predicting his defeat, Chess Master Bobby Fisher was asked "How could you foresee the success of that apparently disastrous move? What was your reasoning?" Fisher responded, "I don't know. It just felt right."[3]

This typifies the mystery of cognitive functioning that creates the Amadeus Myth. The reality is more prosaic. Psychologists William Chase and Herbert Simon estimate that chess masters acquire about 50,000 patterns of four or five pieces in each, from which they analyze most positions faced in most games.[4] At first glance these numbers seem miraculously large. It is more easily understood when one considers that over the course of ten years of intense play, one can easily accumulate 25,000 hours playing time. At two moves per hour, the individual can acquire the necessary knowledge and skills to become a chess master. There is nothing miraculous about it. Anyone in any field, in the course of ten years of practice, will naturally put in this much time. I recall joining a chess club as a young teenager and, for a few months anyway, playing four to five hours every day. (Then the baseball season started—for the aspiring chess master, however, the whole year is chess season.) One day a chess master came to our club and played 15 of us simultaneously, winning every game. He would spend no more than a couple of seconds per move at each board, almost as if he had seen all the patterns before. I thought at the time that he must be some kind of genius; now I know how he did it. Of course, the difference between an average player and master, and master and champion, is where the quantitative difference becomes so great that it becomes, de facto, a qualitative difference—Bobby Fisher is wholly different from the rest of us.

I witnessed further evidence of this quantitative analysis of genius when I coauthored the book *Mathemagics* with lightning calculator Arthur Benjamin.[5] The subtitle is revealing: *How To Look Like a Genius Without Really Trying*. When you watch Art perform his mathematical magic he looks like a genius born with a special gift, an ersatz Rainman plucking numbers out of thin air. But Art was not born with it. He learned it over years of practice—from early childhood

Art has been doing mental calculations. When you observe him computing a multiplication problem of two different 5-digit numbers, the answer appears to come to him in a flash. But when he explains how he does it you realize he is not doing anything more than most people could do with considerable practice. That is, if you know your basic multiplication tables and you practice (and practice and practice) the techniques in Art's book, you could master the skill. Where Art becomes a mathematical Other—a mathemagician as he is called— is in the speed and adaptability he has acquired over the years. There is virtually no problem anyone can give him that holds any surprises. He can take any 3-, 4-, or 5-digit number and almost instantly reduce it to a simpler multiplication problem. He uses a mnemonic system to convert numbers into words to store them in memory while he computes another problem, then converts the words back into numbers in a later step in the problem. He has done this so often that the conversion process is second nature.

When we were working on the book together Art used me as his test subject to see if he really could teach mathematical magic to a relative neophyte, and if a novice really could learn his system. It had been over a decade since I had done any serious mathematics (and that was mostly statistics, which has little application to what Art does), so it was a fair test of his program. It was a little anxiety-producing for me, for failure would not only be personally embarrassing, it would signal to us that our book was a waste of time and that Art really was an Amadeus-like mathematical wizard. Fortunately for Art's program (and my ego), I was able to master the techniques to the point where I could even square three-digit numbers through a variety of tricks. (One of Art's favorites is to round a number up or down to simplify the problem, then add or subtract the difference at the end. For example, to square 108 Art would round down to 100 and up to 116, multiply 100×116, with is simply 11,600, then add back the square of the number rounded up and down, or $8 \times 8 = 64$, giving a total of 11,664. Simple!) And for a couple of months I challenged people to test my mental calculation skills. But after the book was published and I moved on to other projects, someone asked me to compute a simple problem and I couldn't do it. Without constant practice the skills slip away like gossamer wings, further evidence of the fleeting nature of such geniuslike abilities that are crucially dependent on routinized algorithms perfected through practice.

Cognitive psychologist Dean Keith Simonton's research on genius, creativity, and leadership[6] has opened up new vistas into the nature of this elusive human trait, culminating in a Darwinian perspective on the problem in his 1999 work *Origins of Genius.*[7] Simonton argues that creative genius is best understood as a

Darwinian process of variation and selection. Just as nature produces a vast array of variation upon which natural selection to select for those variations most conducive to survival and reproductive success, creative geniuses generate a massive variety of ideas from which they select only those most likely to survive and reproduce. As the two-time Nobel laureate and scientific genius Linus Pauling observed, one must "have lots of ideas and throw away the bad ones. . . . You aren't going to have good ideas unless you have lots of ideas and and some sort of principle of selection."[8] For Simonton, qua contra Forest Gump, genius is as genius does: "these are individuals credited with creative ideas or products that have left a large impression on a particular domain of intellectual or aesthetic activity. In other words, the creative genius attains eminence by leaving for posterity an impressive body of contributions that are both original and adaptive. In fact, empirical studies have repeatedly shown that the single most powerful predictor of eminence within any creative domain is the sheer number of influential products an individual has given the world."[9]

In science, for example, the number one predictor of receiving the Nobel Prize is the rate of journal citation. As well, Simonton notes, Shakespeare is a literary genius not just because he was good, but because "probably only the Bible is more likely to be found in English-speaking homes than is a volume containing the complete works of Shakespeare." And in music, ápropos my thesis, Simonton notes that "Mozart is considered a greater musical genius than Tartini in part because the former accounts for 30 times as much music in the classical repertoire as does the latter. Indeed, almost a fifth of all classical music performed in modern times was written by just three composers: Bach, Mozart, and Beethoven."[10]

I saw a splendid example of this variation-selection process while conducting a content analysis of the curriculum vitae of eminent scientists for a research project on Carl Sagan and quantitative biography (see chapter 10). When going through the c.vs of Carl Sagan, Stephen Jay Gould, and Jared Diamond, hundreds of essays, articles, reviews, and commentaries in each one reveal how they created their greatest works. When reading Jared Diamond's Pulitzer Prize-winning *Guns, Germs and Steel*, for example, one is almost dumbfounded by the scope and depth of this exceptional book. Drawing on findings from the sciences of paleoanthropology, archaeology, history, history of technology, anthropology, sociology, linguistics, biogeography, zoology, botany, evolutionary biology, evolutionary psychology, and sociobiology, Diamond synthesizes them into a radical new theory about the origin and development of civilization. It is hard to believe that one person, even a genius, could have written such a book . . . until you look through Diamond's c.v. Virtually every chapter in *Guns, Germs and Steel* appeared as articles, essays, and papers published in *Nature, Natural History, Discover*, and other journals and magazines. Through a remarkable variation of articles Diamond not only honed and refined his thinking and writing on the subject from these different fields, he selected from hundreds of his own works those most likely to survive and reproduce in the marketplace of ideas. Genius is as genius does, and Diamond did it; but he did it through Darwinian hard work, not Mozartian miraculous musings.

Of course, quantity alone does not make one a genius. The *quality* of the quantity matters. And how the selection process is applied by both the genius, and the marketplace for the expression of that genius, determines which products survive and which go extinct. But how does one come to produce a high quantity of quality products that constitute genius?

To get at an answer, Robert and Michele Root-Bernstein address "the thirteen thinking tools of the world's most creative people" in their book *Sparks of Genius*.[11] They begin with the "aha" experience of geneticist Barbara McClintock and her sudden insight into why a particular field of corn generated only a third sterile pollen when the expected figure was half. Retreating to the top of a hill for half an hour of contemplative thought, she suddenly "jumped up and ran down to the field. At the top of the field (everyone else was down at the bottom) I shouted 'Eureka, I have it! I have the answer! I know what this 30 percent sterility is.'" When asked later to explain how she arrived at her explanation, she reflected: "When you suddenly see the problem, something happens that you have the answer—before you are able to put it into words. It is all done subconsciously. This has happened many times to me, and I know when to take it seriously. I'm so absolutely sure. I don't talk about it, I don't have to tell anybody about it. I'm just sure this is it."[12] McClintock described this process as having a "feeling for the organism."

But what does it mean to have an insight, or to intuit an answer, to have a "feeling" about something? Insight, intuition, and feeling are just words to describe a thinking process we as yet know little about. In reality, psychologist James Greeno has shown that people respect others more if they appear to solve problems with "real problem solving abilities" versus "mere knowledge."[13] Much of this respect, however, was due to an inability on the part of the observer to identify the "mere" knowledge used by the "real" problem solver. Greeno says that experts are able to solve problems better than novices, not so much because of their problem-solving abilities (the genius of their thinking) as because of their knowledge and expertise in that area. Once again, endless hours of practice trumps strap-it-on-and-go genius.

Simonton summarizes the research on so-called "intuitive cognition" in five points: (1) "the human mind can acquire a vast set of expectations in the absence of any awareness of the basis for those expectations," (2) "unconscious associations can greatly influence the course of thought in the absence of any conscious intervention," (3) "the unconscious material in the mind is often stored in multiple ways, including in manners that are unusual, if not illogical," (4) "unconscious mental operations may be particularly useful to the solution of problems that require creative insight," and (5) "unconscious mental processes can even support 'feeling of knowing' states that are comparable to the unjustified 'hunches' that are so often reported by creative individuals."[14]

One of the appealing aspects of the Amadeus Myth is the mystery that surrounds those who appear wholly Other. Ignorance of the step-by-step mental operations and the year-by-year accumulation of knowledge creates awe and wonder. But when you know how it's done the mystery disappears and the genius is humanized, becoming quantitatively, not qualitatively different. As cognitive scientists learn more about the mental processes of creative genius, the curtain of mystery will be pulled back to reveal the inner workings that are not so different from what the rest of us do. This is one reason why magicians have a sanction against revealing their secrets, and why they never do the same trick twice for the same audience. Without the mystery there is no magic. Magic is an example of the foiling of the Amadeus Myth. Skilled professional magicians are so adept at their craft that they appear almost divine in their ability. It looks like a miraculous gift, not an acquired skill. Yet as I got to know a number of world-class magicians such as Penn and Teller, Jamy Ian Swiss, Banachek, and Randi, I came to realize that the gift of magic comes only through countless hours of practice, and in most cases it starts at a very early age. It is not at all uncommon for professional magicians, like their counterparts in almost any other profession, to put in thirty to forty hours of practice a week, in addition to shows and performances, which for active professionals can range from 150–300 a year. How do you get to Carnegie Hall? The answer is trite, but true nontheless, and it quashes the Amadeus Myth—practice, practice, practice.

Thus far, in studies by physiological, cognitive, and personality psychologists, no characteristic has been found in the brain, mind, or person of the genius that does not exist in the rest of us. Does this mean the genius is no different from the ordinary person? Of course not. The differences are enormous, but quantifiable. The other side of the Amadeus Myth is that since we all have the same qualities as the genius, genius is a matter of degree and thus one's abilities can be improved, limited only by genetic makeup and environmental influences. We can rise to the upper ceiling of our abilities by modeling what some geniuses do. Social scientist Frank Sulloway identified three characteristics in Charles Darwin's personality that accentuated his creative genius: (1) he respected others' opinions but was willing to challenge authorities (Darwin intimately understood the theory of special creation, yet he overturned it with his own theory of evolution by natural selection); (2) he paid close attention to negative evidence (Darwin included a chapter called "Difficulties on Theory" in the *Origin of Species*—as a result his opponents could rarely present him with a challenge that he had not already encountered, let alone addressed); (3) he generously used the work of others (Darwin's collected correspondence numbers over 14,000 letters, most of which include lengthy discussions and question-and-answer sequences about scientific problems). Darwin was constantly questioning, always learning, confident enough to formulate original ideas, yet modest enough to recognize his own fallibility. Sulloway shows why this combination of traits is so special:

> Implicit in these three qualitative aspects of Darwin's genius is his ability to tolerate an unusual degree of what Thomas Kuhn has called 'the essential tension' that is necessary for doing good scientific research. That tension involves a willingness to trust authority (and hence to be guided by the accepted theoretical beliefs of one's age) and a simultaneous ability to engage, whenever necessary, in revolutionary science. Usually, it is the scientific community as a whole that displays this essential tension between tradition and change, since most people have a preference for one or the other way of thinking. What is relatively rare in the history of science is to find these contradictory qualities combined in such a successful manner in one individual.[15]

Following Darwin's methodology does not guarantee the production of creative genius, of course. As Sulloway shows in his book *Born to Rebel*, creativity, openness to new ideas, and the ability to balance that essential tension between conservatism and radicalism, is mitigated by the development of personality, itself shaped by a combination of genetics, family dynamics, peer influences, and cultural forces.

Nevertheless, from this research into genius, there are certain steps we can take to cultivate our creative gardens:

1. Be sensitive to the important problems in a field to be solved, and ignore those that are unimportant or unsolvable.
2. Educate yourself with as much knowledge of a subject as is available, and if a skill is involved, practice, practice, practice.
3. Respect both the knowledge and experts in the field, but do not be afraid to challenge anyone or anything. There are no sacred cows.
4. Look for new ways to solve old problems and try something no one else has ever done. Think outside the box. Do not fear ridicule. Listen to your critics, but do not let them dictate your thinking.
5. Communicate your new ideas with others in the field. Intellect dies in isolation. The conflation of old ideas into new configurations comes from outside stimulation.
6. Generate lots and lots of creative products to give yourself an extensive variety from which to select those most likely to survive and reproduce in the cultural marketplace.[16]

Can one learn to be a genius? Not exactly. One can certainly learn to be more creative, to work harder, to accumulate knowledge about a problem, and model the behavior of known geniuses. But it would be naive not to recognize the inherent limitations that nature and culture sets on most of us. There is a big difference between reaching one's personal upper ceiling and attaining the greatest limits of human achievement. It is the rare individual—the genius—whose personal limitations happen also to be those of humanity as a whole. That is, the methods of geniuses may be ordinary, but the results are spectacular. The difference between Salieri and Mozart may only be quantitative, but what a difference!

THE HISTORY OF GENIUS

Historical evidence for genius is just as important as laboratory research, because history provides a record of what geniuses actually did. As case studies in genius, we will examine the creative lives of Einstein, Kekule, Darwin and Wallace, Galois, Coleridge, Newton, and, of course, Mozart, to peel back the layers surrounding the Amadeus Myth and reveal what lies inside.

Einstein. Arguably the most famous "aha" story about Einstein's genius is his dream about riding on a beam of light. But it is important to realize that not everyone can have this dream, or be inspired by it to create a new model of the universe. On December 14, 1922, Einstein gave a talk at Kyoto University in Japan on "How I Created the Theory of Relativity."[17] In it you can see the step-by-step process of reasoning and problem solving that is anything but the stereotypical lightbulb going "ping" in the head. "It is not easy to talk about how I reached the idea of the theory of relativity," Einstein recalled, because "there were so many hidden complexities to motivate my thought, and the impact of each thought was different at different stages in the development of the idea." He then took himself back seventeen years to confess "I cannot say exactly where that thought came from," but he is "certain that it was contained in the problem of the optical properties of moving bodies." Looking for experimental evidence for the propagation of light through the ether, "I came to know the strange result of Michelson's experiment. Soon I came to the conclusion that our idea about the motion of the earth with respect to the ether is incorrect. This was the first path which led me to the special theory of relativity."[18]

Einstein then explained that he read Lorentz's monograph of 1895 on electrodynamics but was puzzled by how electrons move within a frame of reference that is itself moving and, it seemed to him, "the concept of the invariance of the velocity of light . . . contradicts the addition rule of velocities used in mechanics." Did Einstein *then* suddenly hit upon a solution? No. "I spent almost a year in vain trying to modify the idea of Lorentz in the hope of resolving this problem." Then, after a year, he visited his friend Michele Besso in Bern:

> We discussed every aspect of this problem. Then suddenly I understood where the key to this problem lay. Next day I came back to him again and said to him, without even saying "hello," "Thank you. I've completely solved the problem." An analysis of the concept of time was my solution. Time cannot be absolutely defined, and there is an inseparable relation between time and signal velocity. With this new concept, I could resolve all the difficulties completely for the first time. Within five weeks the special theory of relativity was completed.[19]

Notice that the "aha" came only after a lengthy period of time spent working on the problem. The general theory of relativity has a similar pathway of discovery. Though Einstein says "the idea occurred suddenly," if you read further you see that it really did not. First he says, "I was sitting on a chair in my patent office in Bern. Suddenly a thought struck me: If a man falls freely, he would not feel his weight. I was taken aback. This simple thought experiment made a deep impression on me. This led to the theory of gravity." Einstein then explained how a falling man does not feel his weight because in his frame of reference there is a new gravitational field that cancels the Earth's gravitation. But then he confesses that "it took me eight more years until I finally obtained the complete solution. During these years I obtained partial answers to this problem."[20] In other words, continuity, not discontinuity; Beethoven, not Mozart.

Kekule. With Friedrich August von Kekule's discovery of the benzene ring of six carbon atoms, we see once again that the history and mythistory of genius do not match. Kekule described his now famous dream:

> I turned my chair to the fire and dozed. Again the atoms were gamboling before my eyes. This time the smaller groups kept modestly to the background. My mental eye, rendered more acute by repeated vision of this kind, could now distinguish larger structures, of manifold conformation; long rows, sometimes more closely fitted together; all twining and twisting in snakelike motion. But look! What was that? One of the snakes had seized hold of its own tail, and the form whirled mockingly before my eyes. As if by a flash of lightning I awoke. Let us learn to dream, gentlemen.[21]

There are two problems with this description. First, something may have been lost in the translation from the German, in which the word "Halbschlaf" is translated as "doze" but may more accurately mean "half-sleep." This is more likely a case of daydreaming, where one is lost in thought, not asleep and dreaming. And by "dream" Kekule may very well have meant a type of thinking, as in "I'm dreaming of winning the Nobel prize." In the sentence subsequent to the above quote, in fact, Kekule warns his audience to "take care not to make our dreams known before they have been worked through by the wakened understanding." Of course, working through a problem is a far less romantic story than an inspirational dream.

Second, the reference to a snake biting its own tail follows Kekule's description of a "snakelike" motion, meaning he is most likely speaking figuratively, drawing an analogy between something known and something unknown, a process everyone uses in solving problems—the solar system is like a giant clock; memory is like a hologram. Metaphor and analogy are useful problem-solving tools, but they can cloud our understanding of how the creative mind works, and there is nothing miraculous about them.

Darwin and Wallace. The myth of Darwin discovering evolution while in the Galapagos has now been thoroughly shattered by Frank Sulloway, who shows conclusively that even after the five-year voyage of the *Beagle*, it was not until he returned to England that Darwin's creationism was slowly displaced by his evolutionism. It is true, he did notice the differences among both birds and tortoises on the different islands of the Galapagos, but at the time he did not consider it important enough to carefully log those differences. In fact, it was not until he returned home that Darwin realized the importance of such geographical variation, and he had to turn to notes and samples taken by others on the ship including, ironically, the staunch creationist captain of the *Beagle*, Robert Fitz-Roy. The conversion happened gradually, not suddenly. Sulloway explains: "Contrary to the legend, Darwin's finches do not appear to have inspired his earliest theoretical views on evolution, even after he finally became an evolutionist in 1837; rather it was his evolutionary views that allowed him, retrospectively, to understand the complex case of the finches."[22] Stephen Jay Gould finds a lesson on genius in Sulloway's debunking of the Amadeus Myth of Darwin's discovery, by comparing the myth with the actual history of Darwin:

> The first version upholds the romantic and empirical view that genius attains its status from an ability to see nature through eyes unclouded by the prejudices of surrounding culture and philosophical presupposition. The vision of such pure and unsullied brilliance has nurtured most legends in the history of science and purveys seriously false views about the process of scientific thought. Human beings cannot escape their presuppositions and see "purely"; Darwin functioned as an active creationist all through the *Beagle* voyage. Creativity is not an escape from culture but a unique use of its opportunities combined with a clever end run around its constraints.[23]

The case of Alfred Russel Wallace's discovery of the mechanism of evolution is similarly fraught with myth and mystery. Forty years after the event, Wallace recalled that during a feverish attack of malaria in the Malay Archipelago in 1858, "something led me to think of Malthus's *Essay on Population* and the 'positive checks' which he adduced as keeping all savage populations nearly stationary." He continued:

> It then occurred to me that these checks must also act upon animals, and keep down their numbers; While vaguely thinking how this would affect any

species, there suddenly flashed upon me the idea of the *survival of the fittest* —
that the individuals removed by these checks must be, on the whole, inferior
to those that survived. Then, considering the variations continually occurring
in every fresh generation of animals or plants, and the changes of climate,
of food, of enemies always in progress, the whole method of specific modification became clear to me, and in the two hours of my fit I had thought
the main points of the theory.[24]

Selective memory, however, can be misleading. This is why historians are
leery of autobiographical sources and late-in-life reconstructions of earlier
events. Wallace's reference to the "survival of the fittest" is a phrase coined by
Herbert Spencer in 1861. Also, by 1858 Wallace had already spent four years in
the Amazonian jungle and another four in Malaya, all the while considering the
transformation of species in relation to their geographical variation. The "aha"
came only after many years of painstaking collecting and exhaustive thought on
the problem. (See chapter 13 for details on the Darwin-Wallace codiscovery of
natural selection.)

Galois. The tragic story of the French mathematician Evariste Galois (1812–
1832), killed at the age of twenty in a duel over "an infamous coquette," is
legendary in the annals of the history of mathematics. A precociously brilliant
student, Galois lay the foundation for a branch of mathematics known as group
theory. Legend has it that he penned his theory the night before the duel,
anticipating his demise and wanting to leave his legacy to the mathematical
community. Hours before his death, on May 30, 1832, Galois wrote to Auguste
Chevalier: "I have made some new discoveries in analysis. The first concerns the
theory of equations, the others integral functions." After describing these he
asked his friend: "Make a public request of Jacobi or Gauss to give their opinions not as to the truth but as to the importance of these theorems. After that,
I hope some men will find it profitable to sort out this mess."[25]

As we have seen, however, romantic legend and historical truth do not always
match. What Galois penned the night before his death were corrections and
editorial changes to papers that had been accepted by the Academy of Sciences
long before. Further, Galois's initial papers on the subject had been submitted
three years prior to the duel, when he was only seventeen! It was after this that
Galois became embroiled in political controversy, was arrested, spent time in a
prison dungeon and, ultimately, found himself mixed up in a dispute over a
woman that led to his demise. Aware of his own precocity, Galois noted: "I
have carried out researches which will halt many savants in theirs." For over a

century that proved to be the case, but the researches were years in the making, not a fear-of-death-inspired dash of brilliance.

Coleridge. The genius of artists may seem different than scientists, but the same mythological themes are found in both, and the story of Samuel Taylor Coleridge's creation of his famous poem *Kubla Khan* is a classic example of the Amadeus Myth. One day, as the story goes, Coleridge was feeling "indisposed" and so took his prescribed dose of two grams of opium, after which he fell asleep in a chair while reading *Purchas's Pilgrimage*. The book is about exotic locals, in which one passage reads "Here the Khan Kubla commended a palace to be built, and a stately garden thereunto. And thus ten miles of fertile ground were enclosed with a wall."[26] Then, Coleridge recalls, he fell asleep and the poem appeared in his mind, as he describes it in the third person:

> The Author continued for about three hours in a profound sleep, at least of the external senses, during which time he had the most vivid confidence that he could not have composed less than from two to three hundred lines; if that indeed can be called composition in which all the images rose up before him as things, with a parallel production of the concurrent expressions, without any sensation or consciousness of effort. On awakening he appeared to himself to have a distinct recollection of the whole, and taking his pen, ink, and paper, instantly and eagerly wrote down the lines that are here preserved. At this moment he was unfortunately called out by a person on business from Porlock, and detained by him above an hour, and on his return to his room, found, to his no small surprise and mortification, that though he still retained some vague and dim recollection of the general purport of the vision, yet, with the exception of some eight or ten lines and images, all the rest had passed away like the images on the surface of a stream into which a stone has been cast, but alas! without the after restoration of the latter![27]

Before challenging the authenticity of the poet's own recollection, let us examine the statement itself. First, while Coleridge does say he was asleep, he says nothing about dreaming; rather that he had "the most vivid confidence," whatever that is. Further, when he says that "images rose up before him as things, with a parallel production of the concurrent expressions," does he mean just fuzzy images or does he mean actual lines of poetry, or both? When he awoke he wrote down what he recalled, but was he then creating poetry from images, or was he, in essence, taking dictation from his unconscious mind? This too is not clear.

As a plausible scenario, it would not be unreasonable to expect that vague images could give inspiration to poetry. Again, consider what Coleridge read in *Purchas's Pilgrimage:* "Here the Khan Kubla commended a palace to be built, and a stately garden thereunto. And thus ten miles of fertile ground were enclosed with a wall." As he falls asleep the image presents itself in his mind—a normal process of incorporating external stimuli into dreams, as when you dream of a song only to awaken and discover that the song is playing on the radio. Coleridge awakens, recalls the imagery, then writes:

> In Xanadu did Kubla Khan
> A stately pleasure-dome decree:
> Where Alph, the sacred river, ran
> Through caverns measureless to man
> Down to a sunless sea.
> So twice five miles of fertile ground
> With walls and towers were girdled round:
> And here were forests ancient as the hills,
> Enfolding sunny spots of greenery.

Could the lines themselves have appeared word-for-word in his dreams, recalled hours later to perfection then copied down? From what we know of human memory and the dreaming process, this seems unlikely. Further, according to Elisabeth Schneider, another version of the poem has been discovered that is slightly different from the final version in such a way that she feels it was written earlier. If written before the dream, then the above story is a fabrication. If written after, then it did not appear to him in a final form, attenuating the importance of the dream. In addition, Schneider thinks Coleridge's dream was more like "a sort of Reverie," not induced by opium, which does not have this effect on mental processes; rather more like a daydream. Finally, Schneider reports that Coleridge had a reputation for exaggeration and falsehoods, especially regarding his own work.[28]

For Coleridge, *Kubla Khan* was not complete and, says Schneider, he might have failed in his attempt to bring closure to the project. If an incomplete fragment is part of a larger whole forever lost inside the mind of the poet, however, mystery and wonder surround it and it then takes on greater significance. But why would Coleridge, or anyone else for that matter, want to fabricate the origin of his or her work? One reason, and a compelling one at that, is that the Amadeus Myth lofts the individual into genius status. The mystery surrounding one's original ideas and creations makes it seem like they came from the Muses, or gods of inspiration, making one godlike.

Newton. Everyone knows the story of Newton and the apple. Most physics students have at least heard a rough outline of Newton's *annus mirabiles* — his miracle year from 1665–1666, when, escaping plague-stricken London, he returned home to Woolsthorpe where his mind was allowed to reflect and then generate his most brilliant ideas. The myth of the miracle year originates, not surprisingly, from Newton himself, and most later writers quote Newton's own manuscript description:

> In the beginning of the year 1665 I found the Method of approximating series & the Rule for reducing any dignity of any Binomial into such a series. The same year in May I found the method of Tangents of Gregory & Slusius, & in November had the direct method of fuxions & the next year in January had the Theory of Colours & in May following I had entrance into y[e] inverse method of fluxions. And the same year I began to think of gravity extending to y[e] orb of the Moon & (having found out how to estimate the force with w[ch] [a] globe revolving within a sphere presses the surface of the sphere) from Keplers rule of the periodical times of the Planets being in sesquialterate proportion of their distances from the center of their Orbs, I deduced that the forces w[ch] keep the Planets in their Orbs must [be] reciprocally as the squares of their distances from the centers about w[ch] they revolve: & thereby compared the force requisite to keep the Moon in her Orb with the force of gravity at the surface of the earth, & found them answer pretty nearly. All this was in the two plague years of 1665–1666. For in those days I was in the prime of my age for invention & minded Mathematicks & Philosophy more then at any time since.[29]

Much has been made of this passage, but it must be kept in mind that it was written fifty years after the fact, and we have already seen how easily the memory of a creation can be altered by the creator. What actually happened is that intellectually he "departed from Cambridge more than a year before the plague drove him away physically," says Newton's biographer, Richard Westfall, who has demonstrated conclusively that Newton had already taken critical steps toward the calculus by the spring of 1665, before the plague, in addition to penning two important papers in May 1666, after he had returned. "If we focus our attention on the record of his studies," says Westfall, "the plague and Woolsthorpe fade in importance in comparison to the continuity of his growth." In actual fact, Westfall argues, "when 1666 closed, Newton was not in command of the results that have made his reputation deathless, not in mathematics, not in mechanics, not in optics. What he had done in all three was to lay foundations, some more extensive than others, on which he could build with assurance, but nothing was complete at the end of 1666, and most were not even close to complete."[30]

But there is a critically important lesson to learn from proper contextual history of science. It is, in Westfall's words, this: "Far from diminishing Newton's stature, such a judgment enhances it by treating his achievement as a human drama of toil and struggle rather than a tale of divine revelation."[31] Newton himself invoked the metaphor of the room slowly illuminated from above when he said: "I keep the subject constantly before me and wait 'till the first dawnings open slowly, by little and little, into a full and clear light."[32]

On a larger scale, the Newton story brings us back to the definition of genius as an individual so quantitatively different as to be, de facto, qualitatively different, even if sharing common features. Westfall, himself a brilliant scholar and renowned historian, opened his biography of Newton with this enlightening observation:

> The more I have studied him, the more Newton has receded from me. It has been my privilege at various times to know a number of brilliant men, men whom I acknowledge without hesitation to be my intellectual superiors. I have never, however, met one against whom I was unwilling to measure myself, so that it seemed reasonable to say that I was half as able as the person in question, or a third or a fourth, but in every case a finite fraction. The end result of my study of Newton has served to convince me that with him there is no measure. He has become for me wholly other, one of the tiny handful of supreme geniuses who have shaped the categories of the human intellect, a man not finally reducible to the criteria by which we comprehend our fellow beings.[33]

The genius, so far ahead and thus so far removed from the rest of us, appears wholly Other and thus a mystery that invokes the Muses of the gods.

Mozart. Finally we come to Amadeus himself—the myth that best characterizes this theme. There is a telling incident at the beginning of Peter Shaffer's and Milos Forman's film *Amadeus*, when the Italian composer Antonio Salieri, now aged and decrepit in a mental institute, asks his consulting priest: "Do you know who I am?" The priest responds: "It doesn't matter. All men are equal in

God's eye." With a contradictory smile Salieri retorts with the simple question, "Are they?," knowing they are not. For Salieri, Mozart was a vehicle for divine providence. "It seemed to me I was hearing the voice of God." This was further confirmed when Salieri first laid eyes on Mozart's works, brought to him by Mozart's wife Constanza in an appeal to obtain for her husband gainful employment. Salieri could not believe his eyes. "These were first and only drafts of music but they showed no corrections of any kind. Not one. He had simply written down music already finished in his head, page after page, as if he were just taking dictation." Reinforcing the myth, Salieri drops Mozart's music in shock. "Is it not good?," Constanza inquires. "It is miraculous," gasps Salieri.

Is it? What do we really know about Mozart's composing genius? Well, his music is, by almost everyone's standard, magnificent and ingenious. But *how* did he compose? An oft-cited passage from Mozart himself is revealing:

> When I feel well and in a good humor, or when I am taking a drive or walking after a good meal, or in the night when I cannot sleep, thoughts crowd into my mind as easily as you could wish. Whence and how do they come? I do not know and I have nothing to do with it. Those which please me, I keep in my head and hum them; at least others have told me that I do so. Once I have my theme, another melody comes, linking itself to the first one, in accordance to the needs of the composition as a whole.[34]

Though many writers claim that Mozart composed complete works without preparation and without further refinement, the fact is Mozart's notebooks contain many fragmentary compositions that were revised, or returned to later. Some were never completed. Further, Mozart admits that he has not a clue where his musical thoughts come from or how they are constructed. He only knows the ones he likes, which he reinforces through humming, discarding the rest. This is a good description of Simonton's Darwinian model of genius and creativity: variety and selection. Humming these melodies before writing them down is a form of practice and editing. A musician with a keen ear and memory for notes could easily construct whole pieces in the mind alone, making alterations along the way, and then write the whole down later, giving the appearance of "dictating" a musical miracle. One of my colleagues at Occidental College, music professor Richard Grayson, in fact, annually gives an improvisional performance in which he requests members of the audience to call out a musical theme, which he then incorporates into several other styles. For example, one year the theme from Sesame Street was to be played in a number of musical styles, including baroque! Grayson would take three or four different requests (with many of which he was unfamiliar, like the theme from *Star Trek*, which the student had to hum or whistle to give him the basic muscial structure), pace back and forth for a minute or so, then sit down at the piano and stare at the keys for about thirty seconds. It was obvious from his actions that he was playing the themes in his mind, after which he would play a complete piece — no mistakes, no editing — as if it just popped into his head.

The point is that *how* Mozart did his composition was (and is) not unusual. *What* he did with his composition is what makes him a genius. We must remember too that as a child prodigy Mozart was exposed to music and musical theory from the earliest age possible by his zealous father. Though Leopold Mozart described his son as the "miracle which God let be born in Salzburg," the fact is the father left nothing to God or chance. Himself a composer, Leopold had the young Amadeus composing minuets by age six, a symphony by age nine, his first oratorio at eleven, and at twelve his first opera. By the time Salieri met Mozart, the young man had already composed many hundreds of pieces. By then composing came almost second nature to him.[35] In addition, musicologist R. Baker estimates that approximately eighty percent of Mozart's melodies also appear in the compositions of his contemporaries, a practice apparently not unusual at the time.[36] Finally, John Hayes has provided powerful evidence that learning plays an important role in composition, and in going through all of Mozart's works he concluded that his early ones were of lower quality than his later ones, basing the criteria of quality on the number of recordings of it that were available (i.e., how much interest there was in the particular work).[37] Since Mozart was well-known throughout Europe from an early age, his fame was held constant (as an experimental variable) throughout his life. Thus, Hayes concludes, Mozart too had to practice and learn.

This conclusion is reinforced by the following excerpt from a letter Leopold Mozart wrote to his son dated February 16, 1778, when Wolfgang was twenty-two years old. It is at once illuminating about the intensity and concentration with which the young Mozart would practice his craft, and prophetic at the tragedy of his early death at age thirty-five: "As a child and a boy you were serious rather than childish and when you sat at the clavier or were otherwise intent on music, no one dared to have the slightest jest with you. Why, even your expression was so solemn that, observing the early efflorescence of your talent and your ever grave and thoughtful little face, many discerning people of different countries sadly doubted whether your life would be a long one."[38]

Was Mozart a genius? Of course, but not because of the Amadeus Myth, but because he really was so much better than almost everyone else, and worked so much harder, and from an earlier age that he appears, to them and to us, wholly Other, a genius.

13
A GENTLEMANLY ARRANGEMENT
Science at its Best in the Great Evolution Priority Dispute

WHEN THE NINETEENTH-CENTURY British naturalist Alfred Russel Wallace returned home from eight years in the jungles of the Malay Archipelago in the Spring of 1861, he boasted an almost unbelievable collection of 125,660 total specimens, including 310 mammals, 100 reptiles, 8,050 birds, 7,500 shells, 13,100 butterflies, 83,200 beetles, and 13,400 "other insects." In addition to collecting, Wallace also wanted to put the seemingly infinite variety of nature's pieces together into a puzzle so that as a historical scientist he could solve the riddle of what his friend and colleague Charles Darwin called the "mystery of mysteries"—the origin of species. It was this combination of broad observational scope and penetrating theoretical depth that set Wallace apart from most of his contemporaries and led him to his discovery about the mutable nature of species and the interdependency of organisms in their geographical location. Wallace was demonstrating the practice of science at its best—the blending of process and product into an art form described by Sir Peter Medawar as "the art of making difficult problems soluble by devising means of getting at them."[1]

The story of how Charles Darwin and Alfred Russel Wallace discovered the mechanism of evolution demonstrates how the progress of science is often an interaction between steady historical trends and serendipitous flashes of insight and discovery. As Thomas Kuhn and others have shown, the history of science is not an asymptotic curve of progress toward Truth, or a steady withdrawal of the shroud cloaking Reality. Rather, it consists of long periods of paradigmatic status quo, occasionally interrupted by shifts in the shared worldview, resulting in a new and different way of interpreting nature. The particulars of a specific historical event, however, do not always fit Kuhn's universal concept, as each is unique to itself (in fact, Ernst Mayr and others argue that Kuhn's general model of how science changes fits almost no particular science).[2] Because of the contingent nature of history, no two paradigms or paradigm shifts are ever the same. The history of the independent discovery of natural selection by Darwin and Wallace, and the resolution of the ensuing priority dispute, provides a case study in the scientific process, how an idea shifts from heretical science to normal science, and the interactive nature of contingency and necessity in history.

AN EARLY SKEPTIC

The life of Charles Darwin is well known, documented, and presented in countless narrative biographies. The life of Alfred Russel Wallace lies in Darwin's shadow and is correspondingly less well known. To understand the generalities of science and history we must study the particulars, so I wish to spend a few moments with Wallace because his life (as we saw in chapters 7 and 8) was so remarkable, made even more so when it intersected with the lives of the other intellectual giants of his age, including and especially Darwin.

Alfred Russel Wallace was born in Usk, Monmouthshire, Wales, the eighth child of working-class parents who were devout members of the Church of England. Though his father was apprenticed to a solicitor (preparing legal cases), and for a brief time earned a very respectable 500 pounds a year, his unsuccessful entrepreneurial ventures left the family in difficult economic circumstances. Wallace remembered that "we were all of us very much thrown on our own resources to make our way in life."[3] Education at Hertford Grammar School for seven years turned out to be Wallace's only formal schooling, which he remembered as essentially valueless. Wallace did not let this fact interfere with his self-education, as he explored the natural world on his own. It was during this period, he later recalled, that he "began to feel the influence of nature and to wish to know more of the various flowers, shrubs, and trees I daily met."[4]

When Wallace was thirteen the family finances collapsed. The now six-foot-tall, brown-haired, blue-eyed "little Saxon" was sent to London to live and work with his older brother John, who was apprenticed to a master builder. Wallace's earliest significant intellectual pursuits came from evenings spent with his brother "at what was then termed a 'Hall of Science' . . . really a kind of club or mechanics' institute for advanced thinkers among workmen, and especially for the followers of Robert Owen, the founder of the Socialist movement in England."[5] The Mechanics' Institutes, founded in London and Glasgow the year he was born, were custom-designed for intellectually restless working-class adults such as Wallace. It was here that the intellectual foundation was laid for a lifetime of science, borderlands science, and pseudoscience. His religious and philosophical leanings and political-economic theorizing also found their roots at the Institutes, where all manner of radicals, heretics, and embracers of fringe beliefs congregated. "Here we sometimes heard lectures on Owen's [socialist] doctrines, or on the principles of secularism or agnosticism, as it is now called." The intellectual explorations knew no bounds, as this was when "I also received my first knowledge of the arguments of sceptics, and read among other books Paine's *Age of Reason*."[6]

Wallace's participation in the Mechanics' Institutes was emblematic of a growing class of individuals without money or formal education who were determined to enter the technological and scientific trades. Science periodicals of the time reflected this new "Republic of Science," where anyone could become his own scientist (sadly it was still mostly "he"), adding a brick here or a stone there in the overall edifice of knowledge and truth. Amateurs could participate in the unsullied joy of discovery previously monopolized by an elite minority. These working-class affiliations were radically different from the privileged upbringing and education experienced by many of Wallace's later colleagues, such as Darwin and Lyell, and would play an influential role in the development of his ideas, particularly as they diverged from these other thinkers. An early interest in astronomy, for example, in the twilight of his career, was conflated with his lifelong interest in the evolution of life in his book *Man's Place in the Universe*.[7] Anticipating the arguments made by "intelligent design" creationists a century later, Wallace argued that only a higher spiritual intelligence could account for the complexity and delicately balanced state of the cosmos. Many of Wallace's earliest interests hatched from these societies and institutes and were carried through to the end of his life.

Wallace called this time of scientific exploration "the turning point of my life." While he grew intellectually rich, however, he remained economically poor. In 1843 his father died and the family scattered. The following year Alfred applied for and received a teaching post at Reverend Abraham Hill's Collegiate School at Leicester, where he met the soon-to-be famous entomologist Henry Walter Bates. Both of modest means, Bates and Wallace took a liking to each other, and developed a close friendship that would culminate in a joint venture to South America. Bates introduced Wallace to the importance of variety in nature, particularly the abundant diversity of insect species locally—an estimated 10,000 varieties in a circle of only ten miles! Wallace added literary discoveries to his entomological ones, including Darwin's *Voyage of the Beagle* and "perhaps the most important book I read," Malthus's *Essay on Population*, the "main principles" of which "remained with me as a permanent possession."[8]

Like most naturalists, Wallace read the anonymously written *Vestiges of the Natural History of Creation* (by Robert Chambers) and was intrigued by the author's hypothesis that "the simplest and most primitive type, under a law to which that of like-production is subordinate, gave birth to the type next above it, that this again produced the next higher, and so on to the very highest, the stages of advance being in all cases very small." The highest, of course, was man, placed there ultimately by "Providence."[9] No pure materialist was Chambers, who stated unequivocally: "God created animated beings, as well as the terraqueous theatre of their being, is a fact so powerfully evidenced, and so universally received, that I at once take it for granted."[10] The idea stuck in Wallace's mind.

Wallace's religious views in these early years, almost conspicuous by their absence in his later ventures into spiritualism, were already nontraditional. "What little religious belief I had," he recalled, "very quickly vanished under the influence of philosophical or scientific scepticism." While living with his brothers, particularly William, "though the subject of religion was not often mentioned, there was a pervading spirit of scepticism, or free-thought as it was then called, which strengthened and confirmed my doubts as to the truth or value of all ordinary religious teaching."[11] Wallace's skepticism had two sources: (1) His experiences at the Mechanics' Institutes and his reading of the socialist Robert Owen led him to believe that "the only true religion is that which preaches service to humanity and brotherhood of man." (2) His involvement in science sounded the death knell of his already weak religious beliefs. "In addition to these influences my growing taste for various branches of physical science and my increasing love of nature disinclined me more and more for either the observances or the doctrines of orthodox religion, so that by the time I came of age I was absolutely non-religious, I cared and thought nothing about it, and could be best described by the modern term 'agnostic'."[12]

THE EVOLUTION OF A HERETIC

In 1847, while in London to wrap up some business affairs of his brother, Wallace visited the Insect Room of the British Museum of Natural History. He was so impressed with the collection that he approached Bates with a joint venture proposal for an expedition up the Amazon to, in Bates's words, "gather facts, as Mr. Wallace expressed it in one of his letters, 'towards solving the problem of the origin of species,' a subject on which we had conversed and corresponded much together."[13]

On April 20, 1848, Bates and the 25-year-old Wallace left England, the latter pocketing his entire life savings of £100, gambling that it would carry him through until he could begin selling his anticipated Amazonian specimens to collectors back home. It did. The first shipment back to England consisted of 400 butterflies, 450 beetles, and 1,300 other assorted insects. The pecuniary reward soon followed, and the expedition was given its first of many financial boosts. By 1850 the men were a thousand miles up the Amazon and on March 26 they split up, Bates to explore the Solimoens, or Upper Amazon, Wallace the Rio Negro and the unknown Uaupes. Alfred was then joined by his brother Herbert and, among other activities, they occasionally "mesmerized" willing native subjects, a skill Wallace had learned in Leicester in 1844 from Spencer Hall (and one he would find useful in later odysseys into the world of spiritualism).

In time, malaria and other diseases overcame Wallace and in the summer of 1852 he headed for home, but the real adventure was about to begin. Wallace recounted the dramatic events that then unfolded on his return voyage, to the readers of the *Zoologist*.[14] On August 6 at 9:00 a.m., "smoke was discovered issuing from the hatchways." The ship had caught fire and "the smoke became more dense and suffocating, and soon filled the cabin, so as to render it very difficult to get any necessaries out of it."[15] These necessaries included his notes, journals, and collections. By noon the flames had spread on deck and the crew scrambled to launch the lifeboats which, "being much shrunk by exposure to the sun, required all our exertions to keep them from filling with water."[16] The following morning they hoisted a small sail and steered for Bermuda, a full 700 miles away. After ten days they were rescued by the London-bound *Jordeson* from Cuba, but with her complement now doubled she began to run short of food and water. Finally, after eighty days on the open ocean, they made port, Wallace having managed to save "my watch, my drawings of fishes, and a portion of my notes and journals." The tragedy was that "most of my journals, notes on the habits of animals, and drawings of the transformation of insects were lost," not to mention ten species of river tortoises, 100 species of Rio Negro fishes, skeletons and skins of an anteater and cowfish (Manatus), and living monkeys, parrots, macaws, and other birds, all "irrecoverably lost."[17]

Despite this disaster, Wallace and Bates would boast of a remarkable assortment of 14,712 species of insects, birds, reptiles, and other variegated biological items, approximately 8,000 of which had never before been seen in England or Europe. There is no way no know how much sooner Wallace might have discovered the mechanism of species transformation had he had his detailed notes and specimens upon his return. On the other hand, perhaps he would not thus have been properly motivated to travel to the Malay Archipelago where he did, in fact, make his discovery. How contingent life can be, making it ever so difficult to predict what might happen next (or what might have happened if contingencies had unfolded differently). In any case, the trip was emblematic of Wallace's event-filled life of heretical science, adventure, and unrestricted speculation.

A STRIKING COINCIDENCE

At the age of thirty-one Wallace left for the Malay Archipelago, where he invested eight years of intense observation and reflection in a variety of locales. In 1855, for example, he went to Sarawak, the northwest region of Borneo, where he collected 320 different species of beetles in fourteen days. (Many of these species still carry his name, such as *Ectatorphinus wallacei* and *Cryiophalpus wallacei*.) More important, it was here that Wallace formulated the *Sarawak Law*, presented in a paper entitled "On the Law which has Regulated the Introduction of New Species." Using a series of arguments based on geographical and geological evidence, Wallace concluded: *"Every species has come into existence coincident both in time and space with a pre-existing closely allied species."*[18] The paper appeared in *The Annals and Magazine of Natural History* in September 1855, and was read by Charles Darwin, who told Wallace that he agreed with "almost every word."

During this period of ambitious collecting, cataloging, and synthesizing, Wallace barely survived, often weak, sick and starving, not to mention poor. In early 1858 he caught a ship to Ternate and Gilolo, in the Moluccas, where there was, as he told Bates on January 25, "perhaps the most perfect entomological *terra incognita* now to be found. I think I shall stay in this place two or three years, as it is the centre of a most interesting and almost unknown region."[19] One night, while stricken with malaria, trembling, delirious, and fearing for his life, it occurred to Wallace that death would befall individuals unequally throughout a species. The stronger, healthier, and faster would be more likely to survive while the less favored would die. Thus, there was a selection for and against certain individuals, depending on the variety of their characteristics. That eve-

Figure 45. Alfred Russel Wallace

ning Wallace "sketched out the draft of a paper," and in two nights penned his complete theory in an essay entitled "On the Tendency of Varieties to Depart Indefinitely from the Original Type." He promptly "sent it by the next post to Mr. Darwin," which apparently was on March 9, 1858. Most likely Darwin received Wallace's essay on June 18. It contained, for Darwin, an all too familiar theme that "there is a general principle in nature which will cause many varieties to survive the parent species, and to give rise to successive variations, departing further and further from the original type. . . ."[20] Darwin was stunned. "I never saw a more striking coincidence," he told Lyell in a letter dated the 18th (also

Figure 46. Linnean Society Meeting Room

presumed to be June), "if Wallace had my manuscript sketch written out in 1842, he could not have made a better short extract! Even his terms now stand as heads of my chapters."[21] At the Linnean Society meeting of July 1, 1858, Darwin and Wallace were both awarded credit for the discovery.

A QUESTION OF PRIORITY

The matter of who was first in the discovery and description of natural selection has essentially remained unresolved for two reasons: (1) the letter and essay from Wallace to Darwin in the spring of 1858 is missing, making empirical resolution impossible; (2) the pugnacious zero-sum game (win-lose) model of priority held by scientific communities does not recognize the interactive and social nature of the scientific process.

Wallace's codiscoverer status with Darwin is generally accepted by most biologists and historians. The question some raise, however, is this: should Wallace be given even more credit? In his maximally tendentious 1980 work, *A Delicate Arrangement*, journalist Arnold Brackman makes an emotional appeal for Wallace's case.[22] Brackman suggests that Charles Lyell and Joseph Hooker, with Darwin's knowledge (but not his direction), conspired to negate Wallace's credit, while simultaneously boosting Darwin's. Specifically, Brackman claims that Darwin received Wallace's letter and essay earlier than the announced June 18 date, and that he might have spent that time fleshing out the missing pieces of his theory from Wallace's essay, then feigned distress at Wallace's parallel ideas. Motive, of course, is virtually impossible to prove, but the chronological sequence can be analyzed.

The strongest associative evidence we have is another letter sent by Wallace to Frederick Bates, the younger brother of his naturalist colleague and Amazon companion Henry Walter Bates. The letter is assumed to have been sent the same day, March 9 (mail ships were not a daily occurrence), and appeared in London on June 3. The clearly-dated, postmarked letter (no envelope—the letter itself was addressed and postmarked), is in possession of Wallace's grandson, Alfred John Russel Wallace, and available for study. In the letter Wallace tells Bates of the seemingly incoherent diversity of insect coloration in the Malay, and notes that "such facts as these puzzled me for a long time, but I have lately worked out a theory which accounts for them naturally."[23] That theory, "lately worked out," was the essay sent to Darwin, the original autograph manuscript and cover letter of which is, tragically, nowhere to be found.

A DELICATE ARRANGEMENT OR A GENTLEMANLY ONE?

Thomas Huxley's son Leonard called the Wallace situation "a delicate arrangement."[24] Arnold Brackman argues that since Darwin had been working on his theory for twenty years, and that because he was an established scientist with a recognized role within the scientific community, when this young amateur naturalist appeared on the verge of scooping his senior, Lyell and Hooker determined that if Darwin was not given the lion's share of the credit, no one would accept Wallace's theory. The "delicate arrangement" was, according to Brackman, as follows: Wallace was not part of the traditional scientific community in England (owing to his working-class background and lack of formal university training), and since he spent most of his professional life (thus far) outside the country, it was necessary for there to be an organized conspiracy by the intellectual elite surrounding Darwin to lessen the value of Wallace's contribution.

Wallace, with a working-class mentality, deferred to his superior. Brackman explains:

> No matter how heinous is a conspiracy, the participants—especially if it is successful—are apt to develop a plausible rationale for gilding it. "I do not think that Wallace can think my conduct unfair in allowing you and Hooker to do whatever you thought fair," Darwin wrote to Lyell. The message was clear: Lyell and Hooker bore historical responsibility for the cover-up. Darwin did not "allow" Lyell and Hooker to act independently. In this instance, he appeared helpless, informed powerful friends of his impending doom, pointed subtly in the direction of a solution, let his friends solve the problem by dubious means, and went along with the solution—claiming it, of course, as theirs.[25]

There is no doubt that the Darwin-Wallace situation was a "delicate" one. Any time there is a question of scientific priority—and in this case the priority is over one of the half dozen most important ideas in the history of Western civilization—the situation could be nothing but delicate. But the respect and deference shown by both Darwin and Wallace toward each other provides us with evidence that though the arrangement may have been a delicate one, it was worked out between the two men in a gentlemanly way. In a letter dated January 25, 1859, for example, Darwin wrote to Wallace:

> I owe indirectly much to you and them [Lyell and Hooker] for I almost think that Lyell would have proved right and I should never have completed my larger work, for I have found my abstract hard enough with more poor health. Everyone whom I have seen has thought your paper very well written and interesting. It puts my extracts (written in 1839, now just twenty years ago!), which I say in apology were never for an instant for publication, in the shade.[26]

Wallace was equally generous in giving credit to Darwin, as this passage from a letter written on May 29, 1864, shows:

> You are always so ready to appreciate what others do and especially to overestimate my desultory efforts, that I can not be surprised at your very kind and flattering remarks on my paper. I am glad however that you have made a few critical observations and am only sorry you were not well enough to make more, as that enables me to say a few words in explanation.[27]

DARWIN'S SURPRISE OR CHAGRIN?

What is surprising, if anything, is Darwin's apparent surprise at the receipt of Wallace's essay. A clipping of a letter from Wallace to Darwin dated (in Darwin's hand) September 27, 1857, clearly shows that Wallace was continuing work on the problem of the origin of species that he had begun with the publication of his 1855 paper ("On the Law which has Regulated the Introduction of New Species"), for which he voices to Darwin his disappointment in a lack of response: "[cut off] . . . of May last, that my views on the order of succession of species were in accordance with your own, for I had begun to be a little disappointed that my paper had neither excited discussion nor even elicited opposition. The mere statement and illustration of the theory in that paper is of course but preliminary to an attempt at a detailed proof of it, the plan of which I have arranged, and in part written, but which of course requires much [cut off] and collections, a labor which I look [cut off]."[28]

It seems clear from this passage, albeit truncated through deliberate cutting, that the only thing Darwin could have been surprised about was how quickly Wallace completed the promised "detailed proof" of the theory that did, in fact, parallel Darwin's and result in the 1858 essay sent to Down House in the spring. (The deliberate cutting up of letters, manuscripts, notes, and various forms of correspondence by Darwin was his regular, rather disjointed method of organizing his major publishing projects. When one requests original manuscript and papers of Darwin's at the Cambridge Library, for example, one receives a box filled with clippings, snippets, notes cribbed on the backs of envelopes, and the like. Darwin collected these and labeled them as to their source, date of receipt, the chapter into which they would fit, etc. The above letter clipping from Wallace to Darwin, labeled by Darwin, fits this pattern.) But it leaves one to wonder what plan Wallace was working on that he had already written part of, since the 1858 essay was composed in the course of two nights in late February, a full five months after this letter to Darwin. Did his feverish discovery overturn the ideas he was developing in this 1857 plan? If not, what happened to this manuscript? If so, then why did Wallace not expand the 1858 essay into a longer book-length manuscript? One possible answer may be found in a letter written to Bates between these two dates, on January 4, 1858, in which Wallace discusses what appears to be this same "plan" or "work.":

> To persons who have not thought much on the subject I fear my paper on
> the succession of species [the Sarawak Law of 1855] will not appear so clear

as it does to you. That paper is, of course, only the announcement of the theory, not its development. I have prepared the plan & written portions of an extensive work embracing the subject in all its bearings & endeavouring to prove what in the paper I have only indicated. I have been much gratified by a letter from Darwin, in which he says that he agrees with *"almost every word"* of my paper. He is now preparing for publication his great work on *Species & Varieties,* for which he has been collecting information 20 years. He may save me the trouble of writing the 2nd part of my hypothesis, by proving that there is no difference in nature between the origin of species & varieties, or he may give me trouble by arriving at another conclusion, but at all events his facts will be given for me to work upon. Your collections and my own will furnish most valuable material to illustrate & prove the universal applicability of the hypothesis.[29]

Here a plausible scenario presents itself. Wallace, after years of collecting and observing, formed a hypothesis — "On the Law which has regulated the Intro- duction of New Species" (the 1855 "Sarawak Law"). Lacking further supportive evidence for a mechanism to drive evolutionary change, coupled to the fact that he perceived his paper to be largely ignored by the scientific community, Wallace continued about his business of collecting in relative anonymity, but never aban- doned his ultimate quest to understand the origin of species. He knew that Darwin had been working on the problem for twenty years and was currently writing his "big species book"(originally entitled *Natural Selection,* later changed to *The Origin of Species*). Wallace, in no position (either logistically in his travels, or scientifically in his research) to complete a work thorough enough to be well received, decided to sit back and wait to see what Darwin would produce. If Darwin was successful (that is, if Wallace agreed with his arguments), then he would have no need to repeat what had already been done ("He may save me the trouble of writing the second part of my hypothesis"). If Darwin was not successful ("he may give me trouble by arriving at another conclusion"), then Wallace could respond accordingly with his own theory and data. It seems clear that Darwin's *Origin* satisfied the first set of criteria, and Wallace never did write his own "big species book" until he published *Darwinism* in 1889, the very title of which indicates his own leanings on the priority question.

The September 27, 1857 clipping also indicates that, if anything, Darwin should have been a little chagrined instead of surprised, having already been warned by Lyell that he should publish. Darwin's response to this portent in- dicates his dislike of publishing solely for the sake of priority, yet states his own fear of being forestalled. On May 3, 1856, Darwin wrote to Lyell: "I rather hate the idea of writing for priority, yet I certainly should be vexed if anyone were to publish my doctrine before me."[30] His hand forced by Wallace in 1858, Darwin found a solution to his apparent dilemma (i.e., publish for priority sake only, or be completely scooped) by writing a book that was midway between a brief sketch and a magnum opus — *The Origin of Species.*

WHAT WE SHALL NEVER KNOW

With the primary evidence missing in this historical mystery, we can only speculate on what really happened at Down. The extreme interpretation of a conspiratorial cover-up is not supported by the evidence. If Darwin were going to rig (or allow to be rigged) the editorial presentation of the papers to award him priority; or worse, plagiarize from Wallace certain needed ideas (such as the divergence of species), why announce the arrival of Wallace's essay and submit it for publication in the first place? Why not either just take what was needed, or, if Wallace's essay added nothing new to the theory, just destroy the essay and letter and blame the loss on an inefficient postal service, or the mishandling of his mail at Down, or whatever? If one is going to accuse Darwin of such devious finagling as delicate arrangements or plagiarism, then would not the same guileful and scheming personality think of complete elimination of Wallace's essay as a successful strategy?

There is no question that much confusion surrounds the critical period of the spring and summer of 1858, and biologist John Langdon Brooks's epilogue, "What Really Happened at Down House?" draws the pieces together and he wisely concludes: "The simple answer is that no one knows."[31] But then Brooks proceeds "to sketch an alternative reconstruction" in which he concludes that Darwin's letter to Lyell, dated "Down, 18th" and assumed by most to be June 18, was "probably written May 18, 1858" but "it is my view, however, that Darwin did not mail the letter then. Probably after much soul-searching, he restudied Wallace's Ternate manuscript and, with recourse again to Wallace's 1855 paper, wrote the material on [divergence] and inserted it into the text of his chapter on 'Natural Selection'."[32]

Brooks's subsequent analysis of various manuscripts and letters after that incident, then, are all based on the assumption that Darwin received Wallace's letter and essay on May 18. But his analysis is inconsistent. Earlier in the book Brooks says that "the evidence indicates that Darwin must have received Wallace's manuscript on either of two dates in May. Receipt on May 18 would leave 25 days for completion of those folios [on divergence] by June 12 [the date

Darwin noted his thoughts on divergence in the manuscript]; May 28–29 would leave scarcely two weeks. But it must be conceded that desperation will make the pen move quickly."[33] Conspiratorial historicism also makes the pen move quickly, perhaps too quickly. First Brooks gives us the dates of May 18 or May 28–29 for the arrival of Wallace's letter and essay, then he tells us he thinks the "Down 18th" letter to Lyell announcing the arrival of Wallace's letter and essay was actually written on May 18, thus completely negating the May 28–29 option. But even worse, Brooks assumes the Wallace-Bates letter that arrived in London (and postmarked) June 3, was in the same batch as the Wallace-Darwin letter and essay. This is not a historical fact. But an inference, but even if true, this makes both May dates impossible, and, assuming Darwin did not lie in the letter to Lyell about the arrival of the Wallace material on the same day (the 18th), then the arrival date must be June, not May.

Wallace biographer H. L. McKinney has consistency problems as well. He first concludes that the mail from Malaya to London averaged ten weeks in transition, and thus "ten weeks from 9 March, when the communication was mailed, is precisely 18 May, one month before Darwin acknowledged receiving it." McKinney then points to the Wallace-Bates June 3 letter and concludes: "It is only reasonable to assume that Wallace's communication to Darwin arrived at the same time and was delivered to Darwin at Down House on 3 June 1858, the same day Bates's letter arrived in Leicester." To account for the delay from May 18 to June 3, McKinney explains: "Knowing the numerous delays in such matters, we should perhaps allow some leeway, although one month appears to be an excessive allowance."[34] Fine, but then why no "delays" and "leeway" for the Bates letter? And what was Darwin doing with Wallace's manuscript in that time? McKinney wisely ends his discussion "with a series of question marks," but then hints that Darwin might have been filling in the gaps "on divergence in his long version of the *Origin;* he finished that section on 12 June."[35]

So which is it? Either the Bates letter is damning evidence, or it is not. Brooks and McKinney cannot have it both ways. They cannot use the Wallace-Bates letter as evidence that the Wallace-Darwin materials arrived on June 3, and then have Darwin writing Lyell announcing the same on May 18 (as Brooks does); or that the Darwin letter was delayed while the Bates letter was not (in McKinney's case). Either way, to accuse one of the greatest scientists in history of committing one of the most heinous crimes in science on one of the most important aspects of his theory, one better have compelling evidence.[36] Modern skeptics are fond of saying that extraordinary claims require extraordinary evidence. These claims against Darwin are truly extraordinary, but the evidence is not.

In addition, Darwin's contribution to the joint Linnean Society papers did not include materials developed in 1858; rather, he included a letter to the American scientist Asa Gray, written in September 1857. If Darwin had cribbed the concept of divergence from Wallace, why submit this older version? And why was divergence listed in the table of contents for *Natural Selection* in March 1857? And how could he have explained divergence to Asa Gray, almost a year before the Wallace essay?

In 1858, Darwin was knee-deep in producing a massive, multivolume work entitled *Natural Selection*. He planned on taking several more years to complete it, and without outside pressure to publish, he was in no hurry. He had seen the fallout of other theorists who had published prematurely (Chamber's *Vestiges* being the most prominent), and he was not about to be subjected to that kind of criticism. But Wallace's 1858 letter and essay, whether they arrived in May or June, changed all that. Darwin was forced to publish a "shorter" (490-page) book the following year—*The Origin of Species*. Unless the Wallace letter miraculously turns up, we shall never know what really happened. The most logical conclusion is that under the circumstances the delicate arrangement was handled in the most gentlemanly way possible, not as a zero-sum exchange but as a plus-sum relationship.

THE ZERO-SUM MODEL

In 1947, the mathematician John von Neumann published *Theory of Games and Economic Behavior,* in which he described the zero-sum model: the gain of one participant means the loss of the other, and the more one gains the more the other loses.[37] If I win six games of tennis, my opponent must lose six games, and thus they sum to zero $(6 + -6 = 0)$. This antagonistic win-lose model, however, misses the interdependent, sometimes cooperative, and always social nature of the scientific process. Wallace's priority credit and recognition for scientific achievement can and should be significantly enhanced without taking anything away from Darwin. Richard Dawkins, in *The Selfish Gene,*[38] describes these symbiotic relationships—called by Robert Trivers "reciprocal altruism"[39]— as common throughout plant and animal relationships, including human interactions. To describe these reciprocal relationships, Dawkins adopts the Prisoner's Dilemma (PD) model developed by Axelrod and Hamilton,[40] a game in which two prisoners have several options: (1) they can cooperate with each other and get light sentence terms; (2) if one defects while the other cooperates the defector is freed while the cooperator gets an even longer jail sentence; or (3) both can defect, in which case both receive longer jail stays. When this game is reiterated, or repeated, the majority of responses produced are cooperative, as this strategy leads in the long run to "the greatest good for the greatest number." In the short run—that is, in a noniterated or one-trial game—defection is the rule. But over time consistent defectors lose out. Dawkins demonstrates that for animals and humans, those who adopted the zero-sum model were losers.

The zero-sum model is at the heart of most disputes of scientific priority because it assumes that the only way one scientist can profit is through the loss of another scientist. Clearly Newton and Newtonian supporters saw Newton's gain in the priority of the invention of the calculus to be Leibniz's loss, and vice versa, leading to centuries of contentious debate and bitter disagreement. Likewise, some perceive Darwin's gain as Wallace's loss, and Wallace's gain as Darwin's loss. Because of this, it becomes difficult in most of these debates to tease out the facts from the emotion, the information from the rhetoric. This disputatious posturing on both sides wedges historians into a defensive stance that compels an attack-or-be-attacked response. Thus, the antagonism between scholars and historians in both camps could be attenuated by the rejection of the zero-sum model. Darwin, and especially Wallace, clearly rejected the zero-sum model, as they recognized the gain to be had through cooperative interaction. Consider this exchange of letters between the two men. The first, an April 6, 1859, letter from Darwin to Wallace, reveals a man paying the highest respect for a fellow winner in this game of scientific cooperation:

> You cannot tell how I admire your spirit, in the manner in which you have taken all that was done about establishing our papers. I had actually written a letter to you, stating that I could not publish anything before you had published. I had not sent that letter to the post when I received one from Lyell and Hooker, urging me to send some ms. to them, and allow them to act as they thought fair and honourably to both of us. I did so.[41]

Wallace responded with an equally generous dose of recognition in this passage from a May 29, 1864 letter:

> As to the theory of Natural Selection itself, I shall always maintain it to be actually yours and yours only. You had worked it out in details I had never thought of, years before I had a ray of light on the subject, and my paper would never have convinced anybody or been noticed as more than an ingenious speculation, whereas your book has revolutionized the study of Natural History, and carried away captive the best men of the present age.[42]

DARWINISM

AN EXPOSITION OF THE

THEORY OF NATURAL SELECTION

WITH SOME OF ITS APPLICATIONS

BY

ALFRED RUSSEL WALLACE

LL.D., F.L.S., ETC.

WITH A PORTRAIT OF THE AUTHOR, MAP AND ILLUSTRATIONS

London

MACMILLAN AND CO.

AND NEW YORK

1889

Figure 47. Wallace's *Darwinism* title page

(It is interesting to note that not only Alfred Wallace, but his grandson John, were and are satisfied with the historical priority outcome. After a lengthy conversation on this question, John Wallace told me: "I can't understand what all the fuss is about. Grandfather was satisfied with the arrangement, none of us desire to call it 'Wallace's theory of natural selection,' but many of the Darwin people seem defensive about it." There is no doubt about the latter, but it is understandable because the aforementioned Wallace defenders have embraced the zero-sum model, causing them to give more credit to Wallace while simultaneously taking credit away from Darwin. Darwin scholars, in turn, adopt the zero-sum model in defense, as they feel Wallace's gain is Darwin's loss.)

THE PLUS-SUM MODEL

A plus-sum model—the gain of one is the gain of another—recognizes the contingent, cooperative, and interdependent nature of scientific discovery. Both Darwin and Wallace profited by the profit of the other. Both were winners in the game to understand the origin of species. An 1870 letter of "reflection" from Darwin to Wallace shows the special win-win nature of their relationship: "I hope it is a satisfaction to you to reflect—and very few things in my life have been more satisfactory to me—that we have never felt any jealousy towards each other, though in one sense rivals." In the most gentlemanly fashion Wallace always politely addressed Darwin in virtually every letter written, and Darwin nearly always responded in kind. "I was much pleased to receive your note this morning," reads a typical letter opening from Wallace to Darwin. "Hoping your health is now quite restored," "I sincerely trust that your little boy is by this time convalescent," and so on.[43] Darwin and Wallace used each other and each other's ideas to their mutual benefit, and the world of science is better off for it.

One of the other problems with the zero-sum model is an assumption that the ideas under priority dispute are identical, leading to the conclusion that only one individual can be first in discovery. But a law of nature is the product of both the *discovery and description* of a phenomenon. Two individuals may make the same discovery, but they may not make the same description. This is the case with Darwin and Wallace, where their theories of evolution by means of natural selection are similar and complementary, but not identical. Through their numerous intellectual exchanges in letters, papers, and books, they stimulated one another in both knowledge and theory, with a net gain profit for both, making them genuinely codiscoverers.

The historical record has been read differently, however, beginning with the ranking of the joint papers presented at the July 1, 1858 Linnean Society meeting that placed Darwin's 1844 extract and his 1857 Asa Gray letter ahead of Wallace's 1858 essay. If considered by dates of ideas alone, then the ranking is chronologically correct. (It is also alphabetically correct, which was how the names were listed.) But, in fact, what has happened is that Darwin has become a household name and Wallace all but forgotten. This historical reality, of course, was not caused by the ranking of their names at this meeting. In actual fact, according to the Linnean Society president, Thomas Bell, in a reflection of the year's activities, nothing of significance happened in 1858: "The year which has passed . . . has not, indeed, been marked by any of those striking discoveries which at once revolutionize, so to speak, the department of science on which they bear."[44] Obviously Bell and his colleagues did not grasp the significance of the theory of natural selection at its time of presentation. Darwin's fame and importance accrued over many decades of sound scientific work, not through a "delicate arrangement" and clandestine priority ranking of his name over Wallace's. Besides, other than later noting that his paper "was printed without my knowledge, and of course without any correction of proofs," Wallace was most delighted to finally gain the recognition of the scientific community he had desired for so many years, as he indicated to his mother on October 6, 1858, while still in the Malay Archipelago: "I have received letters from Mr. Darwin and Dr. Hooker, two of the most eminent naturalists in England, which has highly gratified me. I sent Mr. Darwin an essay on a subject on which he is now writing a great work. He showed it to Dr. Hooker and Sir C. Lyell, who thought so highly of it that they immediately read it before the Linnean Society. This assures me the acquaintance and assistance of these eminent men on my return home."[45]

Consider Wallace's position at this time. He was a relatively unknown 35-year-old whose only theoretical work—the 1855 Sarawak Law paper—was largely ignored (or, at least, so he thought). He had been away from England and the center of scientific activity already four years, and was, by all rights, still cutting his teeth on such weighty theoretical matters. Darwin, by contrast, was 49 years old, fairly well-known in scientific circles, had already published numerous important scientific articles, and had shared his theoretical ideas with the most important scientists in England. Wallace did not feel the loser because he was not. An essay written in two nights, sent to the right place at the right time, put him in the scientific inner circle and into the historical record—his name next to Darwin's—forever. Anyone who thinks that this was Wallace's loss should reconsider the circumstances in light of the plus-sum model of scientific priority. The gain of Darwin was the gain of Wallace, not the loss.

Consider also Wallace's reception and review of Darwin's *Origin of Species*. In only seven months, he told his boyhood friend George Silk on September 1, 1860, "I have read it through 5 or 6 times & each time with increasing admiration. It is the *'Principia'* of Natural History. It will live as long as the *'Principia'* of Newton."[46] Silk was not someone Wallace needed to impress with false praise, as he continued the comparison: "The most intricate effects of the law of gravitation, the mutual disturbances of all the bodies of the solar system are simplicity itself, compared with the intricate relations & complicated struggle which has determined what forms of life shall exist & in what proportions. Mr. Darwin has given the world a *new science* & his name should in my opinion stand above that of every philosopher of ancient or modern times. The force of admiration can no further go!!!"[47]

THE PLUS-SUM MODEL AND THE NATURE OF HISTORY

Wallace, perhaps better than most, understood the plus-sum model of scientific interaction, and provides us with a brilliant example of this interpretation in an article entitled "The Origin of the Theory of Natural Selection," published in *The Popular Science Monthly* as a reply to his being honored with the Darwin-Wallace medal of the Linnean Society of London on the 50th anniversary of the July 1, 1858 joint reading of the papers. The 1908 celebration rekindled interest in reconstructing the events that led to the theory of natural selection, and in the popular media in particular, there was much historical confusion. It had become apparent to Wallace that there was much misunderstanding of what actually happened in the years leading up to 1858. An analysis of his article on this subject not only supports the plus-sum model, it offers us insight into the interdependent nature of scientific progress in particular, and historical change in general.

In this article we see Wallace's generosity in offering more of the share of the credit to Darwin (whom he refers to as "my honored friend and teacher"), while at the same time firmly reestablishing what he did and did not do. The paper also contains a certain amount of the obligatory modesty that is usually elicited when one is being so honored, such as when Wallace states that the share of the credit should be allocated "proportional to the time we had each bestowed upon it . . . that is to say, as twenty years is to one week."[48]

Wallace did discover and describe natural selection all in the course of a week in late February 1858, but his four years in the Amazonian tropical rain forest and another eight in the Malay Archipelago hardly represent one week to Dar-

DARWIN-WALLACE MEDAL.
1st July, 1908.

**Figure 48. Darwin-Wallace
Medal, 1st July, 1908.**

win's twenty years (in fact, Darwin was five years on his voyage, Wallace a total
of twelve in two voyages). It is true, however, that had Darwin published, "after
ten years—fifteen years—or even eighteen years" instead of the twenty following
the opening of his notebook in 1838, Wallace "should have had no part in it
whatever, and he would have been recognized as the sole and undisputed dis-
coverer of 'natural selection'."[49] The fact is, however, Darwin waited twenty

years, and would have likely waited longer had Wallace not triggered Darwin's productive burst.

In addition, to the modern historian interested in the relative historical role of contingency (unplanned conjunctures of events) and necessity (constraining forces and trends compelling certain actions), it is interesting to note Wallace's recognition of the role of both sets of historical forces in the development of scientific discoveries. For example, after first clarifying that he and Darwin independently, not simultaneously, discovered natural selection ("the idea occurred to Darwin in October, 1838, nearly twenty years earlier than to myself in February, 1858,"), Wallace recognizes the role of contingency in scientific discovery, when he notes: "It was really a singular piece of good luck that gave to me any share whatever in the discovery."[50] He then turns to an analysis that shows how a number of contingencies in the lives of both men led to the necessary discovery of natural selection: "we find a curious series of correspondences, both in mind and in environment, which led Darwin and myself . . . to reach identically the same theory," including:

1. Being "ardent beetle-hunters, [a] group of organisms that so impresses the collector by the almost infinite number of its specific forms, the endless modifications of structure, shape, color and surface-markings that distinguish them from each other, and their innumerable adaptations to diverse environments."

2. Having "an intense interest in the mere variety of living things . . . which are soon found to differ in several distinct characteristics."

3. A "superficial and almost child-like interest in the outward forms of living things, which, though often despised as unscientific, happened to be the only one which would lead us towards a solution of the problem of species."

4. Both "were of a speculative turn of mind [and] constantly led to think upon the 'why' and the 'how' of all this wonderful variety in nature."

5. "Then, a little later (and with both of us almost accidentally) we became travellers, collectors and observers, in some of the richest and most interesting portions of the earth" (Darwin's five-year global circumnavigation and Wallace's four years in the Amazon and eight in the Malay Archipelago). "Thence-forward our interest in the great mystery of *how* species came into existence was intensified."

6. Both men on their voyages and in their home lives enjoyed "a large amount of solitude . . . which, at the most impressionable period of our lives, gave us ample time for reflection on the phenomena we were daily observing."

7. Both men carefully read Lyell's *Principles of Geology* and Malthus's *Principles of Population,* the latter "at the critical period when our minds were freshly stored with a considerable body of personal observation and reflection bearing upon the problem to be solved," that acted on both like "that of friction upon the specially-prepared match, producing that flash of insight which led us immediately to the simple but universal law of the 'survival of the fittest.' "⁵¹

All of these contingencies created necessities (what Wallace calls "the combination of certain mental faculties and external conditions") that drove Darwin and Wallace down parallel paths that became cut ever deeper until they finally crossed in the spring of 1858. This historical tension between what might be and what must be—the contingent and the necessary—for an eighty-five-year-old Wallace reflecting back on a life of science, explains why it was Darwin and himself who finished first and "a very bad second," in the "truly Olympian race" to discover the mechanism of evolutionary change; and why it was not the "philosophical biologists, from Buffon and Erasmus Darwin to Richard Owen and Robert Chambers." For Wallace, the explanation is simple. These "great biological thinkers and workers" were on different paths at different times that made it impossible for them to "hit upon what is really so very simple a solution of the great problem." An adequate explanation of a historical development requires a healthy balance of the internal and the external, individual thought and collective culture, or "the combination of certain mental faculties and external conditions that led Darwin and myself to an identical conception."⁵²

Finally, Wallace applies his model to the large picture of the development of ideas in general, and comes to the conclusion that "no one deserves either praise or blame for the ideas that come to him, but only for the actions resulting therefrom." Wallace is suggesting that the vagaries and nuances of our life and thoughts lead us down certain paths toward conclusions that can only be reached by way of that particular road. Wallace and Darwin shared nearly parallel paths for a time (which later diverged on other issues), and Wallace acknowledges the role of such historical contingencies and necessities in the larger scale of the discovery of scientific ideas: "They come to us—we hardly know how or whence, and once they have got possession of us we can not reject or change them at will." Wallace also addresses the even larger role of human freedom within historical trends by explaining that it is not the development of ideas but in the "actions which result from our ideas" that individuals have the most say in their historical context. Here we catch a glimpse of Wallace, the

CHARLES DARWIN
AND ALFRED RUSSEL WALLACE
MADE THE FIRST COMMUNICATION
OF THEIR VIEWS ON
THE ORIGIN OF SPECIES
BY NATURAL SELECTION
AT A MEETING OF THE LINNEAN SOCIETY
1ST JULY 1858   1ST JULY 1908

Figure 49. A facsimile of the 50th anniversary plaque commemorating the reading of the joint Darwin-Wallace papers.

hard-working, common man who made a most uncommon discovery: "it is only by patient thought and work, that new ideas, if good and true, become adopted and utilized; while, if untrue or if not adequately presented to the world, they are rejected or forgotten."[53] Such is the nature of science and history.

14

THE GREAT BONE HOAX

Piltdown and the Self-Correcting Nature of Science

IN THE PHYSICAL SCIENCES the uncovering of new facts takes one down a reasonably straight and clear path toward support of one theory or rejection of another (where Karl Popper's falsifiability criteria so well applies). There is a fairly straight one-to-one correspondence between theory and reality. The problems to solve, while difficult, are not nearly so complex as those in the biological and especially the social sciences (where falsifiability criteria and operational definitions of measured variables become clouded in fuzzy complexities). When these sciences bump up against national prejudices, as in the race-IQ debates, and religious preferences, as in the evolution-creationsm controversies, emotions run high, theories are clouded in bias, scientists and their critics exchange salvos of prejudicial accusation, and the science seems far removed from whatever the reality might be.

To take one example of a scientific controversy that erupted in 1859 and has not abated to this day, the theory of evolution continues to stir emotions in people who wouldn't know the difference between a pongid and a hominid (but if enlightened would unhesitatingly express their preference for the latter). When the creationists were defeated in the 1987 Supreme Court decision that prohibited Louisiana teachers from teaching so-called creation-science in public schools, it appeared that this peculiarly American social movement had finally heard its death knell that should have been rung in the 1925 Scopes trial.[1] And when Pope John Paul II released his 1996 encyclical that gave sanction to the theory of evolution and admitted that evolution was "more than a theory" (in the standard vernacular meaning of more than a hypothesis and less than a law), it seemed as if our troubles were over with the political and religious intonations of Darwin's theory.[2] But then came the Kansas decision of 1999 to make optional the teaching of evolution in public schools, on the heals of the Templeton Foundation-sponsored revival of attempts to link science and religion, coupled to the rise of "intelligent design" creationism and such sophisticated-sounding arguments as "irreducible complexity." It would appear that the reports of creationism's death were premature, with science's hegemony challenged on every front.[3]

Even within purely scientific circles the theory of evolution remains controversial. Not *whether* it happened, of course, but *how* it happened. Did life evolve gradually or in punctuated bursts?[4] Does life evolve from simple to complex as a result of some inherently progressive tendencies within the laws of nature, or does life become more complex because it simply cannot become any simpler than it was when it began?[5] Did all the races of humans evolve in one place and spread out to the four corners of the world (the "Out of Africa" hypothesis), or did different human groups take different evolutionary tracks after they arrived in various locales (the "Candelabra" hypothesis)?[6] Did big brains evolve as a product of the physical or social environment?[7] Are big brains primarily due to selection factors or are they a contingent accident of nature?[8] Still more controversially, do racial differences in IQ reflect evolutionary (and thus genetic) differences from the races evolving in disperate environments, or do these differences reflect modern cultural biases?[9]

These and many more myths and mysteries surround human evolution today, as they did at the start of the twentieth century when the science was still in its infancy. No doubt that at the end of the twenty-first century these problems will mostly be solved but another set of questions and mysteries will have taken their place. In fact, in her splendid book *Narratives of Human Evolution*, anthropologist Misia Landau shows how scientific theories, especially those purporting to explain human evolution, are narratives, or stories that, like all stories, are very much influenced by the times and cultures of the storytellers themselves.[10] Even though these storytellers are scientists, and their stories are riddled with empirical evidence, the data never just speak for themselves. Evidence must be interpreted through theories and hypotheses, and these are shaped by non-scientific factors. Landau analyzes these theories as stories, and like the scholar of folktales she deconstructs evolutionary narratives into their component parts: "Every paleoanthropological account sets out to answer the question, what really happened in human evolution? Paleoanthropologists generally discuss four main events: a move from the trees to the ground (terrestriality); the development of the upright posture (bipedalism); the development of the brain, intelligence, and language (encephalization); and the development of technology, morals, and society (civilization). Although these events occur in all theories of human evolution, they do not always occur in the same order. Nor do they always have the same significance."[11] Was it bipedalism that gave rise to tool use, which generated big brains? Or was it tool use that led to bipedalism and then big brains? Were early hominids primarily hunters—man the killer ape, warlike in nature? Or were they primarily gatherers—man the vegetarian, pacifist in nature? More importantly, does the narrative change in response to empirical evidence, or does the interpretation of the evidence change as a result of the currently popular narrative? This is a serious problem in the philosophy of science—to what extent are observations in science driven by theory? Quite a bit as it turns out. The scientific method of purposefully searching for evidence to falsify our most deeply held beliefs does not come naturally. Telling stories in the service of a scientific theory does, and there is, perhaps, no better example of this than the discovery of Piltdown man, followed by the revelation decades later that it was all a hoax. It is a bitter and hard-won lesson in the realities of how science really works.

In the annals of evolutionary theories there is one enduring and unsolved mystery that continues to this day to compel writers to speculate in the best of "whodunnit" modes, and that is the Piltdown hoax—the set of ancient hominid bones found in England that would turn out to be the invention of a clever scam artist. There is no finer example of how cultural forces, psychological expectations, and the power of belief led even the most intelligent and educated evolutionary scientists of the first half of the twentieth century to be duped for decades. One recent Piltdown chronicler, John Walsh, calls it "the science fraud of the century."[12] But despite the author's claim to offer a solution, like the JFK whodunnit Walsh's return to a lone perpetrator will be met with scorn from the Oliver Stones of Piltdown. The reason is simple. There is no smoking gun, and the remaining speculative theories will continue being contested as long as intelligent and imaginative people are interested. The list is a long one, including J. S. Weiner's 1955 book *The Piltdown Forgery*, Ronald Millar's 1972 *The Piltdown Men*, Charles Blinderman's 1986 *The Piltdown Inquest*, and Frank Spencer's 1990 *Piltdown: A Scientific Forgery*, to name just a few.[13] Hundreds more articles and essays have been penned in the half century since the hoax was exposed. Why is this such an enduring myth and why is the case still not closed?

As a narrative story, the Piltdown discovery—a big brain atop an apelike jaw—fit the scientific and cultural expectations of the day in that it conveniently supported the prevailing theory (read "hope") that humans first evolved a big brain and only later such features as bipedalism and tool use. Afterall, it was argued, it was our singular ability to think in abstract ways, to plot and strategize and communicate complex ideas, that allowed us, in this progressivist model, to take the great leap forward in evolution above and beyond our simian ancestors. Their bodies may have been similar, but their brains were not. Exceptional encephalization was what set us apart. Further, since it was believed that the most advanced races of today (white) can be found in northern climes (Europe and Asia), fossils of our ancestors would most likely be found here, and certainly not Africa. (This belief prevailed despite Charles Darwin's speculations to the contrary, which he based on the logical inference that since Africa is where our evolutionary cousins, the great apes, presently reside, so too should our fossil ancestors. Half a century later Louis Leakey proved him right and another half century of finds have supported the theory of our African origins.) Indeed, out of Germany came a treasure trove of fossils, starting with the breathtaking finds from the valley of Neander, giving the name to the most famous of all our ancestors. Out of France came our most recent and advanced relatives, the Cro-Magnons, with their cave paintings, clothing, jewelry, and complex tool kits that allowed them to develop what could genuinely be called culture. Additional fossils were discovered in Holland, Belgium, and scattered areas of Asia and Southeast Asia, including significant finds at Peking ("Peking Man") in China and at Java ("Java Man") in southeast Asia.

It seemed everyone was getting in on the great human fossil hunt; everyone except the English, that is. Was it possible that humans did not evolve in England? Were Englishmen nothing more than a recent migration from the continent, a backwater of human evolution? If only an ancient hominid could be found here. And what a coup it would be that if that hominid, unlike many of the finds coming from elsewhere, clearly showed a humanlike brain sitting atop more primitive primate features, especially a jaw. Seek and ye shall find, build it and they will come—pick your metaphor. The British got what they were wishing for in 1912.

On February 15, 1912, a British lawyer named Charles Dawson, who devoted every moment of his spare time to amateur archaeology, presented to the renowned Keeper of Geology of the British Museum of Natural History, Arthur Smith Woodward, several cranial fragments that appeared to be of an ancient hominid. Dawson told Smith Woodward that in 1908 workmen had unearthed the fragments from a gravel pit at Piltdown in Sussex, accidentally smashing them with their pick. The skull fragments were modern in appearance, with a large, thick casing, yet they were found in deep, ancient layers, indicating great antiquity. On June 2, 1912, Smith Woodward, Dawson, and a youthful paleontologist and Jesuit priest named Pierre Teilhard de Chardin (later to become the world-famous author of *The Phenomenon of Man*, the book that attempted a scientific proof of the spiritual nature of humanity), went to the pit to continue the dig. There Dawson made another find—the lower jaw of the skull, including two molars, very ape-like in structure but indicating humanlike wear. Additional digging uncovered stone tools, chipped bones, and fossil animal teeth that placed the ancient hominid well back in evolutionary history. The December 5, 1912 edition of the most respected British science journal, *Nature*, ran a short news item of the find:

> Remains of a human skull and mandible, considered to belong to the early Pleistocene period, have been discovered by Mr. Charles Dawson in a gravel-deposit in the basin of the River Ouse, north of Lewes, Sussex. Much interest has been aroused in the specimen owing to the exactitude with which its geological age is said to have been fixed.

On December 18, 1912, Dawson, under the auspices and endorsement of Smith Woodward, announced his great find at a meeting of the Geological Society of London. A few skeptics voiced their doubts, but one of the crucial pieces of evidence that might have addressed their concerns—the jaw—was, mysteriously (and conveniently as it turns out), broken in just the right places to preclude resolution. In its December 19, 1912 edition, *Nature* ratched up the excitement by proclaiming this as the singular find in the history of British paleontology: "The fossil human skull and mandible to be described by Mr. Charles Dawson and Dr. Arthur Smith Woodward at the Geological Society as we go to press is the most important discovery of its kind hitherto made in England. The specimen was found in circumstances which seem to leave no doubt of its geological age, and the characters it shows are themselves sufficient to denote its extreme antiquity." The authors went on to explain just why this specimen was so important: "At least one very low type of man with a high forehead was therefore in existence in western Europe long before the low-browed Neanderthal man became widely spread in this region. Dr. Smith Woodward accordingly inclines to the theory that the Neanderthal race was a degenerate offspring of early man and probably became extinct, while surviving modern man may have arisen directly from the primitive sources of which the Piltdown skull provides the first discovered evidence."

Newspaper headlines soon followed. On December 19, the *Times* of London announced:

A PALEOLITHIC SKULL
FIRST EVIDENCE OF A NEW HUMAN TYPE

The *New York Times* followed suit, identifying the deeper, theoretical issues involved in the find:

PALEOLITHIC SKULL IS A MISSING LINK
MAN HAD REASON BEFORE HE SPOKE
DARWIN'S THEORY IS PROVED TRUE

The more provincial *Hastings and St. Leonards Observer*, on February 15, 1913, ran this poetic tribute to Piltdown, drawing out the full moral implications of Dawson's now-heroic discovery:

> Now for the moral of my tale:
> We little humans in this vale
> Of joys and fears our short lives pass,
> And then we're blotted out alas!
> Perchance in a thousand years,
> Our skulls may be unearthed—Our fears,
> Our hopes, our aspirations may
> Be analyzed, some future day,
> By some keen-brained geologist,
> Who'll hold our jawbone in his fist . . .
> Dawson, we owe you a debt
> And hope that you will dig up yet,
> In research geological,
> Our tree genealogical.

That summer Pierre Teilhard de Chardin, with a background in paleontology and completing his theological training at a Jesuit seminary, conveniently (some say suspiciously) near Piltdown, found an apelike lower canine tooth, but worn in a very humanlike fashion. The following summer, as the great nations of

Europe cascaded toward their destiny of total war, Dawson added to the trove a fossilized thigh bone from an elephant and what appeared to be a stone tool, and a fairly advanced one at that. In 1915, at another pit two miles from Pilt-down, Dawson uncovered two more hominid skull pieces along with another tooth similar to the previous finds. Now there could be no doubt as to both the authenticity and the significance of the fossil collection, and for four decades the finds went largely unchallenged. Humans evolved not only in Europe and Asia, but in England too, and these ancient hominids first evolved big brains then the other hominid features, as evidenced by the Piltdown fossils, cementing the progressivist narrative plot: big brains—> bipedalism—> tool use. Re-nowned anatomist Grafton Elliot Smith summarized it this way in his 1927 book *The Evolution of Man*:

> The brain attained what may be termed the human rank when the jaws and face, and no doubt the body also, still retained much of the uncouthness of Man's simian ancestors. In other words, man at first, so far as his general appearance and "build" are concerned, was merely an Ape with an over-grown brain. The importance of the Piltdown skull lies in the fact that it affords tangible confirmation of these inferences.[14]

The scientific world was ecstatic that this new find had confirmed what they always believed must be true about human evolution—our large brain had lifted us above our simian ancestors. And it was an Englishman no less, no trivial matter in this jingoistic era on the eve of the Great War where France and Germany had already produced a number of fossils attesting to their role in humanity's rise out of the detritus. It was too good to be true.

Unfortunately for those who allowed their skeptical facilities to be overcome by cultural expectations (although there were some skeptics, they were outgun-ned by such luminaries as Arthur Keith), it *was* too good to be true. In 1953, scientists Kenneth Oakley, J. S. Weiner, and W. E. le Gros Clark announced that new dating techniques had proven the skull fragments to be of modern origin, as was the orangutan jaw, all stained, chipped, and filed to look ancient. The flint artifacts were worked with modern tools, the fossil animal teeth were from locals elsewhere, and everything was carefully placed in the Piltdown pit. It was all a hoax—four decades worth!

Now the mystery begins. Whodunnit? If this hoax happened in the 1960s, Oliver Stone would have a blockbuster. Conspiratorialists have produced an endless variety of combinations, most of which include Dawson (like his JFK counterpart in Lee Harvey Oswald) as everything from patsy to accomplice to orchestrator. There are no less than twenty candidates, including (in addition to Dawson) Pierre Teilhard de Chardin, Professor Arthur Smith Woodward, the Oxford anatomist Sir Arthur Keith, zoologist and curator at the Natural History Museum in London Martin Hinton, anatomist and Piltdown historian Sir Grafton Elliot Smith, Oxford geologist W. J. Sollas, amateur collector Lewis Abbott, and even (for an added twist of mystery) the celebrated author and creator of the Sherlock Holmes mysteries, Sir Arthur Conan Doyle, who lived near the Piltdown pit and had visited the site at least once.

The first full-length treatment of the hoax since its exposé, J. S. Weiner's 1955 *The Piltdown Forgery,* concluded that Dawson either acted alone or was at the very least the key figure in a conspiracy. In what has to be one of the most understated indictments in history, Weiner concluded (with a double negative thrown in for good measure): "We have seen how strangely difficult it is to dissociate Charles Dawson from the suspicious episodes of the Piltdown history. We have tried to provide exculpatory interpretations of his entanglement in these events. What emerges, however, is that it is not possible to maintain that Dawson could not have been the actual perpetrator." Yet Weiner lets down those who were taken in: "Let it be said, however, in exoneration of those who have assumed the Piltdown fragments to belong to a single individual, or who, having examined the original specimens, either regarded the mandible and ca-nine as those of a fossil ape or else assumed (tacitly or explicitly) that the prob-lem was not capable of solution on the available evidence, that the faking of the mandible and canine is so extraordinarily skillful, and the perpetration of the hoax appears to have been so entirely unscrupulous and inexplicable, as to find no parallel in the history of palaeontological discovery."[15]

In 1972, Ronald Millar, in his book with the pluralistic title (and suspects) *The Piltdown Men*, made the case for the involvement of the anatomist Grafton Elliot Smith based on circumstantial evidence (he was involved in a similar controversy over a skull in his home country of Australia) and the fact that he preferred the human evolution theory that big brains came first. But most evo-lutionary biologists supported this theory, and Smith did not arrive in England until 1915–1916, so he could not have been involved in the preparing or planting of the original fossil finds. Charles Blinderman, in his 1986 book *The Piltdown Inquest*, argued for amateur collector and scientist W. J. Lewis-Abbott, since Abbott claimed to have directed Dawson to the Piltdown pit in the first place. Blinderman reprints a letter from F. H. Edmunds of the Geological Survey and Museum to one of the hoax busters, Kenneth Oakley, in which he implicated

Abbott: "While I was in the district I made the acquaintance of a jeweller who owned a shop in Hastings, one W. J. Lewis-Abbott, whose name is not unknown in the geological world of 40 years ago. He himself told me that he had worked with Dawson on the Piltdown skull and that the skull had been in his possession in his house six months before Smith Woodward saw it; and I gathered from him that he soaked it in bichromate to harden it. I have every reason to believe that these statements were matters of fact."[16] Blinderman queried Weiner about his suspicion of Lewis-Abbott "I have material about Abbott which I think makes it difficult to incriminate him."[17] In fact, Dawson never said who it was that tipped him off on the pit, so once again we are left with circumstantial evidence only, not good enough to convict in a court of law.

In the early 1980s a Piltdown brouhaha arose around a 1979 *Natural History* essay by Harvard paleontologist Stephen Jay Gould[18] (followed by two more in 1980 and 1981[19]), who rang in on the conspiratorial side by making a case for Pierre Teilhard de Chardin (admitting that he was not the first to suspect Teilhard and that even the great paleoanthropologist Louis Leakey had begun writing a book on Teilhard's involvement, but died before completing it). Gould notes that Teilhard was well trained professionally in paleoanthropology, had considerable field experience, was intensely interested in the debate about human evolution, was present at the digs during many of the most important finds, and even made a significant Piltdown discovery himself when he found a tooth of a lower jaw on August 30, 1913. When the hoax was exposed in 1953, Teilhard was an old man and one of the few from the original Piltdown cast still alive. Gould points out a number of mistakes and inconsistencies in letters from Teilhard to Kenneth Oakley, confusing dates and locations. Gould admits these literary blunders may be due to "the faulty memory off an aging man," but finds it more than curious that after the initial excitement over the Piltdown discovery and his own involvement, Teilhard shut down and "Piltdown never again received as much as a full sentence in all his published works," which were substantial.[20]

Perhaps Teilhard lost interest as he turned to other studies, but at the end of his life he told Oakley in a letter that Piltdown was "one of my brightest and and earliest paleontological memories." By comparison, Teilhard's later involvement in the Peking Man find in China generated voluminous articles. Since Piltdown was every bit as spectacular a find as Peking, especially since it so keenly supported his progressive and spiritual theory of evolution (in which an early big brain plays a crucial role), we have to ask about Teilhard. Why the silence? Perhaps Teilhard knew Piltdown was a hoax (and maybe was even in on it), but when the discovery became a landmark in human evolution, it was too late to fess up. Better to just become tight-lipped and hope the story breaks before too much damage is done. Years rolled over into decades, careers were built (and subsequently destroyed) on this single find, and too much was at stake. When Teilhard visited Oakley after the exposé, for example, he allegedly was visibly uncomfortable talking about it, passed through the hoax exhibit hurriedly, and informed Oakley that Piltdown was a sensitive subject. But why, Gould asks? "Unless, of course, the embarrassment arose from guilt about another aspect of his silence—his inability to come clean while he watched men he loved and respected make fools of themselves, partly on his account."[21]

Teilhard de Chardin is a tantalizing candidate, but Blinderman, after an extensive analysis of the evidence against Teilhard, concludes: "I don't think that there's any more evidence that Teilhard suffered pain for having collaborated on the hoax or that he paid his debt than there is that he committed it in the first place."[22] Piltdown's latest theorist, John Walsh, also dismisses Teilhard, along with all the others. The case is now closed, he says. Dawson acted alone. His proof? Dawson had both the means and the motive: overweaning ambition to make a name for himself in science, and previously faked antiquities. Walsh carefully documents Dawson's long history of questionable finds, including a Roman horseshoe that he dated before there existed such a thing, an ancient boat with not-so-ancient timbers, some fraudulant Roman Britain bricks he tried to pass off as genuine, several accusations of plagiarism, and more. Dawson had both the knowledge and the skill to hoax Piltdown all by himself. Plus, Walsh argues, there is no evidence beyond a reasonable doubt to convict anyone else, and as another colorful lawyer chimed a few years ago in yet another "trial of the century," if the evidence doesn't fit you must acquit.

This is a fair and reasonable argument. If this were a court of law (especially in Los Angeles), they would all walk. Walsh even dismisses Martin A. C. Hinton, a curator who worked for Smith Woodward at the museum and who was the subject of a May 23, 1996 paper in *Nature* by King's College paleontologist, Brian Gardiner. According to Gardiner, Hinton played the hoax for revenge against his boss who apparently refused to pay his wages in a timely fashion. Gardiner was given a trunk found twenty years prior, allegedly belonging to Hinton, in which he found fossils, teeth, and other remains similar to those found in Piltdown, as well as a mixture of chemicals used in staining said items. "I was able to show that the stains in the teeth and the stains in the trunk were the same as those at Piltdown," Gardiner said, also explaining that Hinton had done some staining for Dawson, knew what Dawson would expect to find, and planted the evidence for his dupe to take to his boss. In other words, Dawson was a patsy, not a perpetrator. Walsh counters with the reasonable conclusion that the staining was done after the hoax was exposed in 1953, and that Hinton was experimenting with different staining techniques to figure out how Dawson did it.

Who faked Piltdown? Who knows? Like JFK, the lone perpetrator is a cleaner, simpler theory than a complex conspiracy, but conspiracies do happen in history so we cannot dismiss them *a priori*. Walsh builds a powerful case against Dawson, but Gardiner claims that "Dawson was too ignorant—an old country solicitor." Lawyers may be many things, including the butt of jokes, but ignorant isn't one of them. Short of the discovery of an unequivocal smoking gun that incriminates one or more men, the Piltdown mystery will not soon end. While I feel no particular compunction to ring in with who I think dunit ("I don't have a dog in that fight," as my friend and colleague Frank Miele likes to say), but if scientific crimes were resolved in courts of law and I was called to serve on a Piltdown jury, I would force a hung jury for everyone but Dawson.

Regardless of the solution to this unsolved mystery, Piltdown provides two valuable lessons in the nature of science:

1. *Science is subject to bias*. The head of the Institute for Creation Research, Duane T. Gish, wrote of Piltdown in 1978 that "The success of this monumental hoax served to demonstrate that scientists, just like everyone else, are very prone to find what they are looking for."[23] Well, sure, that much is true; but this is the very reason that science is constructed to be self-correcting. It has built into it methods to detect not only hoaxes, but both conscious and unconscious biases. This is what sets science apart from all other knowledge systems and intellectual disciplines. If it were not for this self-correcting mechanism, in fact, science could not have made the remarkable progress it has over its 500 year history. Its greatest weakness, in fact, is its greatest strength.

2. *Science is self-correcting*. It is critical to emphasize that in the Piltdown hoax, as with all other scientific hoaxes and errors, it was scientists who exposed the hoax and science that corrected the mistake and moved forward with more and better research. Creationists are only too happy to point to such scams and errors (see FIGURE 50 with the note on Piltdown that "the jawbone turned out to belong to a modern ape") as evidence that science never gets the story right (right defined by scripture, or at least a particular and narrow strand of Christian fundamentalist scriptural interpretation). Dr. Gary Parker, from the Institute for Creation Research in Santee, California, expressed this well when he wrote:

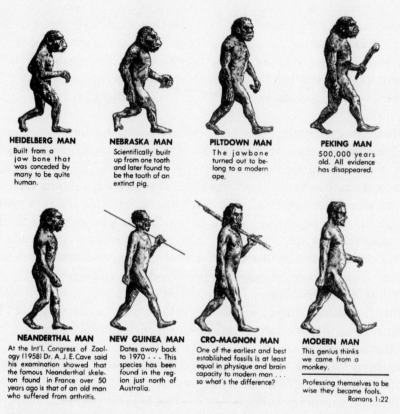

HEIDELBERG MAN
Built from a jaw bone that was conceded by many to be quite human.

NEBRASKA MAN
Scientifically built up from one tooth and later found to be the tooth of an extinct pig.

PILTDOWN MAN
The jawbone turned out to belong to a modern ape.

PEKING MAN
500,000 years old. All evidence has disappeared.

NEANDERTHAL MAN
At the Int'l. Congress of Zoology (1958) Dr. A. J. E. Cave said his examination showed that the famous Neanderthal skeleton found in France over 50 years ago is that of an old man who suffered from arthritis.

NEW GUINEA MAN
Dates away back to 1970 - - - This species has been found in the region just north of Australia.

CRO-MAGNON MAN
One of the earliest and best established fossils is at least equal in physique and brain capacity to modern man . . . so what's the difference?

MODERN MAN
This genius thinks we came from a monkey.

Professing themselves to be wise they became fools.
Romans 1:22

Figure 50. A creationist perspective on paleoanthropology: Piltdown as another in a long line of evolution frauds.

At least Piltdown Man answers one often-asked question: "Can virtually all scientists be wrong about such an important matter as human origins?" The answer, most emphatically is, "Yes, and it wouldn't be the first time." Over 500 doctoral dissertations were done on Piltdown, yet all this intense scrutiny failed to expose the fake. Students may rightly wonder what today's "facts of evolution" will turn out to be in another 40 years.[24]

Beyond the fact that it was evolutionists, not creationists, who exposed the Piltdown hoax, the fundamental flaw in this reasoning is that the theory of evolution is not proved through one fossil, or even a set of fossils. We know evolution happened through what the nineteenth-century British philosopher of science, William Whewell, called a consilience of inductions, or what is better expressed as a convergence of evidence. It is not that evolution is supported by a few or a lot of fossils (the total numbers are irrelevant), it is that these fossils, along with numerous other strands of evidence from genetics, biochemistry, anatomy, physiology, zoology, botany, geology, and many other sciences, independently of one another converge toward the same conclusion. There are so many different lines of evidence in support of the theory of evolution now, in fact, that if half a dozen more Piltdown-like hoaxes were exposed tomorrow it wouldn't shake the theory at all. To do that one would need to propose another theory and mechanism that better accounts for the existing data and lines of evidence. And, it is important to add, the theory "that's just the way God did it" is not a testable hypothesis and thus is not a part of science.

Still, Piltdown is a painful reminder of the fact that intelligence and education is no prophylactic against fraud and flimflam. In Piltdown we saw some of the most highly decorated and respected scientists in the world taken in by someone who was at most an amateur hoaxer. It shows that humans are pattern-seeking, storytelling animals, who seek and find patterns that fit a meaningful story. Once the pattern is found and a story developed around that pattern, additional confirming evidence is sought, and disconfirming evidence (or clues of a hoax) are ignored. It is a testimony to the confirmation bias, one of the most powerful explanatory models of cognitive psychologists who study flaws in critical thinking, and Piltdown shows that scientists—even world-class scientists—are not immune. It is one thing to wonder why people believe weird things, it is quite another, and on one level far more important, to understand why *smart* people believe weird things. One answer is that the belief engine that drives our perceptions is so powerful that, with rare exceptions, it is almost impossible to step outside one's culture, to shed the belief baggage that comes with residence in a community of believers, to filter knowledge through the belief engine in order to see the evidence for what it really is—whether that be truth or hoax.

NOTES

INTRODUCTION: BLURRY LINES AND FUZZY SETS

1. Quoted from the literature of the Western Institute of Remote Viewing.
2. Ibid.
3. Schnabel, 1997.
4. Stories and quotes are from the jacket blurbs and promotional material in the front matter of the book.
5. Quoted in Schnabel, 1997, 368.
6. Ibid., 340.
7. Brown, 1999, 172.
8. Ibid., 200.
9. Ibid., 216.
10. Quoted from the literature of the Western Institute of Remote Viewing.
11. Quoted from the literature of the Western Institute of Remote Viewing, all emphases in original.
12. Quoted in Bloomberg, 1994, 36.
13. Quoted in Lippard, 1994, 31.
14. See also Larue, 1994.
15. See a description and discussion of this in Futuyma, 1989.
16. Kosko, 1993; 1999.
17. Shermer and Grobman, 2000, Chapter 10.
18. Sagan, 1996.
19. Kevles, 1999.
20. Gold, 1999.
21. Taubes, 1993.
22. Park, 2000.

23. See the discussion of both reparations and political motivations in Shermer and Grobman, 2000.

24. For examples of these theories see Hancock, 1995, 1999; Wilson, 1996.

25. See Fagan, 1996 and Lambert 1997 on how archaeology is done professionally, and Feder, 1999, for a discussion on how archaeology is misused.

26. Nickerson, 1998.

27. The most vocal proponent of cold fusion is Gene Mallove, founding publisher of *Infinite Energy* magazine. Mallove, who appeared on my radio show to debate Robert Park, is a tireless crusader for what he sees as blatantly immoral injustices served against the American public by the withholding of such energy sources.

28. Four fair-minded and thoughtful analyses of UFO and alien abductions are Sagan 1996, Matheson 1998, Bartholomew and Howard 1998, and especially Achenbach 1999, who not only goes inside the UFO and alien abduction movement but draws important distinctions between this and the SETI program, NASA, and other scientific organizations with similar goals but radically different methods.

29. For the latest and most thorough analysis of the creationist movement see Pennock, 1999.

30. Miller, 1999, demonstrates the vacuousness of creationists' theories about the origins of life.

31. See the special issue of *Skeptic* devoted to the HIV-AIDS skeptics, vol. 3, no. 2, cited in Harris, 1995.

32. For a strongly skeptical perspective on hypnosis see Baker 1990 and 1992. For the most academically up to date and thorough history of hypnosis see Gauld, 1992.

33. Hilgard, 1968.

34. Hilgard, 1977.

CHAPTER 1. THE KNOWLEDGE FILTER

1. In Campbell, 1992, 18.

2. Dyson, 1997, 20–21.

3. Ibid.

4. A lengthy first-person account of Feynman's involvement in the Challenger disaster is told by Feynman in his posthumously published book *"What Do You Care What Other People Think?": Further Adventures of a Curious Character*. An appendix reprints the report Feynman submitted to NASA.

5. Ibid., 237.

6. Ibid.

CHAPTER 2. THEORIES OF EVERYTHING

1. Originally published in the Winter 1950–51 issue of the *Antioch Review* and reprinted in 1981 in Martin Gardner's *Science: Good, Bad, and Bogus,* 3–14, and includes a postscript.

2. For decades I have had on my shelf the 1957 paperback edition of *Fads and Fallacies*

in the Name of Science. In researching this chapter I wanted to obtain a 1952 first edition, which I was able to do on the net for a modest price of $35.00.

3. From the postscript to the reprinted edition of "Hermit Scientists," in *Science: Good, Bad, and Bogus*, 12.

4. Gardner, 1952, 8.

5. Gardner, 1950, from the 1981 reprint edition, 11.

6. Ibid.

7. Gardner, 1952, 3.

8. Ibid., 4.

9. Ibid., 6.

10. Ibid., 242.

11. Ibid., 7–8.

12. Ibid.

13. Ibid., 8.

14. Ibid., 9.

15. Ibid., 12–13.

16. Ibid., 15.

17. Quoted in Simons, 1993, 3, the original published collection of these letters, sold at the Museum of Jurassic Technology in Los Angeles, California.

18. Ibid., 105.

CHAPTER 3. ONLY GOD CAN DO THAT?

1. Wilmut, 1996.

2. National Bioethics Advisory Commission report, 1.

3. National Bioethics Advisory Commission. 1997. *Cloning Human Beings: Report and Recommendations*. Rockville, Maryland.

4. Ibid., 2.

5. McGoodwin, 1997, 1.

6. Ibid., 2.

7. Kluger, 1997, 67.

8. See Shermer, 1998, B11 and Dixon, 1998, B11.

9. See the interview with Richard Seed by Frank Miele in *Skeptic*, Miele, 1999.

10. See, for example, Rantala and Milgram, 1999; Humber and Almeder, 1998; Kass and Wilson, 1998; Kitcher, 1996; and Kolata, 1998.

11. Quoted in Kluger, 1997, 70.

12. Segal, 1999.

13. Ibid., 314.

14. Quoted in Peters, 1997.

15. Shermer, 1993a, 1997.

16. Marx, 1852, 594.

17. Rosenbaum, 1998.

18. Sulloway, 1996, 286.

19. In Humber and Almeder, 1998, 4.

20. In Kluger, 1997, 71.
21. Ibid., 69.
22. Quoted in the NBAC report.
23. In Rantala and Milgram, 1999, 157.
24. Howard and Rifkin, 1977.
25. Peters, 1997.
26. Woodward, 1997, 60.
27. *Los Angeles Times*, 28 December, 1998, B11.
28. In Rantala and Milgram, 1999, 210.
29. In Shelley, 1965, 205.

CHAPTER 4. BLOOD, SWEAT, AND FEARS

1. Reported in the *Los Angeles Times*, 6 March 2000, Section D.
2. Entine, 2000.
3. Leonard Shapiro, " 'Jimmy the Greek' Snyder Says Blacks are 'Bred' for Sports," *Washington Post*, 16 January 1988, A10.
4. Transcript from *Nightline* quoted in Entine, 2000, 233–234.
5. Quoted in Newhan, 1998. See also Goldstein, 1998.
6. Hoberman, 1998.
7. Quoted in Kane, 1971.
8. Quoted in Almond, 1988.
9. Jackson's quote was taken from a transcription of the original airing of the show made by the author.
10. A copy of the taped interview is available through KPCC, located on the campus of Pasadena City College.
11. Entine, 2000, chapter 5. "Nature's Experiment: The 'Kenyan Miracle'," 43–67.
12. See Shermer, 1985, 1987, 1993.
13. Ritchie, 1988.
14. Personal correspondence.
15. See Halpern, 1996, for a discussion of this and other cognitive biases.
16. See Sarich 2000, Entine 2000, Hoberman 2000, all appearing in *Skeptic*, vol. 8, no. 1, as part of a special issue on race and sports. Sarich's article is in support of Entine's thesis, Entine's is an excerpt from his book, and Hoberman's article is critical of Entine's book.
17. See discussion in Entine, 2000, Chapter 15. "The 'Scheming, Flashy Trickiness' of Jews," 198–203.
18. Ibid.
19. Ibid.
20. For a discussion of this and other forms of extreme Afrocentrism and historical revisionism, see Skeptic, vol. 2, no. 4.
21. See Nickerson, 1998.
22. Sarich, 2000.
23. Complete data available at www.kentuckyderby.com
24. For a discussion of exaptation see Gould and Vrba, 1982.

25. Astrand and Rodahl, 1977.

26. Quoted in Astrand and Rodahl, 1977; see also Sleamaker, 1989; Gross, 1986; Burke, 1986.

27. Personal correspondence.

28. Entine, 2000, 267.

29. For a thorough discussion of this subject see Entine's Chapter 19 "Winning the Genetic Lottery," 246–271.

30. Sarich, 2000.

31. See Astrand and Rodahl, 1977.

32. Personal correspondence.

CHAPTER 5. THE PARADOX OF THE PARADIGM

1. Kuhn, 1962, 10.
2. See, for example, Lakatos and Musgrave, 1970.
3. Kuhn, 1977, 319.
4. Ruse's essay is reprinted in Somit and Peterson, 1992.
5. Quoted in Weaver, 1987, v. ii, 133.
6. Kuhn, 1962, 90.
7. Planck, 1936, 97.
8. Sulloway, 1996.
9. Ruse, 1992.
10. Eldredge, 1971.
11. Eldredge and Gould, 1972.
12. Gould, 1991, 14.
13. Darwin, 1859, 280.
14. Eldredge and Gould, 1972, 205.
15. Ibid., 207–208.
16. Gould, 1991, 16.
17. Ibid.
18. Ibid.
19. Ruse, 1992, 146.
20. Gould, 1991, 16.
21. Prothero, 1992, 40.
22. Mayr, 1992, 25.
23. Mayr, 1954, 179.
24. Shermer and Sulloway, 2000, 79.
25. Ibid.
26. Mayr, 1992, 24.
27. Prothero, 1992, 42.
28. Dennett, 1995; Dawkins, 1998; Ruse, 1999.
29. Dennett, 1995, 262–310.
30. Dawkins, 1998, 197.
31. Ruse, 1996.

32. Ruse, 1999, 50.
33. Ibid., 150–152.
34. Prothero, 1992, 43.
35. Ibid.
36. Miller, 1999.
37. Darwin, 1871, 119–120.
38. Miller, 1999, 115.
39. Darwin, 1859, 113–114.
40. In Gould's Curriculum Vitae.
41. Burkhardt and Smith, 1985–1999.
42. Kohn, 1985, 1–2.
43. Montagu, 1952.
44. Barzun, 1958, 25.
45. Himmelfarb, 1959, 127.
46. Ghiselin, 1969, 1.
47. Ibid., 4.
48. Ibid., 12.
49. Ibid., 243.
50. Quoted in Burkhardt and Smith, 1985–1999.
51. See, for example, Vorzimmer, 1970; Hull, 1973; Glick, 1974; Ruse, 1975; Schweber, 1979; Greene, 1981; Bowler, 1983; and the narrative biographies by Bowlby, 1990 and Desmond and Moore, 1991.
52. Richards, 1987, 46.
53. Hooykaas, 1970, 45.
54. Ghiselin, 1969, 1.
55. Bowler, 1988, ix.
56. Ibid., 3–4.
57. Mayr, 1982.
58. Mayr, 1988.
59. Mayr, 1982, 161.
60. Ibid., 183.
61. Mayr, 1988, 182.
62. Cohen, 1985.
63. Jacob, 1976, 172–174.
64. In Cowen, 1986, 8.
65. In Gould, 1983, 264.
66. Grabiner and Miller, 1974.
67. See Gilkey, 1985, and Shermer, 1991b or 1997 for detailed examples.
68. Rogers, 1992, 86.
69. Campbell, 1988, 123.
70. Campbell, 1949, 30.
71. In Desmond and Moore, 1991, 181–185.
72. Ibid., 186.
73. Ibid, 218.

74. Ibid.
75. Ibid., 341.
76. Ibid., 603.
77. Darwin, 1859, 490.
78. In F. Darwin, 1887, 44–45.
79. In Hull, 1973, 277.

CHAPTER 6. THE DAY THE EARTH MOVED

1. Koestler, 1959, 69.
2. See Koestler, 1959; Toulmin and Goodfield, 1961; and Beer and Beer, 1975, for these quotes and details on this scientific relationship.
3. Popper, 1975, 72–75.
4. Ibid.
5. Snelson, 1993, 44.
6. Planck, 1936, 97.
7. Boring, 1950, 399.
8. Ibid.
9. Mayr, 1982, 835.
10. Cohen, 1985, 35.
11. Quoted in Sulloway, 1996, 539.
12. Ibid.
13. Sulloway, 1990, 15.
14. Ibid., 1.
15. Ibid., 6.
16. Ibid., 10.
17. Ibid., 12.
18. Ibid., 8.
19. Ibid., 7.
20. Sulloway, 1996, 154.
21. Ibid., 178.
22. Cohen, 1985.
23. Koestler, 1959, 284.
24. Ibid, 285.
25. Ibid, 286.
26. Crombie, 1979, 176–177.
27. Kuhn, 1957, 264.
28. Ibid, 1.
29. Cohen, 1985, 106.
30. Ibid.
31. Ibid. 123–124.
32. Personal correspondence.
33. See Munitz, 1957, for a detailed history of this sequence from Aristotle to Copernicus.

34. For a thorough summary of this system see Tillyard, 1944; Koestler, 1959, 51–79; Cohen, 1960, 24–52; and Olson, 1982, 138–141.

35. Tillyard, 1944, 19–33.

36. Ibid., 125.

37. Daly, 1979, 5.

38. Daly, 9.

39. Olson, 1982, 238–241.

40. Cohen, 1985, 183–187.

41. Tillyard, 1944, 25.

42. Ibid., 25.

43. Ibid., 85.

44. Daly, 12–13.

45. Milton, 1948–52, 408.

46. Tillyard, 1944, 82–83.

47. Hobbes, 1651, 47.

48. Prowe, 1883, 232. See also Mason, 1952 for connections between the Protestant Reformation and Copernican cosmology.

49. In Olson, 1991.

50. Hooker, 1594, 104.

51. For defense of these statements see Palter, 1970; Neugebauer, 1975; Rosen, 1971; and Gingerich, 1973.

52. Copernicus, 1978, 9.

53. Ibid., 11.

54. Rosen, 1971.

55. For an elaboration on the development of Copernicus's theory see Duhem, 1969; Gingerich, 1975b; and Swerdlow and Neugebauer, 1984.

56. Olson, 1982, 253–254.

57. Gingerich, 1975a, 85–93.

58. Rosen, 1973, 433.

59. Westman, 1980, 106–107.

60. Olson, 1982, 253–254.

CHAPTER 7. HERETIC-PERSONALITY

1. Marchant, 1916, 451.

2. Quoted in Marchant, 1916, 451.

3. George, 1964, x.

4. *Oxford English Dictionary*, vol. 1, p. 1294.

5. Guilford, 1959, 5–6.

6. For discussions of and data on the Five Factor model of personality see Digman, 1990; Costa and McRae, 1992; Goldberg, 1993.

7. Documents in the archives of the Royal Society, no referencing designation given other than under Wallace's name.

8. Wallace, 1908b, 361–372.

9. Ibid.
10. Ibid.
11. Quoted in Wallace, 1908, 368–369.
12. Letter in the Hope Entomological Collection of Oxford University Museum, no referencing designation given other than under Wallace's name.
13. In Wallace, 1908, 372.
14. Poe, 1966, 5.
15. Ibid., 6.
16. Ibid., 7.
17. Ibid., 8.
18. Ibid., 9.
19. Ibid., 10.
20. Ibid.,
21. Ibid., 11.
22. In Marchant, 1916, 447.
23. Letter in the Hope Entomological Collection of Oxford University Museum, no referencing designation given other than under Wallace's name.
24. Sulloway, 1996.
25. Sulloway, 1990, 10.
26. Sulloway, 1990, 11.
27. Turner and Helms, 1987, 175.
28. Adams and Phillips, 1972.
29. Kidwell, 1981.
30. Markus, 1981.
31. Hilton, 1967.
32. Nisbet, 1968.
33. See also Bank and Kahn, 1982; Dunn and Kendrick, 1982; Koch, 1956; Sutton-Smith and Rosenberg, 1970.
34. Sulloway, 1990, 19.
35. Ibid.
36. Personal correspondence, January 25, 1991.
37. Kurtz, 1986, 477.
38. Ibid., 417.
39. Ibid., 459.
40. British Library, Department of Manuscripts, catalogued volume #46436, folio #299.
41. Ibid.

CHAPTER 8. A SCIENTIST AMONG THE SPIRITUALISTS

1. Tipler, 1994, 3.
2. Wallace, 1869, 391–392
3. Ibid.
4. Ibid., 394.
5. Lyell, 1881, v. ii, 442.

6. In Marchant, 1916, 197.
7. Ibid.
8. Ibid., 199.
9. Ibid., 206.
10. Ibid.
11. Ibid., 200.
12. Ibid., 450.
13. Gould, 1980, 23.
14. Kottler, 1974, 145.
15. Schwartz, 1984, 285.
16. Wallace, 1913, 621.
17. Schwartz, 1984, 285.
18. Ibid., 288.
19. Wallace, 1870, 204.
20. Ibid., 206.
21. Ibid., 209.
22. Ibid., 212.
23. Ibid., 210.
24. Ibid.
25. Ibid., 213.
26. Wallace, 1889, 469.
27. Ibid., 478.
28. Wallace, 1885, 9.
29. Cooter, 1984, 17.
30. Wallace, 1908, 126.
31. Ellenberger, 1970, 84.
32. See Shapin, 1994, for an excellent discussion of the social history of science.
33. Hacking, 1988, 435–437.
34. Wallace, 1866, 10.
35. In Marchant, 1916, 422.
36. Ibid., iii.
37. Ibid., 1.
38. Ibid., 2.
39. Ibid., 3.
40. Ibid., 4.
41. Ibid., 7
42. Ibid., 9.
43. In Marchant, 1916, 423.
44. Huxley, 1900, 1, 419–420.
45. F. Darwin, 1887, v. ii, 364–365.
46. Wallace, 1908, 336–337.
47. Wallace, 1874, 630–657.
48. Letter in the Hope Entomological Collection of Oxford University Museum, no referencing designation given other than under Wallace's name.

49. *Chambers' Encyclopaedia*, 1892, v. ix, 645–649.
50. Wallace, 1875, vii-viii.*

CHAPTER 9. PEDESTALS AND STATUES

1. Quoted in Hook, 1943, 14.
2. Quoted in Mencken, 1987, 533.
3. Ibid.
4. Quoted in Seldes, 1983, 657.
5. Quoted in Hook, 1943, 79.
6. Ibid.
7. Hook, 1943, 20.
8. For an example of a historical matrix, see Shermer, 1988.
9. Hook, 1943, vii.
10. Ibid., 153.
11. Ibid., 155–157.
12. Quoted in Seldes, 1983, 143.
13. Hook, 1943, 77.
14. Boorstin, 1987, 285.
15. Ibid., 285.
16. Ibid., 287.
17. Ibid., 288
18. Ibid., 284.
19. Ibid.
20. Ibid., 289.
21. Quoted in ibid., 284.
22. Sulloway, 1979, 3.
23. Ibid., 445.
24. Quoted in ibid., 276.
25. Ibid.
26. All quotes in this paragraph from Sulloway, 1979, 477–478.
27. Quoted in Sulloway, 1979, 114.
28. Fromm, 1959, 81.
29. Ibid., 82.
30. Sulloway, 1979, 489.
31. Campbell, 1946, 446.
32. Sulloway, 1979, 446–447.
33. Ibid., 21.
34. Ibid., 421.
35. Ibid.
36. Ibid., 443.
37. Quoted in Sulloway, 1979, 4–5.
38. Ibid.
39. Ibid., 479.

40. Ibid., 137.
41. Ibid., 303.
42. Quoted in Korey, 1984, 22.
43. Barzun, 1958, 25.
44. Himmelfarb, 1959, 127.
45. In F. Darwin, 1892, 5.
46. Ibid., 54–55.
47. Ibid., 55.
48. Ibid., 52–53.
49. Quoted in Korey, 20.
50. Quoted in Gould, 1986, 33.
51. Quoted in Korey, 11–12.
52. Ibid., 12–13.
53. Ibid., 18.
54. Ibid., 22.
55. Ibid., 15.
56. Ibid., xvi.
57. Ibid.
58. Ibid., viii.
59. Hitching, 1982, 196.
60. Quoted in Hitching, 210.
61. Ibid., 199.
62. Ibid.
63. Ibid.
64. In Minkoff, 1983, 65.
65. Personal communication, June 18, 1987.
66. Quoted in Mayr, 1982, 495.
67. Mayr, 1982, 495.
68. Quoted in Gould, 1980, 137.
69. Quoted in Korey, 1984, 234.
70. Ibid., 229.

CHAPTER 10. THE EXQUISITE BALANCE

1. Popper, 1959, 27, 34.
2. Ibid., 40–41.
3. Ward and Brownlee, 2000; Broad, 2000.
4. Tuchman, 1978, xvii.
5. Nickerson, 1998.
6. In Davidson, 1999, 374–375.
7. Swift, 1990, 129.
8. Ibid., 42.
9. Comment in survey collected in November, 1999.
10. Davidson, 1999, 113.

11. Ibid., 121.

12. Ibid., 140.

13. Ibid., 395.

14. Ibid., 89.

15. Swift, 1990, 57.

16. Ibid., 236.

17. Ibid., 123.

18. Ibid., 135.

19. Ibid., 28.

20. Ibid., 135.

21. Ibid., 47.

22. Ibid., 105.

23. Ibid., 219.

24. Davidson, 1999, 350.

25. Sagan, 1986, 430–431.

26. Sulloway, 1996, 73.

27. See McCrae and Costa, 1987 and 1990 on the "Big 5," or the Five Factor model of personality.

28. Sulloway, 1996, 213.

29. Davidson, 1999, 89.

30. Ibid.

31. Ibid., 90–91.

32. Kuhn, 1977.

33. Shermer, 1999.

34. Personal correspondence, November 25, 1999.

35. Ibid.

36. Ibid.

37. Personal correspondence, December 7, 1999.

38. Ibid.

39. Ibid.

CHAPTER II. THE BEAUTIFUL PEOPLE MYTH

1. Taussig, 1980.

2. Ibid., 229.

3. Cronon, 1983, 12–13.

4. Merchant, 1980, xvi.

5. Ibid., 2.

6. Ibid., 295.

7. Eisler, 1987, xvi.

8. Ibid., 295.

9. Low, 1996; on the Standard Cross-Cultural Sample see Murdock and White, 1969.

10. Keeley, 1996.

11. Edgerton, 1992.

12. Wrangham and Peterson, 1996.
13. Diamond, 1997.
14. Leaky and Lewin, 1992.
15. Tattersall, 1995.
16. Roberts, 1989.
17. Crosby, 1986.
18. Gellner, 1988.
19. Bronowski, 1973.
20. Diamond, 1997.
21. Boserup, 1988, 31.
22. Crosby, 1994.
23. Boserup, 1988, 29.
24. Flannery, 1969.
25. Cohen, 1977.
26. Crosby, 1986, 20.
27. Worster, 1988.
28. Crosby, 1986.
29. Diamond, 1992.
30. Cassels, 1984.
31. Reed, 1970.
32. Martin and Klein, 1984.
33. Krantz, 1970.
34. Muench and Pike, 1974, 161.
35. Betancourt and Van Devender, 1981.
36. Bingham, 1948.
37. Hemming, 1970.
38. Ibid.
39. Fejos, 1944.
40. Hemming, 1981.
41. Heyerdahl, 1958.
42. Bellwood, 1987.
43. Bahn and Flenley, 1992, 213.
44. Flenley and King, 1984.
45. Shermer, 1993, 1997.

CHAPTER 12. THE AMADEUS MYTH

1. Weisberg, 1986.
2. DeGroot, 1966.
3. In Hardison, 1988, 176.
4. Chase and Simon, 1973.
5. Benjamin and Shermer, 1991.
6. Simonton, 1984, 1988, 1994.
7. Simonton, 1999.

8. Quoted in Simonton, 1999, 28.

9. Ibid.

10. Ibid., 6.

11. Root-Bernstein and Root-Bernstein, 1999.

12. Ibid., 2.

13. Greeno, 1980.

14. Simonton, 1999, 47–48.

15. Sulloway, 1991, 32.

16. The list is my own formulation arrived at through reading numerous books on genius and creativity.

17. Einstein, 1982.

18. Ibid.

19. Ibid.

20. Ibid.

21. In Rothenberg, 1979.

22. Sulloway, 1982.

23. Gould, 1985.

24. Wallace, 1908.

25. In Rothman, 1982.

26. Schneider, 1953.

27. Ibid.

28. Ibid.

29. In Westfall, 1980.

30. Ibid.

31. Ibid.

32. Ibid.

33. Ibid.

34. In Baker, 1982.

35. Gay, 1999.

36. Baker, 1982.

37. Hays, 1981.

38. Turner, 1938.

CHAPTER 13. A GENTLEMANLY ARRANGEMENT

1. Medawar, 1984, 2–3.

2. Mayr, 1982, 1988.

3. Wallace, 1908, 8.

4. Ibid., 61.

5. Ibid., 45.

6. Ibid.

7. Wallace, 1903.

8. Wallace, 1908a, 123–124.

9. Ibid., 222.

10. Ibid., 152.

11. Ibid., 227.

12. Ibid., 228. The term "agnostic" was coined in 1869 by Thomas Huxley, to distinguish himself from theists and gnostics who were certain of their knowledge of God's existence. Huxley was uncertain and believed that the God question was an insoluble one, which is what he meant by the term.

13. Bates, 1863.

14. *Zoologist*, 19 October, 1852, 3641–3643.

15. Ibid.

16. Ibid.

17. Ibid.

18. *The Annals and Magazine of Natural History*, September, 1855, 195, original italics.

19. In Marchant, 1916, 56.

20. Wallace, 1895, 23.

21. Correspondence between Darwin and Wallace on the priority question under discussion here reprinted in *The Correspondence of Charles Darwin*. Vol. 7, 1858–1859. Cambridge University Press.

22. Brackman, 1980.

23. AJRW, l. 40. This is the private collection of Wallace's two grandsons, Alfred John Russel Wallace, and Richard Russel Wallace. Letters are designated by letter number, corresponding to a catalogue of the collection.

24. In Marchant, 1916, 65.

25. Brackman, 1980, 78.

26. In Marchant, 1916, 111–112.

27. Ibid., 128–129.

28. DAR, 47:145 (Darwin Archives, Cambridge University Library, catalogue and number).

29. AJRW, l. 41. Emphasis in original.

30. F. Darwin, 1887, 68.

31. Brooks, 1984, 258.

32. Ibid., 261–263.

33. Ibid., 257.

34. McKinney, 1972, 139.

35. Ibid., 141.

36. See Beddall, 1968 and 1988 for a detailed history of the development of Wallace's ideas.

37. von Neumann, 1947.

38. Dawkins, 1976.

39. Trivers, 1971.

40. Axelrod and Hamilton, 1981.

41. Marchant, 1916, 113.

42. Ibid., 131.

43. DAR:106, 107.

44. Bell, 1859, viii-ix.

45. Marchant, 1916, 57.

46. AJRW, l. 46. Emphasis in original.

47. Ibid.

48. RES, 397 (Royal Entomological Society. No referencing designation given. Number designates page in the article).

49. Ibid.

50. Ibid., 396–397.

51. Ibid., 398–400, enumeration added.

52. Ibid., 399.

53. Ibid., 400.

CHAPTER 14. THE GREAT BONE HOAX

1. See Shermer, 1997, for a complete history of the Louisiana creationism trial.

2. See Gould, 1997, 1999; Ruse, 1997; Scott, 1997; Shermer, 1999, for discussions on the Pope's statement on evolution.

3. See Robert Pennock's brilliant analysis of the new creationism in *Tower of Babel: The Evidence Against the New Creationism,* providing both refutations of their arguments and a social history of their movement.

4. This is the debate over gradualism versus punctuated equilibrium—see Chapter 10 for a longer discussion.

5. Compare, for example, Stephen Jay Gould's *Full House,* where he argues that life had to get larger and more complex simply by pushing away from a left wall of minimal size and complexity, to Robert Wright's *Nonzero,* where he claims that life gets larger and more complex even when it is well away from the left wall of minimal size and complexity.

6. For a good general discussion accompanied by the finest photographs ever presented to the public of fossil hominids see Johanson and Edgar, 1996. See Tattersal, 1995 for a scholarly but highly readable account. The best general textbook on the subject of human evolution is Klein, 1999.

7. Almost everyone has argued that the increase in brain size is a result of changes in the physical environment. Evolutionary psychologists, however, make the case for the role of the social environment. See, for example, Pinker, 1997; Stanford, 1999; Boehm, 1999; Jolly, 1999; Leakey and Lewin, 1992; Leakey, 1994; Dawkins, 1996. A fascinating study on the relationship between grooming, gossip, language, and the brain, see Dunbar, 1996. And, of course, the classic work in the field is Trivers, 1985.

8. Once again, most evolutionary biologists believe that the brain's evolution must be due entirely to selection factors, but Gould, 1989, shows how contingency might have had a role.

9. See the special issue of *Skeptic* devoted to the race and IQ debate in vol. 3, no. 3.

10. Landau, 1991. Landau follows the narrative structure set out by Propp, 1928. See also Shermer, 1999, Chapter 7, "The Storytelling Animal," for a discussion on the evolution of the propensity to tell stories.

11. Ibid.

12. Walsh, 1996.

13. Weiner, 1955; Millar, 1972; Blinderman, 1986; Spencer, 1990.

14. Smith, 1927.

15. Weiner, 1955.

16. Quoted in Blinderman, 1986, 193; original letter in the British Museum of Natural History, dated November 24, 1953.

17. Quoted in Blinderman, 1986, 193. Letter dated May 14, 1981.

18. Gould, 1979.

19. Gould, 1980, 1981.

20. Gould, 1979.

21. Ibid.

22. Blinderman, 1986, 142.

23. In Gish, 1978.

24. Parker, 1981.

BIBLIOGRAPHY

Achenbach, J. 1999. *Captured by Aliens: The Search for Life and Truth in a Very Large Universe*. New York: Simon and Schuster.

Adams, R. L. and B. N. Phillips. 1972. "Motivation and Achievement Differences Among Children of Various Ordinal Birth Positions." *Child Development* 43: 155–164.

Almond, E. "Debate Over Whether Black Athletes Are Superior to Whites is Not a New One," *Los Angeles Times*, III 9.

Astrand, P. O. and K. Rodahl. 1977. *Textbook of Work Physiology*. New York: McGraw Hill.

Axelrod, R. and Hamilton, W. D. 1981. "The Evolution of Cooperation." *Science* 211: 1390–6.

Bahn, P. G. and J. Flenley. 1992. *Easter Island, Earth Island*. New York: Thames and Hudson.

Baker, R. 1982. *Mozart*. London: Thames and Hudson.

Baker, R. 1990. *They Call it Hypnosis*. Amherst: Prometheus Books.

———. 1992. *Hidden Memories: Voices and Visions From Within*. Amherst: Prometheus Books.

Bank, S. and M. D. Kahn. 1982. *The Sibling Bond*. New York: Basic Books.

Bartholomew, R. E. and G. S. Howard. 1998. *UFOs and Alien Contact: Two Centuries of Mystery*. Amherst: Prometheus Books.

Barzun, J. 1958. *Darwin, Marx, Wagner*. New York: Doubleday

Bates, H. W. 1863. *The Naturalist on the River Amazons*. London: Murray.

Beddall, B. G. 1968. "Wallace, Darwin, and the Theory of Natural Selection: A Study in the Development of Ideas and Attitudes." *Journal of the History of Biology* 1: 261–323.

———. 1988. "Darwin and Divergence: The Wallace Connection." *Journal of the History of Biology* 21, 1: 2–68.

Beer, A. and P. Beer (Eds). 1975. *Kepler: Four Hundred Years*. Proceedings of Conferences Held in Honour of Johannes Kepler. New York: Pergamon Press.

Bell, T. 1859. "Presidential Address." *J. Linn. Soc. London, (Zool.)* 4.

Bellwood, P. 1987. *The Polynesians: Prehistory of an Island People*. London: Thames and Hudson.

Benjamin, A. and M. Shermer. 1991. *Mathemagics: How to Look Like a Genius Without Really Trying*. Chicago: Contemporary Books.

Betancourt, J. L. and T. R. Van Devender. 1981. "Holocene Vegetation in Chaco Canyon, New Mexico." *Science* 214: 656–658.

Bingham, H. 1948. *Lost City of the Incas: The Story of Machu Picchu and Its Builders*. New York: Duell, Sloan and Pearce.

Blinderman, C. 1986. *The Piltdown Inquest*. Buffalo: Prometheus Books.

Bloomberg, D. 1994. "The Incredible Mysteries of Sun Pictures." *Skeptic*, Vol. 2, No. 3: 34–37.

Boehm, C. 1999. *Hierarchy in the Forest: The Evolution of Egalitarian Behavior*. Cambridge, Mass: Harvard University Press.

Boorstin, D. 1987. "From Hero to Celebrity." In *Hidden History*. New York: Harper & Row.

Boring, E. G. 1950. *A History of Experimental Psychology*. New York: Appleton.

Boserup, E. 1988. "Environment, Population, and Technology in Primitive Societies." In *The Ends of the Earth: Perspectives on Modern Environmental History*. D. Worster (Ed.). Cambridge: Cambridge University Press.

Bowlby, J. 1990. *Charles Darwin: A New Life*. New York: W. W. Norton.

Bowler, J. 1988. *The Non-Darwinian Revolution*. Baltimore: Johns Hopkins University Press.

———. 1989. *Evolution: The History of an Idea*. Revised Edition (1983). Berkeley: University of California Press.

Bowler, P. J. 1983. *The Eclipse of Darwinism*. Baltimore: John Hopkins University Press.

Brackman, A. 1980. *A Delicate Arrangement: The Strange Case of Charles Darwin and Alfred Russel Wallace*. New York: Times Books.

Broad, W. J. "Maybe We Are Alone in the Univese, After All." *New York Times*, 8, February 2000.

Bronowski, J. 1973. *The Ascent of Man*. Boston: Little, Brown.

Brooks, J. L. 1984. *Just Before the Origin: Alfred Russel Wallace's Theory of Evolution*. New York: Columbia University Press.

Brown, C. 1996. *Cosmic Voyage: A Scientific Discovery of Extraterrestrials Visiting Earth*. New York: Dutton.

———. 1999. *Cosmic Explorers: Scientific Remote Viewing, Extraterrestials, and a Message For Mankind*. New York: Dutton.

Burkhardt, F. and S. Smith (Eds.) 1985–1991. *The Correspondence of Charles Darwin*. 7 vols. Cambridge: Cambridge University Press.

Burke, E. (Ed.) 1986. *Inside the Cyclist: Physiology for the Two-Wheeled Athlete*. Brattleboro, Vermont: Velo-News.

Butterfield, H. 1931. *The Whig Interpretation of History*. London: G. Bell.

Campbell, J. 1949. *The Hero with a Thousand Faces*. Princeton: Princeton University Press.

———. 1972. *Myths to Live By*. New York: Bantam.

———. 1988. *The Power of Myth*. New York: Doubleday.

Cassels, R. 1984. "Faunal Extinction and Prehistoric Man in New Zealand and the Pacific Islands." In *Quaternary Extinctions*. P. S Martin and R. G. Klein (Eds.) Tucson: University of Arizona Press, 741–767.

Chambers, R. 1844. *Vestiges of the Natural History of Creation*. London: J. Churchill.

Chase, W. G., and H. A. Simon 1973. "Perception in Chess." *Cognitive Psychology* 4: 55–81.

Chyba, C. 1999. "An Exobiologist's Life Search." *Nature* 28, October: 857–858.

Cohen, M. N. 1977. *The Food Crisis in Prehistory: Overpopulation and the Origins of Agriculture*. New Haven: Yale University Press.

Cohen, I. B. 1960 (1985 revision). *The Birth of a New Physics*. New York: W. W. Norton.

———. 1985. *Revolution in Science*. Cambridge, Mass.: Harvard University Press.

Combe, G. 1835. *Essay on the Constitution of Man and Its Relation to External Objects*. Boston: Phillips and Sampson.

Cooter, R. 1984. *The Cultural Meaning of Popular Science. Phrenology and the Organization of Consent in Nineteenth-Century Britain*. Cambridge: Cambridge University Press.

Copernicus, N. 1978. *De Revolutionibus Orbium Coelestium*. Jerzy Dobrzycki (Ed.), translation and commentary by Edward Rosen. Baltimore: Johns Hopkins University Press.

Costa, and R. McRae. 1992. "Four Ways Five Factors are Basic." *Personality and Individual Differences*, 13, 6: 653–645.

Cowen, R. 1986. "Creationism and the Science Classroom." *California Science Teacher's Journal* 16, 5: 8–15.

Crombie, A. C. 1979. *Augustine to Galileo*. Cambridge, Mass.: Harvard University Press.

Cronon, W. 1983. *Changes in the Land: Indians, Colonists, and the Ecology of New England*. New York: Hill and Wang.

Crosby, A. W. 1986. *Ecological Imperialism: The Biological Expansion of Europe 900–1900*. Cambridge: Cambridge University Press.

———. 1994. *Germs, Seeds, and Animals: Studies in Ecological History*. London: M. E. Sharpe.

Daly, J. 1979. "Cosmic Harmony and Political Thinking in Early Stuart England." *Transactions of the American Philosophical Society*. Vol. 69.

Dante Alighieri. 1952. *The Divine Comedy*. Translation by Charles Eliot Norton. Chicago: University of Chicago Press. Great Books of the Western World, Vol. 21.

Darwin, C. 1859. *On the Origin of Species by Means of Natural Selection, or the Preservation of Favoured Races in the Struggle for Life*. London: Murray.

———. 1862. *On the Various Contrivances by Which British and Foreign Orchids are Fertilized by Insects*. London: Murray.

———. 1864. Darwin Papers Nos. 82, 106, 107. Handlist of Darwin Papers in University Library, Cambridge.

———. 1868. *The Variation of Animals and Plants Under Domestication* (Vols. I & II). London: Murray.

———. 1871. *The Descent of Man*. London: Murray.

Darwin, C. and A. R. Wallace. 1858. "On the Tendency of Species to Form Varieties and on the Perpetuation of Varieties and Species by Natural Means of Selection." *Journal of the Proceedings of the Linnean Society (Zoology)* 3: 53–62.

Darwin, F. (Ed.). 1892. *The Autobiography of Charles Darwin and Selected Letters*. New York: Dover Publications.

———. 1887. *The Life and Letters of Charles Darwin*. Vols. 1–3. London: John Murray.

Davidson, K. 1999. *Carl Sagan: A Life*. New York: Wiley.

Dawkins, R. 1976. *The Selfish Gene*. Oxford: Oxford University Press.

———. 1996. *Climbing Mount Improbable*. New York: W. W. Norton.

———. 1998. *Unweaving the Rainbow: Science, Delusion and the Appetite for Wonder*. Boston: Houghton Mifflin.

Dennett, D. 1995. *Darwin's Dangerous Idea*. New York: Simon and Schuster.

DeGroot, A. 1966. "Perception and Memory Versus Thought: Some Old Ideas and Recent Findings." In *Problem Solving: Research, Method, and Theory*. B. Kleinmuntz (Ed.) New York: John Wiley.

Desmond, A. and J. Moore. 1991. *Darwin. The Life of a Tormented Evolutionist*. New York: Warner Books.

Diamond, J. 1992. *The Third Chimpanzee*. New York: HarperCollins.

———. 1997. *Guns, Germs, and Steel: The Fates of Human Societies*. New York: W. W. Norton.

Digman, J. 1990. "Personality Structure: Emergence of the Five-Factor Model." *Annual Review of Psychology* 41: 417–440.

Dixon, P. 1998. "Animals Are One Thing, Humans Quite Another." Perspectives on Cloning. *Los Angeles Times*, 28 December: B11.

Donne, J. 1912. *An Anatomie of the World: the First Anniversarie*. In *The Poems of John Donne*. Herbert Grierson (Ed.). Oxford University Press.

Duhem, P. *To Save the Phenomena*. Trans. E. Dolan and C. Maschler. Chicago: University of Chicago Press.

Dunbar, R. 1996. *Grooming, Gossip, and the Evolution of Language*. New York: Oxford University Press.

Dunn, J. and C. Kendrick. 1982. *Siblings: Love, Envy and Understanding*. Cambridge, Mass.: Harvard University Press.

Dyson, F. 1997. *Imagined Worlds*. Cambridge, Mass.: Harvard University Press.

Edgerton, R. 1992. *Sick Societies: Challenging the Myth of Primitive Harmony*. New York: Free Press.

Einstein, A. 1982. "How I Created the Theory of Relativity." *Physics Today*, vol. 35, no. 8: 45–47.

Eisler, R. 1987. *The Chalice and the Blade: Our History, Our Future*. San Francisco: Harper & Row.

Eldredge, N. 1971. "The Allopatric Model and Phylogeny in Paleozoic Invertebrates." *Evolution* 25: 156–167.

———. and S. J. Gould. 1972. "Punctuated Equilibria: An Alternative to Phyletic Gradualism," in *Models in Paleobiology*. T. J. M. Schopf (Ed.). San Francisco: W. H. Freeman.

Ellenberger, H. 1970. *The Discovery of the Unconscious: The History and Evolution of Dynamic Psychology*. New York: Basic Books.

Englert, S. 1970. *Island at the Center of the World: New Light on Easter Island*. (Trans. W. Mulloy.) New York: Charles Scribner's Sons.

Entine, J. 2000. *Taboo: Why Black Athletes Dominate Sports and Why We're Afraid to Talk About it*. New York: Public Affairs.

———. 2000. "Breaking the Taboo: Why Black Athletes Dominate Sports and Why We're No Longer So Afraid to Talk About it." *Skeptic*, vol. 8, no. 1: 29–33.

Fagan, B. (Ed.). 1996. *The Oxford Companion to Archaeology*. New York: Oxford University Press.

Feder, K. 1999. *Frauds, Myths, and Mysteries: Science and Pseudoscience in Archaeology*. Mountain View, California: Mayfield.

Fejos, P. 1944. *Archeological Explorations in the Cordillera Vilcabamba, Southeastern Peru*. New York: Viking.

Feynman, R. 1988. *"What Do You Care What Other People Think?": Further Adventures of a Curious Character*. New York: W. W. Norton.

Flannery, K. V. 1969. "Origins and Ecological Effects of Early Domestication in Iran and the Near East." In *The Domestication and Exploitation of Plants and Animals*. P. J. Ucko & G. W. Dimbleby (Eds.) Chicago: Aldine.

Flenley, J. R. and S. M. King. 1984. "Late Quaternary Pollen Records From Easter Island." *Nature* 307: 47–50.

Fromm, E. 1959. *Sigmund Freud's Mission: An Analysis of His Personality and Influence*. New York: Harper & Row.

Futuyma, D. J. 1989. *Evolutionary Biology*. Sunderland, Mass.: Sinauer Associates.

Gardner, M. 1952. *In the Name of Science*. New York: Putnam.

———. 1981. *Science: Good, Bad, and Bogus*. Amherst: Prometheus Books.

Gasparini, G. and L. Margolies. 1980. *Inca Architecture*. (Trans P. Lyon.) Bloomington: Indiana University Press.

Gauld, A. 1992. *A History of Hypnotism*. Cambridge: Cambridge University Press.

Gay, P. 1999. *Mozart*. New York: Simon and Schuster.

Gellner, E. 1988. *Plough, Sword and Book: The Structure of Human History*. Chicago: University of Chicago Press.

George, W. 1964. *Biologist Philosopher: A Study of the Life and Writings of Alfred Russel Wallace*. London: Abelard Schuman.

Ghiselin, M. T. 1969. *The Triumph of the Darwinian Method*. Berkeley: University of California Press.

Gilkey, L. (ed.). 1985. *Creationism on Trial*. New York: Harper & Row.

Gish, D. J. 1978. *Evolution: The Fossils Say No!* San Diego: Creation-Life.

Grabiner, J. V., and D. Miller. 1974. "Effects of the Scopes Trial." *Science*. 185: 832–836.

Gingerich, O. 1973. "From Copernicus to Kepler: Heliocentrism as Model and as Reality." *Proceedings of the American Philosophical Society*. Vol.117.

———. 1975a. " 'Crisis' Versus Aesthetic in the Copernican Revolution." *Vistas in Astronomy*. Vol. 17.

———. 1975b. *The Nature of Scientific Discovery*. Washington DC: Smithsonian Institution Press.

Glick, T. F. 1988. *The Comparative Reception Of Darwinism*. Chicago: University of Chicago Press.

Gold, T. 1999. *The Deep Hot Biosphere*. New York: Copernicus.

Goldberg, L. 1993. "The Structure of Phenotypic Personality Traits." *American Psychologist* 48, 1: 26–34.

Goldstein, R. "Al Campanis is Dead at 81; Ignited Baseball Over Race," *New York Times*, 22 June 1998, C11.

Gould, S. J. 1979. Piltdown Revisited. *"Natural History"* 88: 86–97.

———. 1980. "Natural Selection and the Human Brain: Darwin vs. Wallace." In *The Panda's Thumb*. New York: W. W. Norton.

———. 1980. "The Piltdown Conspiracy." *Natural History* August: 8–28.

———. 1981. "Piltdown in Letters." *Natural History* 90: 12–30.

———. 1983. "A Visit to Dayton." In *Hen's Teeth and Horse's Toes*. New York: W. W. Norton.

———. 1985. *The Flamingo's Smile*. New York: W. W. Norton.

———. 1986. "Knight Takes Bishop?" *Natural History* 5, 33–37.

———. 1989. *Wonderful Life*. New York: W. W. Norton.

———. 1991. "Opus 200." *Natural History* 9: 12–19.

———. 1992. "Punctuated Equilibrium in Fact and Theory." In *The Dynamics of Evolution*. A. Somit and S. A. Peterson Ithaca: Cornell University Press.

———. 1997. "Nonoverlapping Magisteria: Science and Religion Are Not in Conflict, for Their Teachings Occupy Distinctly Different Domains." *Natural History* 3: 10–18.

———. 1999. *Rocks of Ages: Science and Religion in the Fullness of Life*. New York: Ballantine Books.

Gould, S. J. and E. S. Vrba. 1982. "Exaptation—a Missing Term in the Science of Form." *Paleobiology* 8: 4–15.

Greene, J. C. 1982. *Science, Ideology, and World View*. Berkeley: University press.

Greeno, J. G. 1980. "Trends in the Theory of Knowledge for Problem Solving." In *Problem Solving and Education: Issues in Teaching and Learning*. D. T. Tuma and R. Reif (Eds.). Hillsdale, New Jersey: Erlbaum.

Gross, A. C. 1986. *Endurance: The Events, The Athletes, The Attitude*. New York: Dodd, Mead, and Co.

Guilford, J. 1959. *Personality*. New York: McGraw-Hill.

Hacking, I. 1988. "Telepathy: Origins of Randomization in Experimental Design," *Isis* 79, no. 298: 427–451.

Halpern, D. F. 1996. *Thought and Knowledge: An Introduction to Critical Thinking*. Mahwah, New Jersey: Lawrence Erlbaum Associates.

Hancock, G. 1995. *Fingerprints of the Gods: The Evidence of Earth's Lost Civilization*. New York: Three Rivers Press.

———. 1999. *The Mars Mystery: The Secret Connection Between Earth and the Red Planet*. New York: Crown.

Hardison, R. 1988. *Upon the Shoulders of Giants*. New York: University Press of America.

Harris, S. 1995. "Does HIV Really Cause AIDS?: A Case Study in Skepticism Taken Too Far." *Skeptic*, vol. 3, no. 2.

Hayes, J. R. 1981. *The Complete Problem Solver*. Philadelphia: Franklin Institute Press.

Hemming, J. 1970. *The Conquest of the Incas*. New York: Harcourt Brace Jovanovich.

———. 1981. *Machu Picchu*. New York: Newsweek Book Division.

Heyerdahl, T. 1958. *Aku-Aku: The Secret of Easter Island*. New York: Rand McNally.

Hilgard, E. R. 1968. *The Experience of Hypnosis*. New York: Harcourt Brace Jovanovich.

———. 1977. *Divided Consciousness: Multiple Controls in Human Thought and Action*. New York: Wiley-Interscience.

Hilton, I. 1967. "Differences in the Behavior of Mothers Toward First and Later Born Children." *Journal of Personality and Social Psychology* 7: 282–290.

Himmelfarb, G. 1959. *Darwin and the Darwinian Revolution*. New York: Doubleday.

Hitching, E. 1982. *The Neck of the Giraffe*. New York: New American Library.

Hobbes, T. 1651 (1968). *Leviathan*. C. B. Macpherson (Ed.). New York: Penguin Books.

Hoberman, J. 1998. *Darwin's Athletes: How Sports Has Damaged Black America and Preserved the Myth of Race*. New York: Houghton Mifflin.

———. 2000. "Totum and Taboo: The Myth of Race in Sports." *Skeptic*, vol. 8, no. 1: 42–45.

Hook, S. 1943. *The Hero in History: A Study in Limitation and Possibility*. Boston: Beacon Press.

Hooker, R. 1594. *Of the Laws of Ecclesiastical Polity. Books I–V*. Christopher Morris (Ed.). London.

Hooykaas, R. 1970. "Historiography of Science, its Aim and Methods." *Organon* 7: 37–49.

Howard, T. and J. Rifkin. 1977. *Who Should Play God?: The Artificial Creation of Life and What it Means for the Future of the Human Race*. New York: Delacorte Press.

Hull, D. L. 1973. *Darwin and His Critics*. Chicago: University of Chicago Press.

Humber, J. M. and R. F. Almeder (Eds.). 1998. *Human Cloning*. Biomedical Ethics Reviews. Totowa, New Jersey: Humana Press.

Huxley, L. 1900. *Life and Letters of Thomas Henry Huxley*. 2 vols. New York: D. Appleton and Co.

Jacob, M. C. 1976. *The Newtonians and the English Revolution: 1689–1720*. Ithaca: Cornell University Press.

Johanson, D. and B. Edgar. 1996. *From Lucy to Language*. New York: Simon and Schuster.

Jolly, A. 1999. *Lucy's Legacy: Sex and Intelligence in Human Evolution*. Cambridge, Mass.: Harvard University Press.

Jones, M. (Ed.). 1990. *Fake? The Art of Deception*. London: The British Museum.

Kane, M. "An Assessment of 'Black is Best.'" *Sports Illustrated*, January 18, 1971, 80.

Kass, L. and J. Q. Wilson. 1998. *The Ethics of Human Cloning*. Washington, D.C.: American Enterprise Institute.

Keeley, L. 1996. *War Before Civilization: The Myth of the Peaceful Savage*. New York: Oxford University Press.

Kevles, D. J. 1999. *The Baltimore Affair*. New York: W. W. Norton.

Kidwell, J. S. 1981. "Number of Siblings, Sibling Spacing, Sex, and Birth Order: Their Effects on Perceived Parent-Adolescent Relationships." *J. Marriage and Family*, May: 330–335.

Kitcher, P. 1996. *The Lives to Come: The Genetic Revolution and Human Possibilities*. New York: Simon and Schuster.

Klein, R. G. 1999. *The Human Career: Human Biological and Cultural Origins*. Chicago: University of Chicago Press.

Kluger, J. 1997. "Will We Follow the Sheep?" *Time*, 10 March: 67–72.

Koch, H. L. 1956. "Attitudes of Young Children Toward Their Peers as Related to Certain Characteristics of Their Siblings." *Psychological Monographs* 70, 19.

Koestler, A. 1959. *The Sleepwalkers*. New York: Macmillan.

Kohn, D. 1985. *The Darwinian Heritage*. Princeton: Princeton University Press.

Kolata, G. 1998. *Clone: The Road to Dolly, and the Path Ahead*. New York: William Morrow and Co.

Korey, K. 1984. *The Essential Darwin: Selections and Commentary*. Boston: Little, Brown.

Kosko, B. 1993. *Fuzzy Thinking: The New Science of Fuzzy Logic*. New York: Hyperion.

———. 1999. *The Fuzzy Future: From Society and Science to Heaven in a Chip*. New York: Simon and Schuster.

Kottler, M. J. 1974. "Alfred Russel Wallace, the Origin of Man, and Spiritualism." *Isis* 65: 145–192.

Krantz, G. S. 1970. "Human Activities and Megafaunal Extinctions." *American Scientist* 58: 164–170.

Krieger, L. 1977. *Ranke: The Meaning of History*. Chicago: University of Chicago Press.

Kuhn, T. 1957. *The Copernican Revolution*. Cambridge, Mass.: Harvard University Press.

———. 1962. *The Structure of Scientific Revolutions*. Chicago: University of Chicago Press.

———. 1977. *The Essential Tension*. Chicago: University of Chicago Press.

Kurtz, P. 1986. *The Transcendental Temptation*. Buffalo: Prometheus Books.

Lakatos, I. and A. Musgrave (Eds.). 1970. *Criticism and the Growth of Knowledge*. Cambridge: Cambridge University Press.

Lambert, J. B. 1997. *Traces of the Past: Unraveling the Secrets of Archaeology Through Chemistry*. Reading, Mass.: Helix Books/Addison-Wesley.

Landau, M. 1991. *Narratives of Human Evolution*. New Haven: Yale University Press.

Larue, G. 1994. "Flood Myths and Sunken Arks." *Skeptic*, vol. 2, no. 3: 38–41.

Leakey, R. 1994. *The Origin of Humankind*. New York: Basic Books.

———. and R. Lewin. 1992. *Origins Reconsidered: In Search of What Makes Us Human*. New York: Doubleday.

Lewin, R. 1987. *Bones of Contention: Controversies in the Search for Human Origins*. New York: Simon and Schuster.

Lippard, J. 1994. "Sun Goes Down in Flames: The Jammal Ark Hoax." *Skeptic*, vol. 2, no. 3: 22–33.

Low, B. S. 1996. "Behavioral Ecology of Conservation in Traditional Societies." *Human Nature* 7, 4: 353–379.

Lyell, C. 1830–1833. *Principles of Geology, Being an Attempt to Explain the Former Changes of the Earth's Surface, by Reference to Causes Now in Operation*. 3 Vols. London: Murray.

Lyell, K. M. (Ed.). 1881. *The Life, Letters, and Journals of Sir Charles Lyell*. 2 Vols. London: John Murray.

Marchant, J. 1916. *Alfred Russel Wallace, Letters and Reminiscences*. New York: Arno Press.

Markus, H. 1981. "Sibling Personalities: The Luck of the Draw." *Psychology Today* 15(6): 36–37.

Martin, P. S. and R. G. Klein (Eds.). 1984. *Quaternary Extinctions*. Tucson: University of Arizona Press.

Marx, K. 1852. "The Eighteenth Brumaire of Louis Bonaparte," in *The Marx-Engels Reader*. 2d ed. (1978). R. C. Tucker (Ed.). New York: W. W. Norton.

Mason, S. F. 1952. "The Scientific Revolution and the Protestant Reformation—I: Calvin and Servitus in Relation to the New Astronomy and the Theory of the Circulation of the Blood." *Annals of Science.*

Matheson, T. 1998. *Alien Abductions: Creating a Modern Phenomenon.* Amherst: Prometheus Books.

Mayr, E. 1954. "Change of Genetic Environment and Evolution." In *Evolution as a Process,* J. Huxley (Ed.). London: Allen and Unwin, 157–180.

———. 1982. *Growth of Biological Thought.* Cambridge, Mass.: Harvard University Press.

———. 1988. *Toward a New Philosophy of Biology.* Cambridge, Mass.: Harvard University Press.

———. 1992. "Speciational Evolution or Punctuated Equilibria." In A. Somit and S. A. Peterson. *The Dynamics of Evolution.* Ithaca: Cornell University Press.

McCrae, R. R. and P. T. Costa, Jr. 1987. "Validation of the Five-Factor Model of Personality Across Instruments and Observers." *Journal of Personality and Social Psychology* 52:81–90.

———. 1990. *Personality in Adulthood.* New York: Guilford Press.

McGoodwin, W. L. 1997. "Position Statement" Issued by the Council for Responsible Genetics, 5 Upland Road, Suite 3 Cambridge, MA 02140; wendy@essential.org

McKinney, H. L. 1972. *Wallace and Natural Selection.* New Haven: Yale University Press.

Medawar, P. 1969. *Induction and Intuition in Scientific Thought.* Philadelphia: American Philosophical Society.

———. 1984. *Pluto's Republic: Incorporating the Art of the Soluble and Induction and Intuition in Scientific Thought.* Oxford: Oxford University Press.

Mencken, H. L. 1987. *A New Dictionary of Quotations on Historical Principles From Ancient and Modern Sources.* New York: Knopf.

Merchant, C. 1980. *The Death of Nature: Women, Ecology, and the Scientific Revolution.* New York: HarperCollins.

Miele, F. 1999. "The Man Who Would Be Cloned: An Interview with Dr. Richard Seed, Director of the Human Cloning Foundation." *Skeptic,* vol., 7, no. 2:54–57.

———. 2000. "Introduction to Special Skeptic Symposium on Race and Sports." *Skeptic,* vol. 8, no. 1:28.

Miller, K. R. 1999. *Finding Darwin's God: A Scientist's Search for Common Ground Between God and Evolution.* New York: Harper Collins.

Millar, R. 1972. *The Piltdown Men.* London: Gollancz.

Milton, J. 1942. *Paradise Lost.* H. C. Beeching (Ed.). Oxford: Oxford University Press.

———. 1948–1952. *Of Reformation in England, Milton's Prose Works.* J. Bohn (Ed.), II, London.

Minkoff, E. C. 1983. *Evolutionary Biology.* Reading, Mass.: Addisen-Wesley.

Montagu, M. F. A. 1952. *Darwin: Competition and Cooperation.* New York: Henry Schuman.

Munitz, M. (Ed.). 1957. *Theories of the Universe.* New York: The Free Press.

Muench, D. and D. G. Pike. 1974. *Anasazi: Ancient People of the Rock.* Palo Alto: American West Publishing Company.

Murdock, G. P. and D. White. 1969. "Standard Cross-Cultural Sample." *Ethnology* 8:329–369.

National Bioethics Advisory Commission. 1997. *Cloning Human Beings: Report and Recommendations.* Rockville, Maryland.

Neugebauer, O. *A History of Ancient Mathematical Astronomy*. 3 vols. New York: Springer-Verlag.

Neumann, J. V. and Oskar Morgenstern. 1947. *Theory of Games and Economic Behavior*. Princeton: Princeton University Press.

Newhan, R. "A Lifetime Destroyed by Own Words," *Los Angeles Times*, 22 June 1998.

Nickerson, R. 1998. "Confirmation Bias: A Ubiquitous Phenomenon in Many Guises." *Review of General Psychology*, vol. 2, no. 2:175–220.

Nisbet, R. E. 1968. "Birth Order and Participation in Dangerous Sports." *J. of Personality and Social Psychology* 8:351–353.

Oakley, K. 1981. "Piltdown Man." *New Scientist* 92: 457–458.

Olson, R. 1982. *Science Deified, Science Defied*. Berkeley: University of California Press.

———. 1991. *Science Deified and Science Deified*. Vol. 2. Berkeley: University of California Press.

Palter, R. 1970. "An Approach to the History of Early Astronomy." *Studies in History and Philosophy of Science*. Vol. 1.

Park, R. 2000. *Voodoo Science: The Road from Foolishness to Fraud*. New York: Oxford University Press.

Parker, G. E. 1981. "Origin of Mankind." Impact Series. #101. San Diego: Creation-Life Publishers.

Pennock, R. 1999. *Tower of Babel: The Evidence Against the New Creationism*. Cambridge, Mass.: MIT Press.

Peters, T. 1997. *Playing God?: Genetic Determinism and Human Freedom*. New York: Routledge.

Pinker, S. 1997. *How the Mind Works*. New York: W. W. Norton.

Planck, M. 1936. *The Philosophy of Physics*. New York: W. W. Norton.

Poe, Edgar Allan. 1966. *Edgar Allan Poe: A Series of Seventeen Letters Concerning Poe's Scientific Erudition in Eureka and His Authorship of Leonainie*. No publisher listed.

Popper, K. 1959. *The Logic of Scientific Discovery*. New York: Basic Books.

———. 1975. "The Rationality of Scientific Revolutions." In *Problems of Scientific Revolution: Progress and Obstacles to Progress in the Sciences*. Rom Harre (Ed.). Oxford: Clarendon Press.

Porter, R. 1982. "The Descent of Genius: Charles Darwin's Brilliant Career." *History Today*. Vol. 32.

Poundstone, W. 1999. *Carl Sagan: A Life in the Cosmos*. New York: Henry Holt.

Propp, V. 1928. *Morphology of the Folktale*. Austin: University of Texas Press; English translation, 1968.

Prothero, D. 1992. "Punctuated Equilibrium at Twenty: A Paleontological Perspective." *Skeptic*, vol. 1, no. 3:38–47.

Prowe, L. 1883. *Nicolaus Copernicus*. Berlin.

Ptolemy, C. 1952. *The Almagest*. Trans. Catesby Taliaferro. Chicago: University of Chicago Press. The Great Books of the Western World. Vol. 16.

Rantala, M. L. and A. J. Milgram (Eds.). 1999. *Cloning: For and Against*. Chicago: Open Court.

Reed, C. A. 1970. "Extinction of Mammalian Megafauna in the Old World Late Quaternary." *BioScience* 20, 284–288.

Richards, R. J. 1987. *Darwin and the Emergence of Evolutionary Theories of Mind and Behavior*. Chicago: University of Chicago Press.

Ritchie, A. 1988. *Major Taylor: The Extraordinary Career of a Champion Bicycle Racer*. Baltimore, Maryland: Johns Hopkins University Press.

Roberts, N. 1989. *The Holocene: An Environmental History*. Oxford: Basil Blackwell.

Rogers, J. 1992. "Darwin, Darwinism, and the Darwinian Culture." *Skeptic*, vol. 1, no. 3:86–89.

Root-Bernstein, R. and M. Root-Bernstein. 1999. *Sparks of Genius: The 13 Thinking Tools of the World's Most Creative People*. Boston: Houghton Mifflin.

Rosen, E. 1971. *Three Copernican Treatises*. New York: Octagon Books.

———. 1973. "Reception of Heliocentrism." In *The Nature of Scientific Discovery*. Gingerich (Ed.). Washington DC: Smithsonian Institution Press.

Rosenbaum, R. 1998. *Explaining Hitler*. New York: Random House.

Rothenberg, A. 1979. *The Emerging Goddess*. Chicago: University of Chicago Press.

Rothman, T. 1982. "The Short Life of Evariste Galois." In *Scientific Genius and Creativity*. New York: W. H. Freeman.

Ruse, M. 1992. "Is the Theory of Punctuated Equilibria a New Paradigm?" In Somit, A. and S. A. Peterson, *The Dynamics of Evolution*. Ithaca: Cornell University Press.

———. 1996. *Monad to Man: The Concept of Progress in Evolutionary Biology*. Cambridge, Mass: Harvard University Press.

———. 1997. "John Paul II and Evolution." *The Quarterly Review of Biology*, vol. 72, no. 4, December: 391–395.

———. 1999. *Mystery of Mysteries: Is Evolution a Social Construction?* Cambridge, Mass: Harvard University Press.

Sagan, C. 1986. *Contact*. New York: Pocket Books.

———. 1996. *The Demon Haunted World: Science as a Candle in the Dark*. New York: Random House.

Sarich, V. 2000. "The Final Taboo: Race Differences in Ability." *Skeptic*, vol. 8, no. 1: 46–50.

Schnabel, J. 1997. *Remote Viewers: The Secret History of America's Psychic Spies*. New York: Dell Books.

Schneider, E. 1953. *Coleridge, Opium, and Kubla Khan*. Chicago: University of Chicago Press.

Schonfield, H. 1966. *The Passover Plot*. New York: Bantam.

Scott, E. C. 1997. "Creationists and the Pope's Statement." *The Quarterly Review of Biology*, vol. 72, no. 4, December: 401–406.

Schwartz, J. 1984. "Darwin, Wallace and the Descent of Man." *Journal of the History of Biology* 17, 2: 271–289.

Segal, N. 1999. *Entwined Lives: Twins and What They Tell Us About Human Behavior*. New York: Dutton.

Seldes, G. 1983. *The Great Quotations*. Secaucus, New Jersey: Citadel Press.

Shakespeare, W. 1952. *Troilus and Cressida*. Chicago: University of Chicago Press. The Great Books of the Western World. Vol. 27.

Shapin, S. 1994. *A Social History of Truth*. Chicago: University of Chicago Press.

Shelley, M. 1965. *Frankenstein: or the Modern Prometheus*. New York: Sigert.

Shermer, M. 1985. *Sport Cycling*. Chicago: Contemporary Books.

———. 1987. *Cycling: Endurance and Speed*. Chicago: Contemporary Books.

———. 1988. "The Historical Matrix Model: A Theory of Historical Contingency." Paper presented at the annual meeting of Interface 88, Atlanta, November.

———. 1990. "Darwin, Freud, and the Myth of the Hero in Science." *Knowledge: Creation, Diffusion, Utilization* vol. 11, no. 3: 280–301.

———. 1991a. "Science Defended, Science Defined: The Louisiana Creationism Case." *Science, Technology & Human Values*, Vol. 16 No. 4: 517–539.

———. 1991b. *Heretic-Scientist: Alfred Russel Wallace and the Evolution of Man*. Dissertation Abstracts. Ann Arbor, Michigan: UMI.

———. 1993a. "The Chaos of History: On a Chaotic Model That Represents the Role of Contingency and Necessity in Historical Sequences." *Nonlinear Science Today* 2,4:1–13.

———. 1993b. *Race Across America: The Agonies and Glories of the World's Longest and Cruelest Bicycle Race*. Waco, Texas: WRS Publishing Group.

———. 1997. "The Crooked Timber of History." *Complexity* 2, 6: 23–29.

———. 1998. "If Only God Can Do It, No More Triple Bypasses." *Los Angeles Times*. 28 December: B11.

———. 1999. *How We Believe: The Search for God in an Age of Science*. New York: W. H. Freeman.

——— and A. Grobman. 2000. *Denying History: Who Says the Holocaust Never Happened and Why Do They Say It?* Berkeley: University of California Press.

——— and F. J. Sulloway. 2000. "The Grand Old Man of Evolution: An Interview with Evolutionary Biologist Ernst Mayr." *Skeptic*, vol. 8, no. 1: 76–81.

Simons, S. 1993. *No One May Ever Have the Same Knowledge Again*. West Covina, California: Society for the Diffusion of Useful Information Press.

Simonton, D. K. 1984. *Genius, Creativity, and Leadership: Historiometric Inquiries*. Cambridge, Mass: Harvard University Press.

———. 1988. *Scientific Genius: A Psychology of Science*. Cambridge: Cambridge University Press.

———. 1994. *Greatness: Who Makes History and Why*. New York: Guilford Press.

———. 1999. *Origins of Genius: Darwinian Perspectives on Creativity*. Oxford: Oxford University Press.

Sleamaker, R. 1989. *Serious Training for Serious Athletes*. Champaign, Illinois: Leisure Press.

Smith, G. E. 1927. *The Evolution of Man*. Oxford: Oxford University Press.

Snelson, J. S. 1993. "The Ideological Immune System: Resistance to New Ideas in Science." *Skeptic*, vol. 1, no. 4: 44–55.

Somit, A. and S. A. Peterson. 1992. *The Dynamics of Evolution*. Ithaca: Cornell University Press.

Spencer, F. 1990. *Piltdown: A Scientific Forgery*. Oxford: Oxford University Press.

Stanford, C. B. 1999. *The Hunting Apes: Meat Eating and the Origins of Human Behavior*. Princeton, NJ: Princeton University Press.

Sulloway, F. J. 1979. *Freud, Biologist of the Mind*. New York: Basic Books.

———. 1982. "Darwin's Conversion: The *Beagle* Voyage and its Aftermath." *J. Hist. Biol.* 15: 327–398.

———. 1990. "Orthodoxy and Innovation in Science: The Influence of Birth Order in a Multivariate Context." Preprint courtesy of author.

————. 1991. "Darwinian Psychobiography." Review of *Charles Darwin: A New Life* by John Bowlby. *New York Review of Books*, 10 October.

————. 1996. *Born to Rebel: Birth Order, Family Dynamics, and Creative Lives*. New York: Pantheon.

Sutton-Smith, B. and B. G. Rosenberg. 1970. *The Sibling*. New York: Holt, Rinehart and Winston.

Swerdlow, N. M. and O. Neugebauer. 1984. *Mathematical Astronomy in Copernicus's* De Revolutionibus. 2 vols. New York: Springer-Verlag.

Swift, D. 1990. *SETI Pioneers: Scientists Talk About Their Search for Extraterrestrial Intelligence*. Tucson: University of Arizona Press.

Tattersall, I. 1995. *The Fossil Trail*. New York: Oxford University Press.

Taubes, G. 1993. *Bad Science: The Short Life and Weird Times of Cold Fusion*. New York: Random House.

Taussig, M. T. 1980. *The Devil and Commodity Fetishism in South America*. Chapel Hill: University of North Carolina Press.

Tillyard, E. M. 1944. *The Elizabethan World Picture*. New York: Macmillan Co.

Tipler F. 1994. *The Physics of Immorality*. New York: Doubleday.

Toulmin, S. and J. Goodfield. 1961. *The Fabric of the Heavens*. New York: Harper and Row.

Trivers, R. L. 1971. "The Evolution of Reciprocal Altruism." *Quarterly Review of Biology* 46: 35–57.

————. 1985. *Social Evolution*. Reading, Mass: Benjamin Cummings.

Tuchman, B. 1978. *A Distant Mirror*. New York: Knopf.

Turner, J. S. and D. B. Helms. 1987. *Lifespan Development*. New York: Holt, Rinehart and Winston.

Vere, F. 1955. *The Piltdown Fantasy*. London: Cassell.

Voyimner, P. 1970. *The Year of Contraversy: The Origin of Species and Its Critics*. Cambridge: Cambridge University Press.

Wallace, A. R. 1852. "Letter." October. 19, 1852. Under "Proceedings of Natural-History Collectors in Foreign Countries." *Zoologist*: 3641–3643.

————. 1855. "On the Law Which has Regulated the Introduction of New Species." *Annals and Magazine of Natural History* II., 16: 184–196.

————. 1858. "On the Tendency of Varieties to Depart Indefinitely From the Original Type." The joint Darwin-Wallace paper was entitled "On the Tendency of Species to form Varieties, and on the Perpetuation of Varieties and Species by Natural Means of Selection." *Journal of the Proceedings of the Linnean Society (Zoology)* 3: 53–62.

————. 1866. *The Scientific Aspect of the Supernatural: Indicating the Desirableness of an Experimental Enquiry by Men of Science Into the Alleged Powers of Clairvoyants and Mediums*. London: F. Farrah.

————. 1869. "Sir Charles Lyell on Geological Climates and Origin of Species." *Quarterly Review* 126: 359–394.

————. 1870. "The Limits of Natural Selection as Applied to Man." In *Contributions to the Theory of Natural Selection*. London: Macmillan and Co.

————. 1874. "A Defense of Modern Spiritualism." *The Fortnightly Review*, N. S., 15: 630–657, 785–807.

———. 1875. *On Miracles and Modern Spiritualism*. Three Essays. London: James Burns.

———. 1885. "Modern Spiritualism. Are its Phenomena in Harmony with Science?" *The Sunday Herald*. 26 April 1885, 9c-d.

———. 1889. *Darwinism: An Exposition of the Theory of Natural Selection, with Some of Its Applications*. London: Macmillan.

———. 1892. "Spiritualism." *Chambers' Encyclopaedia*. Vol. 9: 645–649.

———. 1895. *Natural Selection and Tropical Nature: Essays on Descriptive and Theoretical Biology*. London: Macmillan and Co.

———. 1903. *Man's Place in the Universe: A Study of the Results of Scientific Research in Relation to the Unity or Plurality of Worlds*. New York: McClure Phillips and Company.

———. 1908a. "The Origin of the Theory of Natural Selection." *The Popular Science Monthly*. July.

———. 1908b. *My Life: A Record of Events and Opinions*. New edition condensed and revised. London: Chapman & Hall.

———. 1913. "Alfred Russel Wallace." Interview by W. B. Northrop in *The Outlook*. New York 105: 618–622.

———. 1966. *Edgar Allan Poe: A Series of Seventeen Letters Concerning Poe's Scientific Erudition in Eureka and His Authorship of Leonainie*. New York: Haskell House.

Walsh, J. E. 1996. *Unraveling Piltdown: The Science Fraud of the Century and Its Solution*. New York: Random House.

Ward, P. D. and D. Brownlee. 2000. *Rare Earth: Why Complex Life is Uncommon in the Universe*. New York: Copernicus Books.

Weaver, J. H. 1987. *The World of Physics*. New York: Simon and Schuster.

Weiner, J. S. 1955. *The Piltdown Forgery*. Oxford: Oxford University Press.

Weisberg, R. 1986. *Creativity: Genius and Other Myths*. New York: W. H. Freeman.

Westfall, R. 1980. *Never at Rest. A Biography of Isaac Newton*. Cambridge: Cambridge University Press.

Westman, R. S. 1980. "The Astronomer's Role in the Sixteenth Century: A Preliminary Study." *History of Science*. Vol. xviii: 105–147.

Wilmut, I. 1996. "Sheep Cloned by Nuclear Transfer From a Cultured Cell Line." *Nature* 380: 64–66.

Wilson, C. 1996. *From Atlantis to the Sphinx: Recovering the Los Wisdom of the Ancient World*. New York: Fromm International.

Woodward, K. L. 1997. "Today the Sheep . . . Tomorrow the Shepherd?" *Newsweek*. 10 March.

Worster, D. 1988. "Doing Environmental History." In *The Ends of the Earth: Perspectives on Modern Environmental History*. D. Worster (Ed.). Cambridge: Cambridge University Press.

Wrangham, R. and D. Peterson. 1996. *Demonic Males: Apes and the Origins of Human Violence*. New York: Houghton Mifflin.

Wright, R. 2000. *Nonzero: The Logic of Human Destiny*. New York: Pantheon.

ABOUT THE AUTHOR

Michael Shermer is the publisher of *SKEPTIC* magazine, the director of the Skeptics Society, the host of the Skeptics Lecture Series at Caltech, and he is the author of *Denying History* (coauthored with Alex Grobman, about the Holocaust deniers), *How We Believe* (on God and religion), and *Why People Believe Weird Things* (on pseudoscience and superstitions), the latter of which was nominated as one of the top 100 notable books of 1997 by the *Los Angeles Times*. Dr. Shermer is also the author of *Teach Your Child Science*, and coauthored *Teach Your Child Math* and *Mathemagics* (with Arthur Benjamin). Dr. Shermer is also the producer and host of the Fox Family television series *Exploring the Unknown*, and serves as the science correspondent for the NPR affiliate KPCC radio. According to Stephen Jay Gould: "Michael Shermer, as head of one of America's leading skeptic organizations, and as a powerful activist and essayist in the service of this operational form of reason, is an important figure in American public life."

Dr. Shermer received his B.A. in psychology from Pepperdine University, M.A. in experimental psychology from California State University, Fullerton, and his Ph.D. in the history of science from Claremont Graduate School. Since his creation of the Skeptics Society, *Skeptic* magazine, and the Skeptics Lecture Series at Caltech, he has appeared on such shows as *20/20, Dateline, Good Morning America, Extra!, Charlie Rose, Tom Snyder, Donahue, Oprah, Sally, Leeza, Unsolved Mysteries*, and other shows as a skeptic of weird and extraordinary claims.

He has also appeared on documentaries aired on *A & E, Discovery*, and *The Learning Channel*.

Dr. Shermer has also written numerous cycling books, including *Sport Cycling* (Contemporary Books, 1985), *Cycling for Endurance and Speed* (Contemporary Books, 1987), *The Woman Cyclist* (with Elaine Mariolle, Contemporary Books, 1989), and *The Race Across America: The Agonies and Glories of the World's Toughest Bicycle Race* (WRS Books, 1994) based on a 10-year professional career as an ultra-marathon cyclist and competitor in the 3,000-mile, nonstop, transcontinental Race Across America. During his cycling career Dr. Shermer was featured four times on ABC's *Wide World of Sports*, including an in-depth "Up Close and Personal" piece, and has written and hosted five cycling events for ESPN.

INDEX